DEVELOPMENTS IN
APPLIED
SPECTROSCOPY
Volume 3

A Publication of the Chicago Section of the Society for Applied Spectroscopy

DEVELOPMENTS
IN
APPLIED
SPECTROSCOPY

Volume 3

edited by

J. E. Forrette and E. Lanterman

Roy C. Ingersoll Research Center
Borg-Warner Corporation
Des Plaines, Illinois

Proceedings of the
Fourteenth Annual Mid-America Spectroscopy Symposium
Held in Chicago, Illinois
May 20-23, 1963

Distributed by

⊕

PLENUM PRESS
NEW YORK
1964

ISBN 978-1-4684-8690-2   *ISBN 978-1-4684-8688-9 (eBook)*
DOI 10.1007/978-1-4684-8688-9

Library of Congress Catalog Card No. 61-17720

Preface

With the completion of the 14th Annual Mid-America Spectroscopy Symposium another milestone was passed in the history of a most successful endeavor. The Chicago, Cleveland, Detroit, Indianapolis, Milwaukee, Niagara Frontier, and St. Louis Sections of the Society for Applied Spectroscopy united to sponsor the finest of the regional meetings for the advancement of the science of spectroscopy. Enthusiastic response was expressed by participants, speakers, and interested persons from several foreign countries, as well as coast-to-coast in the United States.

Sessions on X-ray, infrared, Raman, visible, ultraviolet, nuclear-magnetic and electron-magnetic resonance, flame, and arc-spark spectroscopy contained interesting arrays of both applied and theoretical papers. Once again, the closely allied field of gas chromatography was represented by a session arranged by the Chicago Gas Chromatography Discussion Group.

This present volume, which continues a series of Proceedings initiated in 1961, represents a collection of 32 papers from the Symposium's program. Providing this material in book form gives a permanent reference to the attendees and makes the material available to those unable to attend the Symposium. Enough requests for publication of the Proceedings have been received to justify the continued publication.

The presentation of a symposium of this size and stature is accomplished through the combined efforts of many people. It is with pride that we acknowledge the untiring efforts of Stuart Armstrong, Murray Barsky, Vivian Biske, James Burroughs, Audrey Companion, Byron Field, E. L. Grove, Roger Harper, Emmett Kaelble, Robert Krupp, Frank Leahy, Robert Morris, John Murphy, Alfred Perkins, Edward Piotrowski, Robert Scholz, and Joan Westermeyer, all of whom made the Symposium a success.

We extend our sincere gratitude to the many authors contained herein who devoted time and effort to submit their manuscripts so that these Proceedings could materialize. In addition, we are grateful to the Borg-Warner Corporation for permitting us to use their time and facilities for organizing the Symposium and editing these Proceedings.

J. E. Forrette
E. Lanterman

Contents

EMISSION SPECTROGRAPHY

X-Ray Spectroscopy

Industrial Applications of Microprobe Analysis

B. R. Banerjee* and W. D. Bingle†

Research Division
Crucible Steel Company of America
Pittsburgh, Pennsylvania

Electron probe X-ray microanalysis has revealed a new realm of analytical investigations on a microscopic scale. But while the microprobe permits high precision in strictly quantitative analysis, such precision may not be pertinent in many commercial problems, where the microconstituent to be analyzed is itself compositionally nonhomogeneous. In those cases, where better precision is warranted, precise quantitative measurements may be applied, or synthetic empirical reference standards may be used. But in many important industrial problems, a rapid semiquantitative approach plus electron and X-ray scanning photographs provide important compositional insight to light metallography.
Therefore some typical industrial applications of the scanning microanalyzer to the analysis of multiphase structures and of compositional heterogeneity in commercial materials are discussed, along with some considerations of instrumental factors in the scanning microanalyzer, which improve image quality, increase speed of semiquantitative analysis, and permit analysis of rough surfaces, such as fractured metal specimens.

INTRODUCTION

Electron probe X-ray microanalysis is a new analytical method, where a fine-electron probe striking a microconstituent on a polished surface excites an X-ray spectrum containing wavelengths characteristic of the elements which make up the constituent; the intensities of these characteristic wavelengths are proportional to mass concentration of these elements. Therefore, by dispersing the spectrum in an X-ray spectrometer and measuring its X-ray intensities, a quantitative analysis of the microconstituent can be made. Thus electron probe microanalysis can readily analyze a microscopic volume, 10^{-12} cc [$(10^{-4}$ cm$)^3$] with a relative accuracy of 1% and a sensitivity of 0.1%, for elements with atomic numbers Z = 11 to 92. These limitations are not absolute, and may indeed be exceeded, but only with special care and techniques; improvements are rapidly being made in techniques, and in novel methods, which may well remove these limitations in the future.

Thus, electron probe microanalysis [1] has revealed a new realm of analytical investigations on a microscopic scale [2]. To the fundamental researcher the microprobe permits fine-scale diffusion analysis and the study of minute compositional variations on a microscopic scale in laboratory controlled samples. But to the industrial researcher the microprobe is a practical tool, of immense importance, in the detailed study of defects and undesirable constituents found in commercial products, in addition to the study of basic phenomena, such as the characterizing of individual phases — solid solutions, transformation products, precipitates, etc. Inclusions, im-

---

*Supervisor, Basic Research and Applied Physics Sections.
†Staff physicist.

purities, and other dispersed materials may be uniquely characterized. Segregation effects, produced by impurity elements, by alloying elements, and by chemical reactions—forming depleted or enriched zones around grain boundaries, near surfaces, and in and around dendrites in cast materials—may be detected. Incomplete diffusion gradients across an interface, frequently found in industrial applications, may be determined.

Because of its instrumental complexity and the large investment involved, the microprobe is frequently regarded as an "ivory-tower" device applicable only to profound academic problems, but somewhat out of place for day-to-day industrial problem solving. But, in reality the microanalyzer, and particularly the scanning type of microanalyzer is uniquely suited to the solving of many problems frequently encountered in industry [3-5].

Therefore, some typical industrial applications of a scanning microanalyzer to the analysis of multiphase structures and compositional heterogeneity in commercial materials will be discussed, along with some considerations of instrumental factors in the scanning microanalyzer, which provide improved image quality, increase speed of semiquantitative analysis, and permit analysis of rough surfaces, such as fractured metal specimens, etc.

THE MICROANALYZER

Most electron probe microanalyzers consist of three essential parts:

1. An electron probe of 0.3 to 5.0 μ diameter, formed by an electron source and two demagnifying lenses. The probe excites X-ray from a microvolume at variable excitation potential, generally between 20 and 30 kV.

2. A sample positioning device to precisely locate a desired microconstituent under the analyzing probe. This may be done (1) light microscopically through the use of light optics coaxial with the electron lens system; (2) through an external light-optical positioning jig; or (3) electron optically through electron scanning in situ and imaging the scan on a cathode-ray tube display.

3. An X-ray spectrometer (air or vacuum path) to measure characteristic intensities: the spectrometer may be either of the full-Rowland focusing or semifocusing types, employing Geiger or proportional counters and electronic circuitry to provide adequate counting statistics.

OPERATION

In principle, operating a microanalyzer requires only focusing the micron-diameter electron probe on the constituent to be analyzed, and measuring the characteristic X-ray intensity for the element analyzed. This intensity—corrected for counter dead time and background—is proportional to mass concentration of the emitting element, relative to similarly corrected X-ray intensity measured from a pure-element reference standard. Because of the small volume of excited material in microprobe analysis, this simple proportionality of mass concentration to relative X-ray intensity is of surprisingly general validity under suitable conditions, and can readily yield a relative accuracy of 1%.

However, when analyzing for elements having low atomic numbers, or operating at high kilovoltages, when the X-ray emergence angle from the specimen is small, errors due to sample absorption will increase. At low concentrations of the analyzed element (less than 1%) a disproportionate increase in counting time is needed to achieve the same accuracy, because of increased white-radiation background. Strong interelement effects can arise when near-neighbor elements in the periodic table are analyzed, for example in an Fe–Cr alloy the FeK_α characteristic emission can excite CrK_α fluorescent radiation. However, analytical methods have been developed to correct for specimen absorption, interelement effects and mutual fluorescence, in addition to counter dead time and background, so that individual calibration curves are not needed for every system analyzed. Thus, by combining appropriate experimental conditions with analytical correction methods, measured spectral intensities can be accurately converted to mass concentrations.

Sample preparation for the microprobe has been adequately described, and the use of a good diamond metallographic polish and avoiding etched samples are accepted practice.

A convenient and rapid way to prepare $1/4$-in.-diameter by $1/8$-in.-high metal samples from bulky or massive segments involves a small laboratory-scale spark-discharge machine.* A $1/4$-in.-ID cylindrical cutter in this machine, will readily drill (without overheating) into any metal, leaving a small pillar, from which the $1/8$-in. segment can be removed with an abrasive wheel, a hacksaw, or the spark machine with an edge cutter. These sample requirements apply to the Cambridge Electron Probe X-Ray Microanalyzer, which, with the indicated modifications, was used in all the experiments.

QUANTITATIVE ANALYSIS

All of the numerical analyses to be reported were made under "fixed-charge" conditions, i.e., for a constant number of probe electrons absorbed within the specimen.

For exact quantitative microprobe analysis, measured intensities must be corrected for X-ray absorption effects in the sample and in the standard, and for interelement fluorescence effects in the sample—excited both by characteristic X-ray wavelengths from elements in the sample, and by the continuous white radiation. Several approximations for these corrections have been suggested and used. Further corrections for electron deceleration within the sample and electron backscatter from the sample may be computed and applied for improved precision of data.

However, for many industrial problems of the types illustrated here, these corrections are inapplicable because of lack of pertinent data; furthermore, heterogeneity in the constituents being analyzed often exceeds the precision obtainable by Castaing's first approximation alone [1]. Therefore, in most such examples, corrections for background and counter are generally sufficient. Absorption corrections, with the generalized absorption–correction curves, may be used when applicable.

While the importance of precise quantitative analysis through meticulous application of all pertinent corrections is not to be minimized, the purpose here is to highlight the many semiquantitative industrial applications of the microprobe which lend to optical microscopy, a quantitative aspect, and understanding that has been hitherto unachievable.

---

*Servomet, Cambridge, England.

SCANNING MICROANALYSIS

In a scanning microanalyzer, as the probe scans an area up to $\frac{1}{2}$ mm square the backscattered-electron intensity detected by a scintillation counter modulates the brightness of a cathode-ray tube. Contrast in such a point-for-point reproduction of the sample image depends upon: (a) surface topography of the sample, and (b) local variations in atomic number on the sample surface. Thus, the image not only has visual qualities of an oblique-incidence micrograph, but also reveals the mean atomic number of elements in the sample. The detailed distribution of a given element may be similarly imaged by modulating brightness on a cathode-ray tube with the signal from the X-ray spectrometer, which receives X-rays characteristic of a single element at a given spectrometer setting.

The unique advantages of electron beam scanning lie in the pictorial display—similar in character to that of the ordinary light micrograph—of the distribution of a particular element over an area, and in positioning the microconstituent, with a resolution equal to or better than that possible with coaxial-light optics. However, because of the moving X-ray source, a semifocusing spectrometer is generally used, with some—though not serious—performance loss in terms of intensity, resolution, and peak-to-background ratio.

The Cambridge Scanning Microanalyzer (see schematic Fig. 1) was modified in several respects (see Fig. 2) to provide: (1) improved image quality, (2) increased speed of semiquantitative analysis, and (3) microanalysis of rough surfaces.

Image Quality

In order to fully utilize the image resolution potential of a 1-μ-diameter electron probe, the cathode-ray tube must be able to resolve $1\ \mu$ at its lowest magnification. Thus, in the present unit, the lowest available magnification being $200\times$, the cathode-ray tube (CRT) must resolve 50 lines/cm, on the screen image. However, the CRT supplied with the Cambridge instrument had an image resolution of 37 lines/cm [6]; also the optimum brightness level of these tubes was rather inadequate. Therefore, these CRT's were

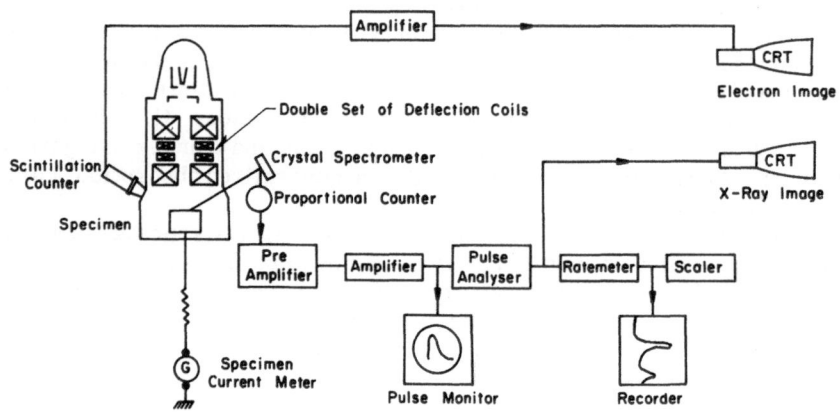

Fig. 1. Original scanning and display system of Cambridge microprobe analyzer.

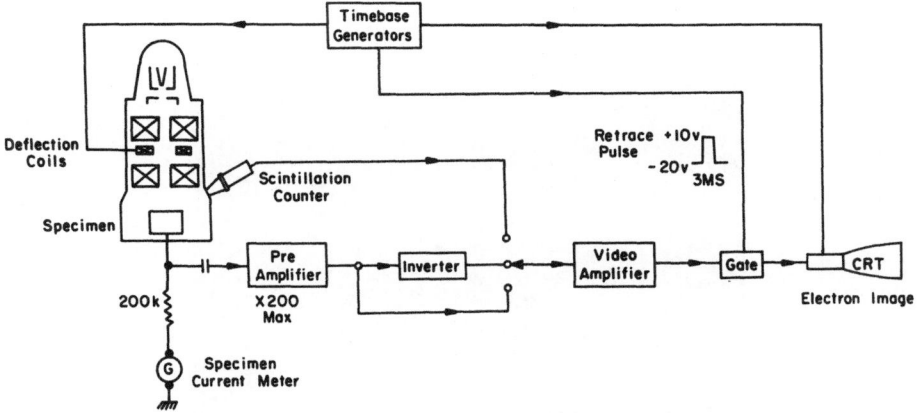

Fig. 2. Modified scanning and display system of electron probe microanalyzer (X-ray imaging not shown).

replaced by RCA-5ADP7-A tubes, operated at 4.5 kV. The power supply and time-base units were appropriately modified for this purpose. In this way, CRT image resolution of 75 lines/cm, and image diameter of 12 cm were achieved.

Blanking the electron-probe signal, during the flyback or retrace period, was done on the original Cambridge instrument (see Fig. 1) by means of additional deflection coils—wound adjacent to the upper set of the Y-deflection coils—which were energized during retrace by a pulsed current. However, this scheme suffered from the inherent disadvantage that during the initial part of the scan, the magnetic field collapsed, causing considerable image distortion in the first 1 to 2 cm of the CRT image. This problem was completely eliminated by disconnecting the blanking coils, and introducing a constant-amplitude blanking pulse into a gating circuit which followed the video amplifier (see Fig. 2). The blanking pulses are introduced during the retrace periods to suppress the video signal during flyback.

Direct Plotting Rate Meter Analysis

Industrial problems involving diffusion gradients due to joining of two metals by welding, etc., surface reactions, or internal inclusions often represent incomplete diffusion and nonequilibrium conditions, where local compositional heterogeneity makes precise quantitative analysis less meaningful. In such applications, microprobe techniques involving scaler counting at various points across the diffusion gradient are not only too time consuming, but redundant.

For analyzing such gradients, the X-shift potentiometer was motorized with a 1-rpm synchronous motor coupled through a two-speed gearbox and slip clutch to a precision potentiometer. The two gear speeds permit a complete scan of the probe across the specimen surface in either 5 min, or 25 min. At these slow scanning rates of the electron probe, the rate meter accumulates sufficient counts of X-ray quanta to obtain adequate semiquantitative statistical accuracy. The rate meter output may be plotted on a strip-chart recorder. The slip clutch permits normal use of the shift control to manually position the beam when the motor drive is not in use.

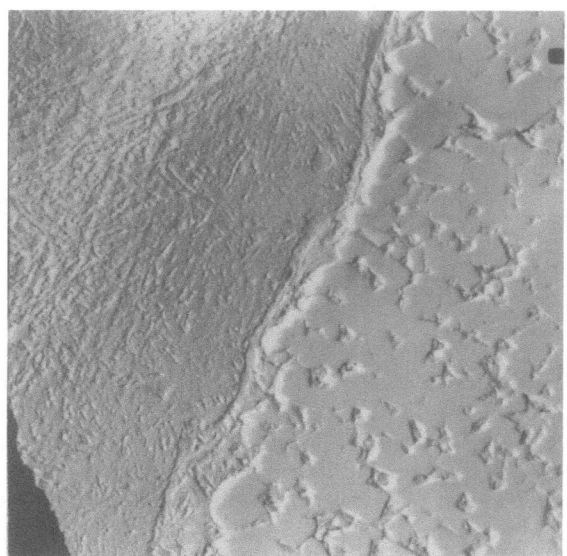

(a) Replica Electron Micrograph

Fig. 3. Electron micrograph, electron image, and X-ray image
illustrating a diffusion boundary (1880 ×).

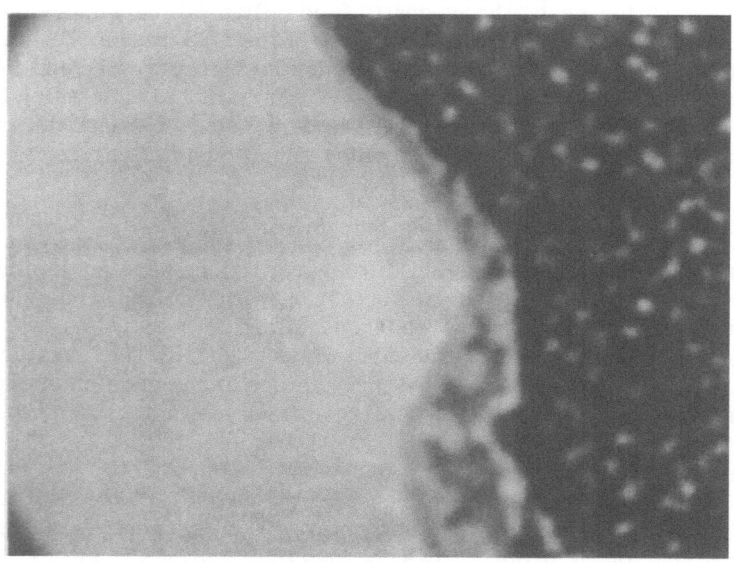

(b) Electron Image

Fig. 3 (continued).

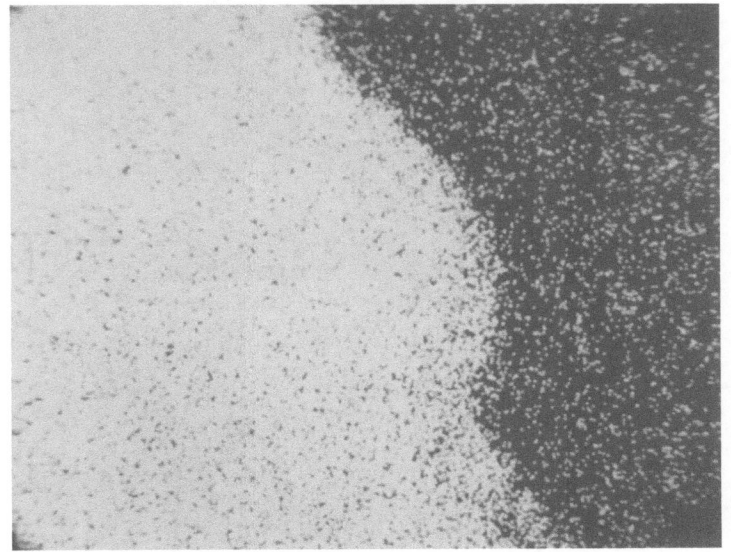

(c) Fe-Kα X-Ray Image

Fig. 3 (continued).

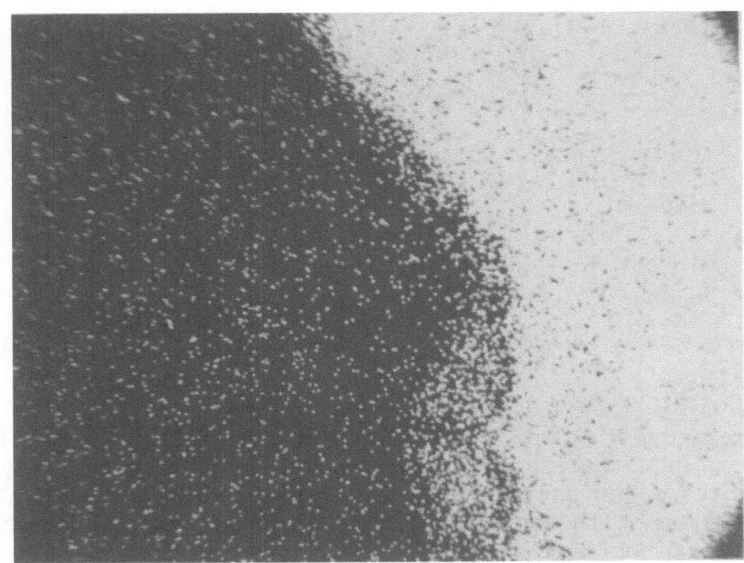

(d) W-Lα X-Ray Image

Fig. 3 (continued).

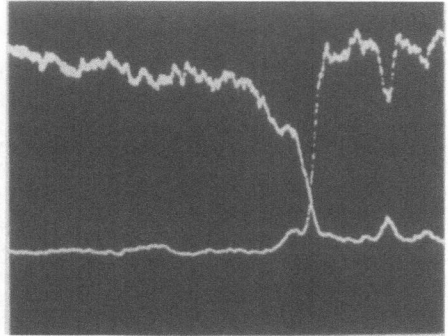

(a) Oscilloscope Line Scan (0.3 min)

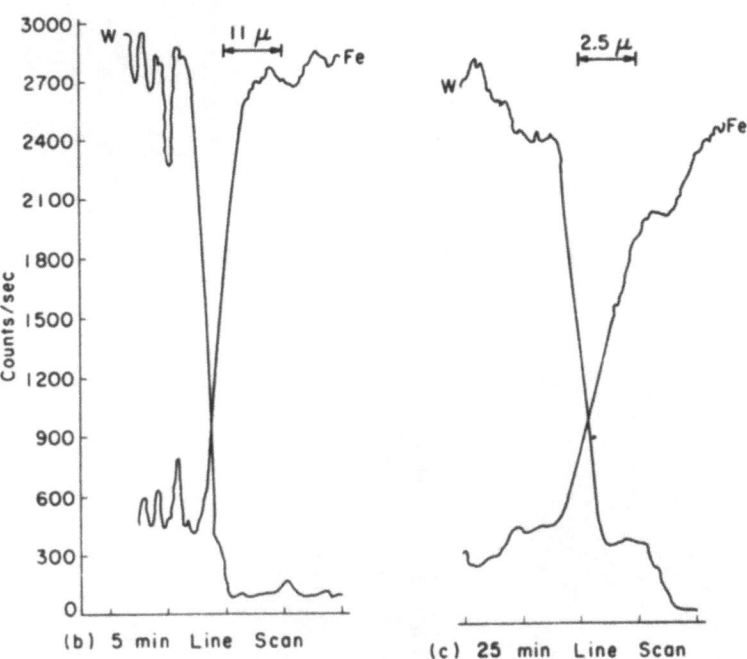

Fig. 4. Directly recorded line scans of X-ray intensity (a) from the cathode-ray tube; (b) and (c) from the strip-chart recorder.

An example of the slow-scan analysis is given in Figs. 3 and 4. Figure 3 shows a replica electron micrograph, as well as electron and X-ray images across the welded interface of a tungsten-carbide tool, bonded to a carbon-steel backing. The electron micrograph (Fig. 3a) shows the cobalt-bonded tungsten-carbide structure on the right, and the martensitic steel structure on the left; the corresponding electron image characteristics are seen in Fig. 3b. Incomplete diffusion of tungsten and iron at the interface is clearly suggested by the X-ray images (Figs. 3c and d).

Figure 4a shows an oscilloscope line scan (about 20-sec duration) while Figs. 4b and c show strip-chart recordings of slow line scans of 5 min and

25 min duration, respectively. The smoother X-ray intensity curve, due to the longer time constant and better statistical accuracy, is clearly apparent.

Specimen Current Electron Image

For analyzing rough surfaces, e.g., fracture surfaces, backscattered electron imaging is unsatisfactory because of extreme shadows created by the low take-off angle of the backscattered electrons. This difficulty was avoided by introducing another mode of CRT imaging, using the specimen current to modulate the brightness of the CRT (see Fig. 2).

The elimination of blanking pulses from the video waveform in the amplifier (previously discussed under image quality), also permits image-polarity choice in the specimen-current pictures; i.e., both positive and negative contrast images may be obtained. Also, increased contrast is possible in these images, due to removal of the gain restriction previously imposed by the retrace-pulse amplitude.

The unique advantages of specimen-current imaging are illustrated in Figs. 5 and 6, which show electron and X-ray images from a fractured surface in an AISI 4340 steel* specimen. The specimen was fractured by wedge impact at −320 F, following a heat-treatment sequence consisting of austenitizing at 2150 F, oil-quenching plus refrigerating to −320 F, and subsequent tempering at 750 F for 2 hr.

The predominantly intergranular fracture outlining the coarse austenite grain boundaries is clearly seen in Fig. 5a, which is the usual backscattered electron image at 25 kV probe excitation. However, the extreme shadows, due to the low take-off angle of the backscattered electrons, obscures many structural details in the shadowed areas. However, these details are clearly revealed in the specimen-current images—also at 25 kV probe excitation in Figs. 5b and c. The advantage of dual polarity in this imaging scheme is also evident through comparison of Figs. 5b and c; areas in dark contrast in Fig. 5b are brought into light contrast in Fig. 5c. Thus structural details within these areas are clearly revealed.

Inclusions on the fracture surface are seen in Fig. 6a, at a higher magnification, and the large inclusion is seen to be MnS, from the X-ray images (6b and c). Thus specimen-current imaging is extremely useful in analyzing rough and irregular surfaces, such as these fracture facets.

A further advantage of specimen-current imaging is in its ability to obtain electron images, with electrons at low-accelerating potentials. In analyzing the very light elements (of low atomic number) greater efficiency of X-ray excitation may be obtained by using lower kilovolt excitation [7] of the electron probe, but the backscattered electron imaging becomes impossible at these low excitation voltages. Therefore the continued effectiveness of specimen-current imaging becomes highly significant.

Figure 7 shows the fractured 4340 sample surface, revealed by specimen current imaging, using 5 kV probe excitation. At this excitation level, the backscattered electron image would show no information whatsoever, and the CRT would register an entirely blank raster.

*Composition of 4340 steel used: C-0.40; Mn-0.72; P-0.02; S-0.015; Si-0.24; Ni-1.80; Cr-0.82; Mo-0.29.

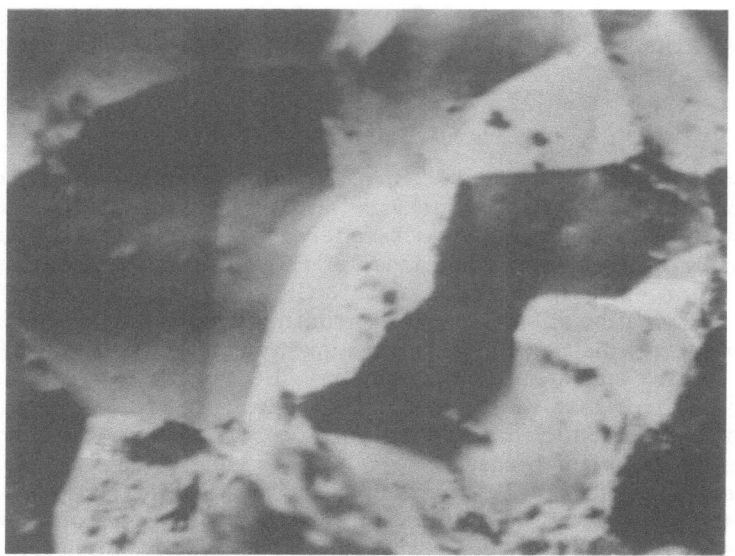

(a) Back Scattered Electron Image

Fig. 5. Electron images from a fractured surface in AISI-4340 steel specimen
(300x).

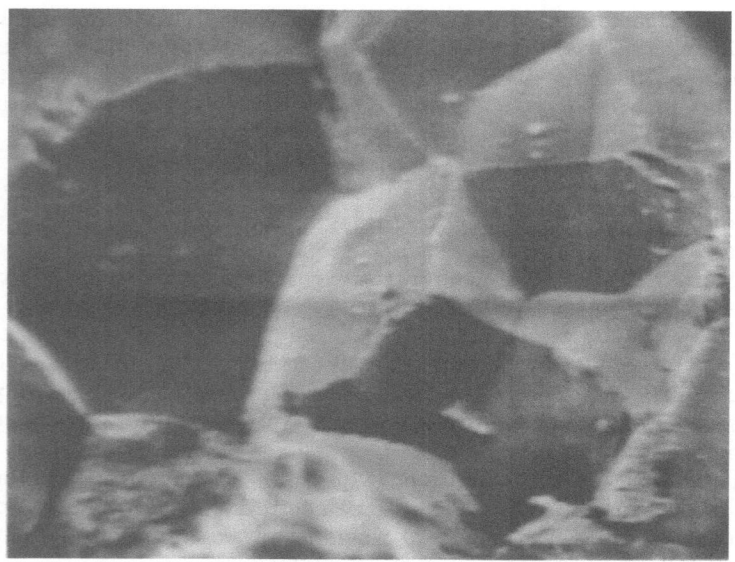

(b) Positive Specimen Current Image

Fig. 5 (continued).

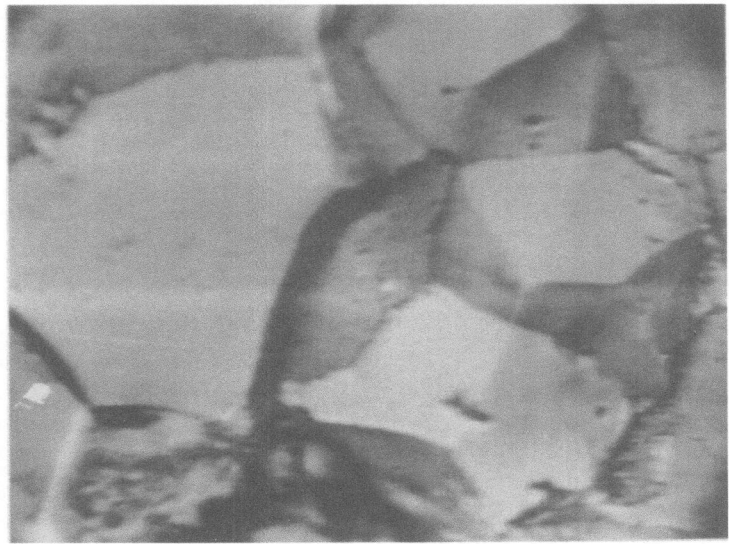

(c) Negative Specimen Current Image

Fig. 5 (continued).

APPLICATIONS

Typical industrial applications of microprobe analysis are illustrated in terms of some inclusions and multiphase structures frequently encountered in industrial materials.

Inclusions

Inclusion analysis by the microprobe is illustrated in terms of its application to resolve a long-standing metallurgical controversy, as well as its application to industrial problems when the inclusion lies on a rough surface, where it must be analyzed in situ.

Sulfide in Free Machining Tool Steel

Ever since the introduction of resulfurized free-machining high-speed tool steels, particularly of the molybdenum-containing (M-2) types, the sulfide inclusions were conjectured to be molybdenum sulfide (MoS_2), which provided lubricity during cutting operations [8]. Though questioned, this hypothesis has never been scientifically disproved.

A specimen of Rex M-2S* (6.4W, 5Mo, 4.15Cr, 1.95V, 0.85C, 0.15S, 0.3Mn) resulfurized high-speed tool steel was sectioned in the longitudinal

*Crucible Registered Trade Mark.

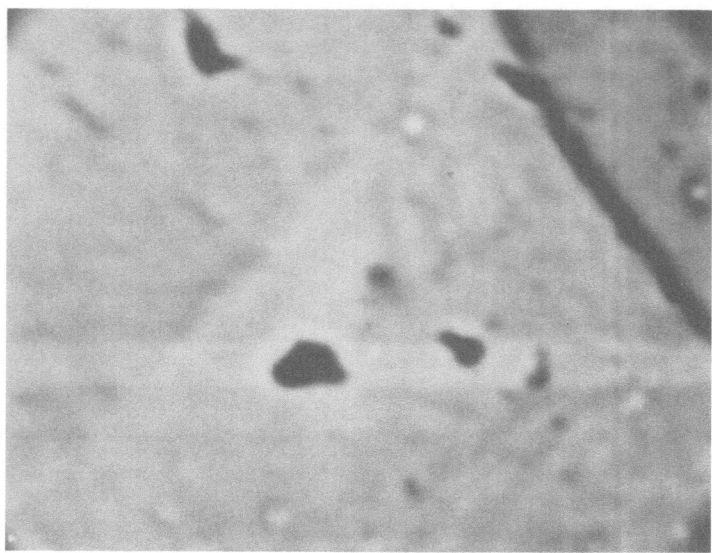

(a) Negative Specimen Current Image

Fig. 6. Electron and X-ray images from a fractured surface (Fig. 5) at higher magnification (2000×).

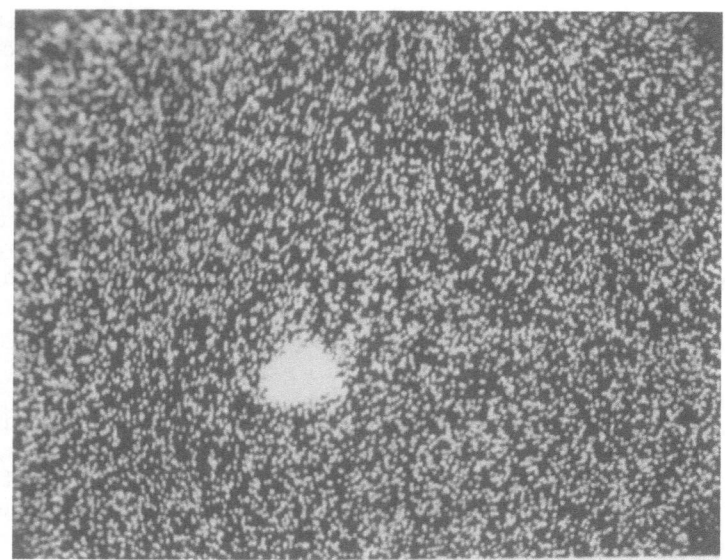

(b) Mn−Kα X−Ray Image

Fig. 6 (continued).

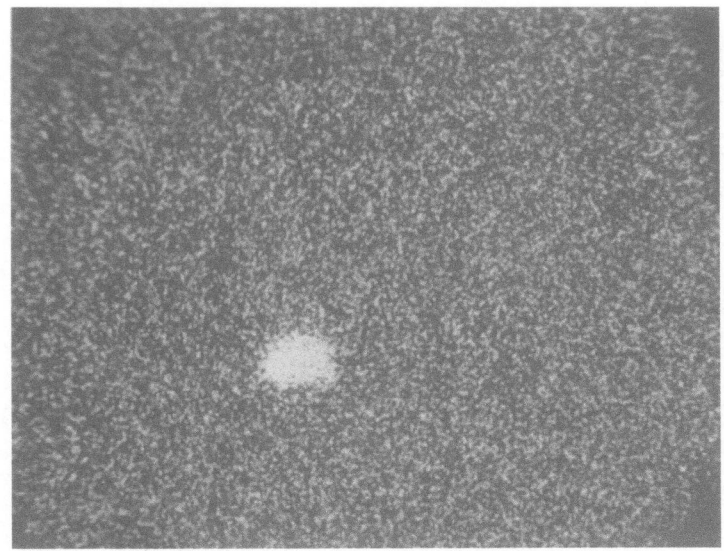

(c) S-Kα X-Ray Image

Fig. 6 (continued).

(working) direction to reveal elongated sulfide stringers. Figure 8a shows the electron-optical image of a sulfide stringer adjacent to a rounded residual carbide. The X-ray images (Figs. 8c and d) clearly establish the sulfide stringer to be mainly manganese sulfide. No evidence of molybdenum or tungsten was found in the stringer. The adjacent residual carbide is seen to be mainly vanadium carbide.

Thus a scanning microprobe analysis resolved this long-standing metallurgical controversy; and the improved machinability of these resulfurized M-2 steels may be understood in terms of easy machinability due to the MnS inclusion stringers, closely adjacent to residual carbide particles.

Inclusions on Rough Surface

While a polished metal surface is optimally desired for electron-probe microanalysis, industrial problems frequently encounter inclusions embedded on a rough surface. Figure 9 illustrates a refractory heavy-metal inclusion in an 85V, 15Al alloy fragment. A molybdenum inclusion particle, of 5-μ diameter, is seen in the electron image. Despite the surface roughness of the bulk-alloy fragment, the molybdenum inclusion was definitely identifiable. Furthermore, quantitative analysis—using scaler counting—proved the particle to be pure molybdenum metal.

Multiphase Structures

In multiphase structures, the electron probe is uniquely suited for analyzing both the dispersed second phase and the matrix. High-temperature reac-

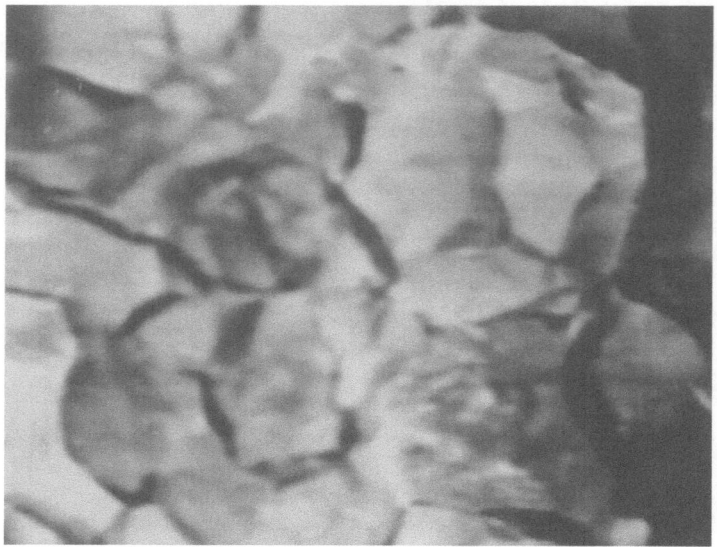

Fig. 7. Negative specimen curve image at 5 kV probe excitation (150X).

tion products and precipitates, such as residual carbides in alloy steels, permit direct microprobe analysis in situ. However, low-temperature reaction products, such as carbides precipitated during the tempering of alloy steels, are frequently smaller than a micron in size. Therefore, special techniques must be combined with microprobe analysis in order to analyze these submicron particles.

Residual Carbide and Matrix Analysis

Alloy element partitioning effects, in a commercial high-carbon Cr–V steel, were precisely determined by microprobe analysis of the carbide-free matrix, as well as the carbides themselves (see Fig. 10). The specimen-current electron images clearly outline the various sized residual carbide particles in the grey matrix. The X-ray image shows the high vanadium content of the carbides.

By reference to specially heat-treated "standard" samples of related composition, where all carbides were put into solution, the matrix in the test material was found by microanalysis to contain only 0.66%V; although the original vanadium content of the alloy was 4.9%. This illustrates the substantial extent of alloy-element partitioning that can occur as a result of a second-phase separation, such as carbide precipitation.

Thus, residual carbides –typical of high-temperature precipitation reactions– were analyzed and combined with an analysis of the matrix.

Low-Temperature Precipitates

Products of low-temperature solid-state reactions, such as carbides precipitating from alloy steels during tempering, cannot be directly micro-

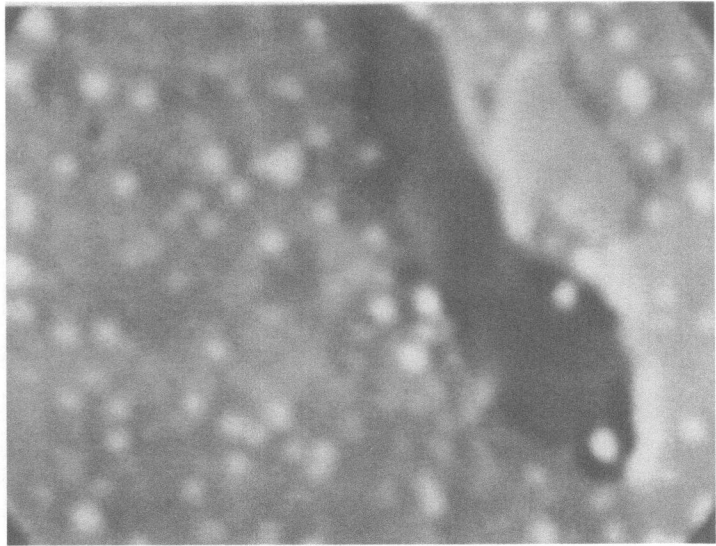

(a) Electron Image

Fig. 8. Electron image and associated X-ray images from MnS inclusion in a
high-speed tool steel (1500x).

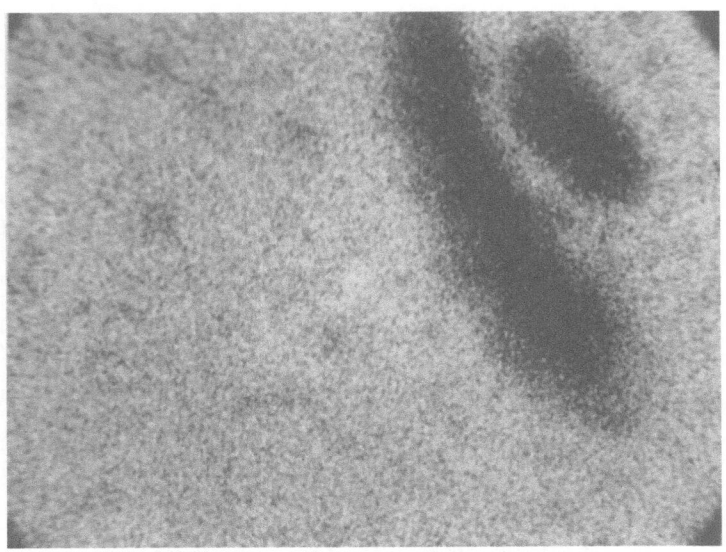

(b) Fe-Kα X-Ray Image

Fig. 8 (continued).

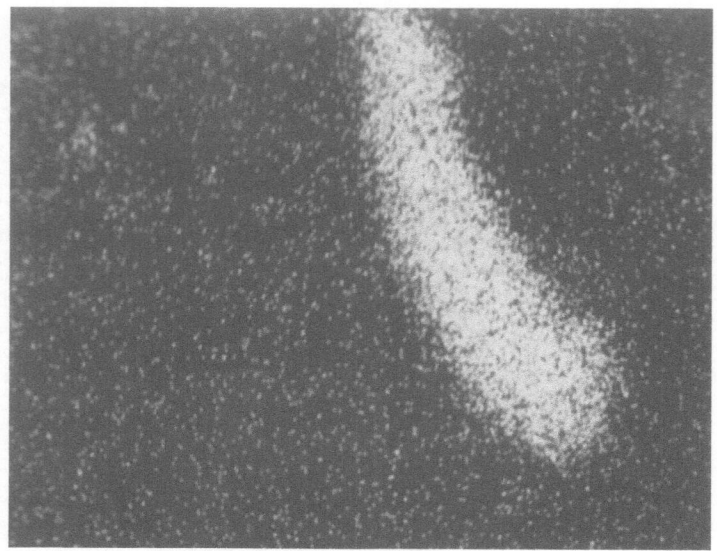

(c) Mn - Kα X - Ray Image

Fig. 8 (continued).

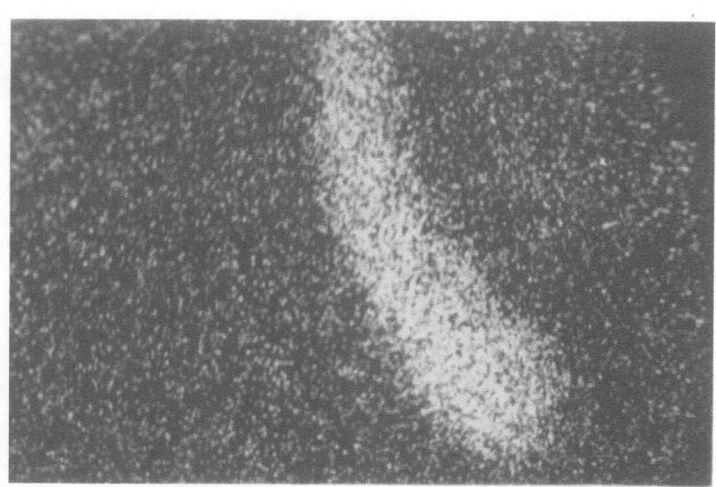

(d) S - Kα X - Ray Image

Fig. 8 (continued).

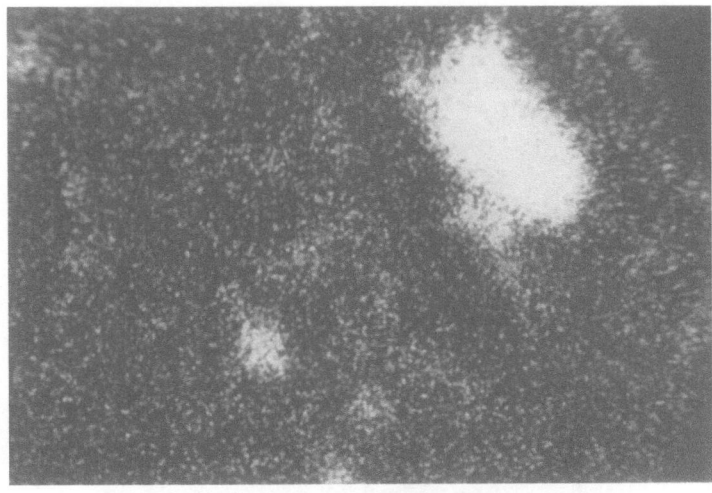

(e) V-Kα X-Ray Image

Fig. 8 (continued).

analyzed because of their submicron size. However, by electrolytically extracting* the carbides through anodic dissolution of the matrix, and collecting the anode residue, these submicron-size carbides can be effectively separated from the matrix. The residue may be compacted into a 1 mm³ volume in a specially constructed die by applying 8000 psi pressure in a compacting press. The surface of such a compacted specimen is seen in Fig. 11, where the electron image shows several cracks in the surface.

Electron probe microanalysis through electron scanning over an area of the compact 0.5 mm square averages over heterogeneities frequently encountered in individual carbide particles. Table I illustrates agreement between microprobe and wet-chemical analyses of these compacts.

An alternative technique for electron probe analysis of submicron precipitates consists of microanalyzing extraction replicas (normally used in electron microscopy). These replicas, containing the precipitates extracted from the etched-metal surface, are placed on a polished light-metal block and microanalyzed by the scanning raster technique.

SUMMARY

While precision quantitative microprobe analysis has been frequently emphasized, many industrial problems involving nonhomogeneous samples may be simply solved through various applications of scanning-microanalysis techniques. Such applications to the analysis of incomplete diffusion gradients, multiphase structures, and other compositional heterogeneity in commercial materials are discussed.

*Electrolyte: ammonium chloride and citric acid, current density—250 mA/in.²

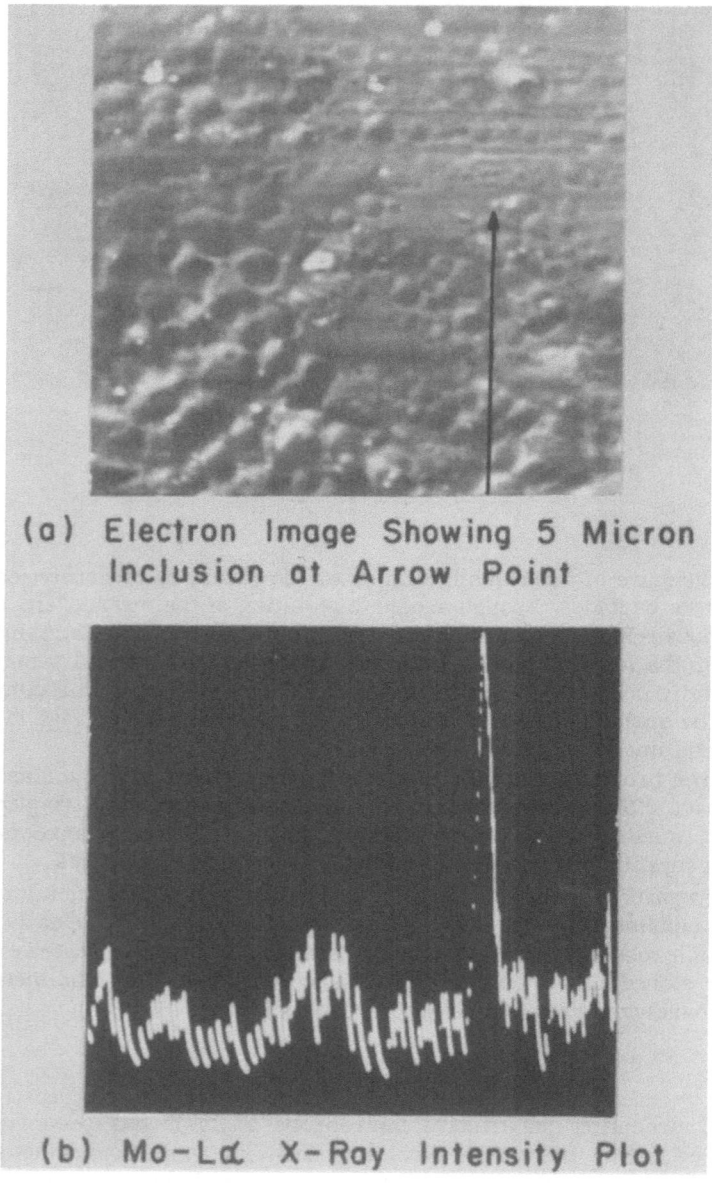

(a) Electron Image Showing 5 Micron
Inclusion at Arrow Point

(b) Mo-Lα X-Ray Intensity Plot

Fig. 9. Refractory heavy metal inclusion in 85 V–15Al alloy fragment (250X).

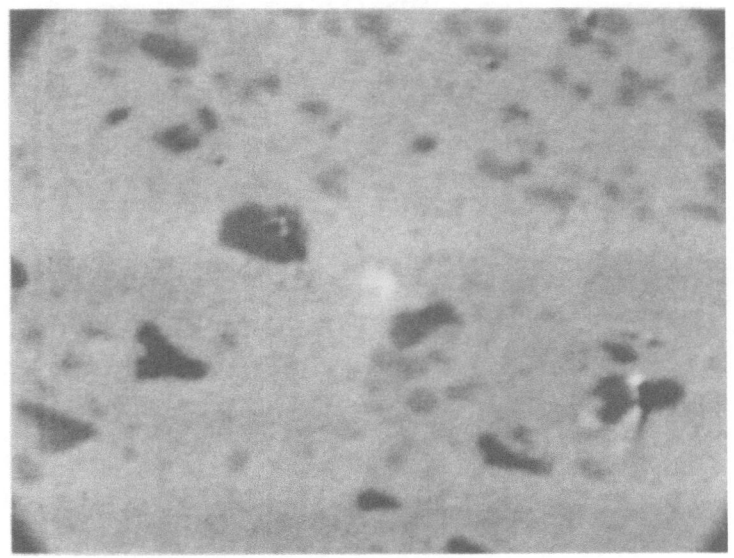

(a) Back Scattered Electron Image

Fig. 10. Electron and X-ray images of vanadium carbides in a high-carbon Cr-V
steel (1000x).

(b) Negative Specimen Current Image

Fig. 10 (continued).

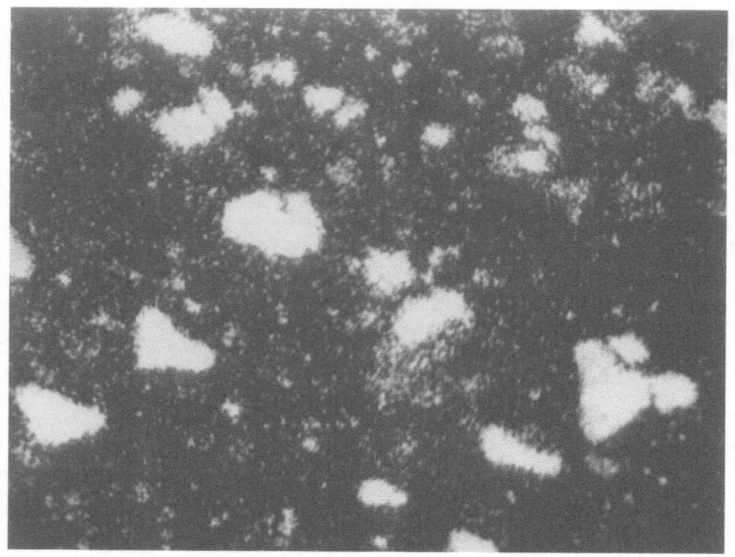

(c) V - Kα X - Ray Image

Fig. 10 (continued).

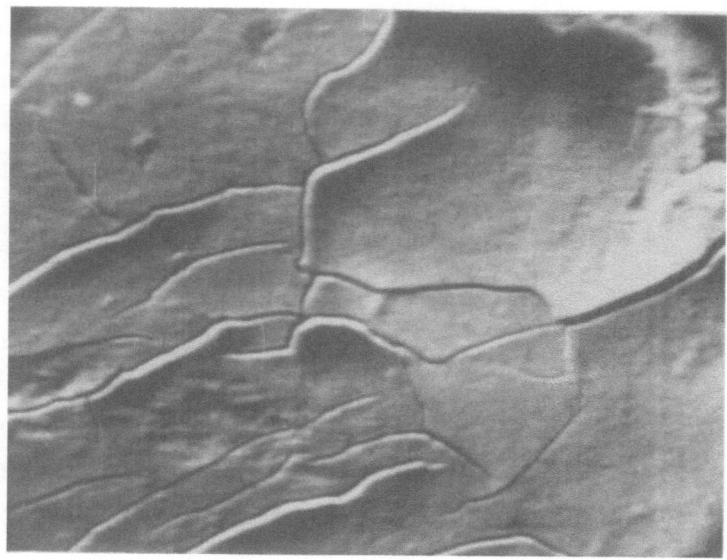

Fig. 11. Electron image of a compacted carbide pellet (300×).

TABLE I
Extracted Residue Compact from Tempered Type H-11 Steel*

Element	Electron probe (wt. %)	Wet chemistry (wt. %)
Fe	17.2	16.65
Cr	51.5	50.4
Mo	3.2	4.95
V	5.7	5.7

*Composition of Type H-11 steel (Crucible 218, hot-work steel): C—0.41; Cr—5.01; Mo—1.29; V—0.48; Mn—0.36; Si—0.99; S—0.012; P—0.015; Ni—0.22.

ACKNOWLEDGMENT

The authors would like to acknowledge the many contributions of their former associate N. S. Blake, in the various instrumental modifications.

REFERENCES

1. R. Castaing, Thesis (1951), Paris University, Paris, France.
2. B. R. Banerjee, "Classified Bibliography on Electron Probe X-Ray Microanalysis," ASTM, STP No. 317 (1962), 190.
3. B. R. Banerjee, "Electron Probe X-Ray Microanalysis," J. of Metals 13:905 (1961).
4. D. A. Melford and P. Duncumb, "The Application of X-Ray Scanning Microanalysis to Some Metallurgical Problems," Metallurgia (1960) 205.
5. D. A. Melford, "Metallurgical Application of the X-Ray Scanning Microanalysis in an Industrial Laboratory," Rev. Univ. des Mines 17:420 (1959).
6. P. Duncumb, "The X-Ray Scanning Microanalyser," Brit. J. Appl. Phys. 10:420 (1959).
7. T. Mulvey, "X-Ray Emission Analysis and the Determination of Gases in Metals," Iron & Steel Institute Special Report No. 68 (1960), 225.
8. G. A. Roberts, J. C. Hamaker, and A. R. Johnson, Tool Steels, ASM, Metals Park, Ohio (1962).

Use of a Multichannel Analyzer for Electron-Probe Microanalysis*

L. S. Birks and A. P. Batt

*U. S. Naval Research Laboratory,
Washington, D. C.*

A transistorized 400-channel multichannel analyzer attached to the electron probe microanalyzer has increased the versatility and decreased the time required for many analyses. Three types of application are illustrated. The multichannel analyzer replaces a scanning crystal spectrometer for quick identification and semiquantitative analysis of components. Quantitative analysis by mathematical unfolding of the energy spectra is made practical by the fast data collection of standard and unknown spectra; comparison of results by unfolding and from crystal spectrometer data shows differences of less than 4% of the amount present. Quantitative two-dimensional topographic display is obtained by multiscaler operation of the 400-channel memory system.

INTRODUCTION

In both X-ray fluorescence and electron-probe microanalysis the use of gas proportional or scintillation detectors has become well known in recent years [1, 3]. Each X-ray quantum detected is converted to an electrical pulse whose amplitude is proportional to the X-ray quantum energy. Most X-ray circuits contain single-channel pulse-height analyzers, so that the characteristic quantum energies and hence the chemical elements may be distinguished from one another, although the resolution is orders of magnitude poorer than with crystal spectrometers. This separation according to energy is often called nondispersive analysis, but energy dispersion is a more descriptive name.

One disadvantage of the usual pulse-height analyzers is that, in order to scan the energy spectrum, one must set a narrow energy window and scan through the energy range one energy increment at a time. More recently, transistorized electronics have made possible the construction of compact multichannel analyzers in which the whole energy spectrum is measured simultaneously. The operation is as follows:

The X-ray quanta are detected by a regular X-ray proportional counter and preamplifier circuit as before. Each quantum is converted to an electrical pulse of amplitude corresponding to the quantum energy. The input stage to the multichannel analyzer examines the pulse and stores it in the memory core according to its amplitude. Examination and storage of a pulse take approximately 30 μsec and then the analyzer is ready to accept the next pulse. In perhaps half a minute the analyzer has collected and stored a sufficient number of pulses to give a very complete picture of the whole X-ray spectrum reaching the detector. Upon command, the analyzer

*Reprinted by permission from Analytical Chemistry 35:778, June, 1963.

displays the energy spectrum on a cathode-ray tube for visual examination and/or prints out the number of pulses of each amplitude numerically or on a digital tape or plots the spectrum automatically on an X-Y plotter. The possible applications to X-ray spectrochemical analyses are unlimited.

This paper describes the initial experiments with a multichannel analyzer attached to the electron probe microanalyzer at the Naval Research Laboratory. Three types of use are discussed: (1) replacement of a scanning crystal spectrometer for quick identification and semiquantitative analysis of unknown precipitates or inclusions; (2) quantitative analysis of multicomponent systems by unfolding of the overlapping energy spectra mathematically or graphically; and (3) quantitative two-dimensional topographic display from beam-scanning operation of the electron probe.

EQUIPMENT AND OPERATION

The multichannel analyzer used is the TMC Model 404, manufactured by Technical Measurements Corp., New Haven, Conn. This is a completely transistorized compact unit with excellent versatility. The memory consists of 400 channels that can be used for detailed coverage of the energy spectrum or divided into four sets of 100 channels each for storage and comparison of four distinct spectra. This latter appears to be the most useful for the electron probe, because the energy range from, for instance, 1 to 20 keV can usually be encompassed with sufficient detail in a 100-channel memory. In fact, it corresponds to the detail achieved with a one-volt window setting in the standard X-ray pulse-height analyzer circuits. The average storage time per pulse is 32 μsec and total counting rates of 20,000 to 30,000 counts/sec are easily accommodated.

Operation of the equipment with the electron probe is completely straightforward. Either a sealed-off or a flow-proportional counter is used with standard preamplifier to read the full X-ray signal (no crystal spectrometer) from any convenient port in the electron probe. The output signal goes directly to the multichannel analyzer, where the gain is adjusted as desired to position the spectrum within the 100-channel memory. For instance, the FeK_α radiation may be peaked at about channel 25 for measurement of the middle range elements. By recording the spectra of several pure elements, relationships between channel number, peak intensity, and atomic number may be obtained for a given set of electron-probe operating conditions.

Figure 1 shows the results for elements from Ti to Ga for 26-keV electrons, 6° take-off angle, and a mica window, xenon-filled proportional counter. With the electron probe striking the desired specimen area, data are collected for some chosen "live" time, usually 1 min or less, and the energy spectrum is displayed on the cathode-ray tube. "Live" time means that the analyzer measures time only during the intervals when it is ready to accept a pulse. Thus, the inconvenient dead-time corrections necessary with ordinary circuitry are eliminated. The first spectrum may be stored in the memory and the instrument switched to the second set of 100 channels as the electron probe is moved to another position on the specimen. After four such spectra are collected, they may be displayed simultaneously for visual comparison as shown in Fig. 2.

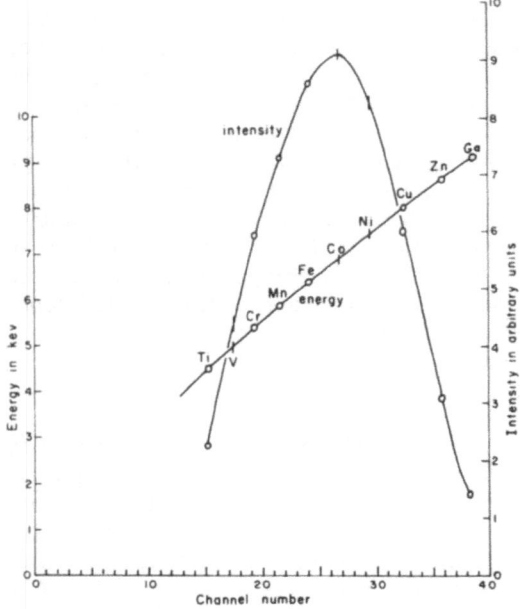

Fig. 1. Peak energy and intensity of X-ray spectra from Ti, Cr, Mn, Ge, Cu, and Zn measured for 100% standards. Values for V, Co, Ni, and Ga interpolated on basis of known excitation potential.

Another mode of operation possible at the turn of a switch is illustrated below and called multiscaler operation. Here the input stage of the instrument acts as a single-channel analyzer and allows the operator to select any energy increment in the spectrum. The memory channels then act as time increments—that is, all the pulses in the selected energy increment are collected in the first memory channel during the first time increment, all those collected in the second time increment are stored in the second memory channel, etc. The sequencing is automatic, so that one may observe the

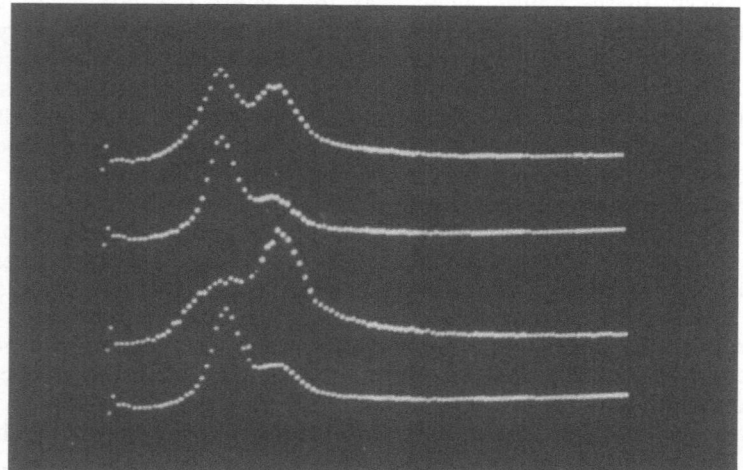

Fig. 2. Cathode-ray tube display of four distinct energy spectra. Spectra from four different precipitates in Type 7075 aluminum alloy, showing varying amounts of Fe, Cu, and Zn.

distribution of, say, FeK_α with time as one scans across the specimen continuously.

APPLICATIONS

Replacement of Scanning Spectrometer

One of the many uses of the electron probe is identification of unknown precipitates or inclusions in metals or minerals. Ordinarily this requires a scanning spectrometer rather than fixed X-ray optics, because one does not know the elements in advance. Even with a rapid scanning rate of 5° 2θ/min for rough identification of major components, it takes about 15 min to cover the usual range of elements from Ti to Mo. With the multichannel analyzer a total counting time of 0.4 min is sufficient for identification of all major constituents simultaneously. It is not that one cannot afford to take 15 min for a determination if he has only a few determinations to make; rather, it is that he can learn so much more if he can make 20 to 25 different identifications in 15 min instead of a single determination.

To illustrate the identification procedure several commercial aluminum alloys containing 2 to 3% added Ga were examined. Table I shows the constituents of the alloys and Fig. 3 shows energy spectra for typical precipitates in each alloy. From the calibration curve of Fig. 1, one can easily identify the major constituents in each case (Mg and Al were not detectable with the operating conditions and detector used, but could be measured with a flow detector) and can make a rough estimate of relative intensity of the constituents to one another. The 24 ST alloy, which contains over 4% Cu, does not show Cu in the precipitates; instead, it shows Mn, which is present only at the 0.5% level, along with Ga, which is present at 2 to 3%. Similarly, the 75 S alloy, which contains about 5% Zn, shows no Zn in the precipitates; instead, it shows Cu along with the Ga. In the 5052 alloy most precipitates showed primarily Ga, although occasionally one was found with Fe in addition to Ga; there is no indicated Fe content for the alloy.

TABLE I

Composition of Precipitates in Aluminum Alloys Containing Added Gallium

Alloy type*	Av. composition, %	Precipitate composition
75S	Zn 5.5, Mg 1.5, Cu 0.3, Cr 0.3, Ga 2 to 3	Ga plus Cu
24ST	Cu 4.5, Mg 1.5, Mn 0.6, Ga 2 to 3	Ga plus Mn
5052	Mg 2.5, Cr 0.25, Ga 2 to 3	Ga or Ga plus Fe

*Gallium added.

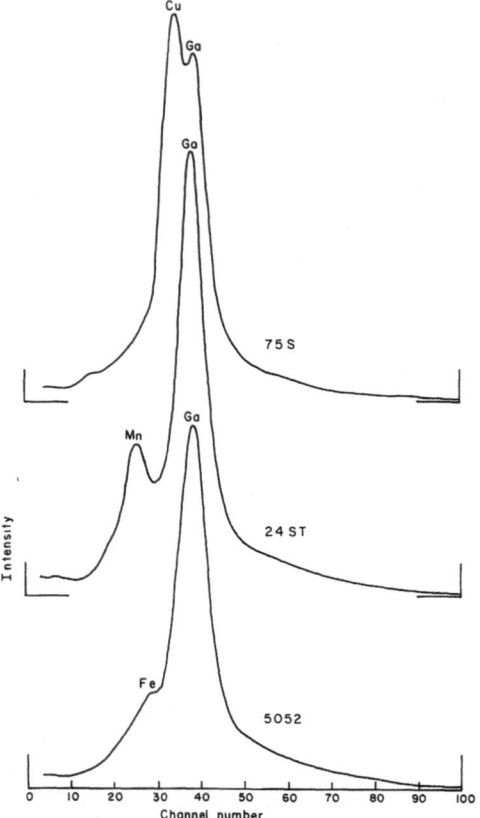

Fig. 3. Energy spectra from precipitates in aluminum alloys containing Ga (see Table I).

Quantitative Analysis by Unfolding of Overlapping Spectra

In the spectra shown in Fig. 2, the pulse amplitude distributions from neighboring elements were not completely resolved (this is always the case when proportional or scintillation detectors are used without crystal spectrometers). Thus the total intensity at the copper position is the sum of the copper intensity plus the fractional contributions from the iron and zinc pulse-amplitude distributions at the copper position. Dolby [2] has shown that a set of linear simultaneous equations may be written for the intensities at the positions corresponding to each element. The set of equations for the individual components is then solved by usual algebraic methods. Each of the simultaneous equations takes the form

$$I_i = I_{ii}R_i + \sum_j R_j I_{jj} P_{ji}$$

where I_i is the total intensity (less background) at the position of element i as measured in an unknown composition); I_{ii} is the intensity from 100% standard of element i at the position of element i; I_{jj} is the intensity from 100% standard of any other element j at the position of element j; P_{ji} is the fractional intensity of element j at the position of element i; R_i, R_j, etc. are the true relative intensities from element i, j, etc., in an unknown compo-

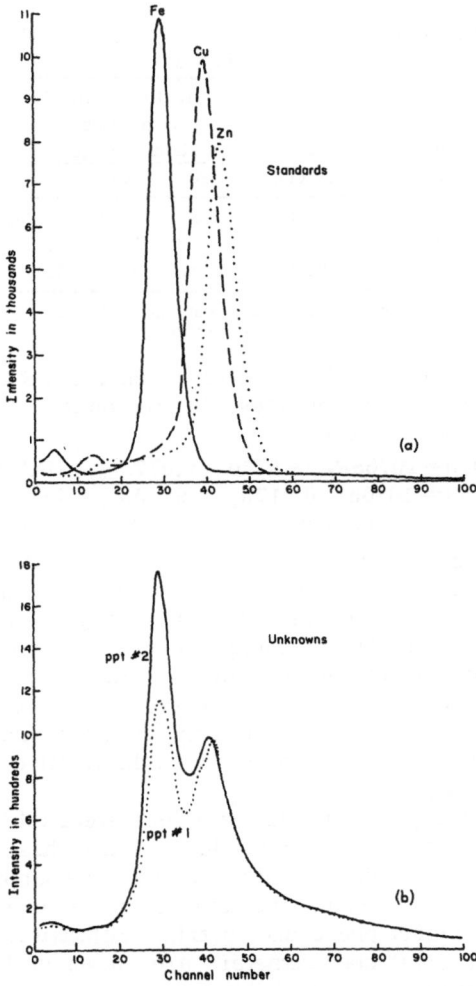

Fig. 4. Quantitative intensity measurements from Fe, Cu, and Zn standards and from two precipitates in 7075 aluminum alloy.

sition. These are the desired values for use in quantitative analysis and are the solutions of the simultaneous equations.

To illustrate practical application of the unfolding procedure we use precipitates in 7075 aluminum alloy (5.8%Zn, 2.3%Mg, 1.7%Cu, 0.2%Fe, 0.2%Cr, balance Al). These contain iron, copper, and zinc, all of which overlap each other. First, individual spectra from 100% standards of the three elements (Fig. 4a) were recorded using the multichannel analyzer and the fractional intensities at the Fe, Cu, and Zn positions were measured (the fractional intensity of an element at its own position is necessarily unity). Table II shows the results. Next, the spectra from two unknown precipitates (Fig. 4b) were recorded and the total intensities were determined at the Fe, Cu, and Zn positions as shown in Table III. The desired

TABLE II

Intensity Distributions from Fe, Cu, and
Zn Standards

Element standard	Peak intensity above background*	Fractional intensity at position of element indicated		
		Fe	Cu	Zn
Fe	10765	1.0	0.017	0.003.
Cu	9730	0.077	1.0	0.559
Zn	7845	0.069	0.506	1.0

*Total counts collected in channel corresponding to
peak of distribution in 0.4-min time.

values to be found are the relative X-ray intensities, R_{Fe}, R_{Cu}, and R_{Zn}, for
the three elements in the precipitates. In precipitate 1 they are found from
the three simultaneous equations:

Iron: $1015 = 10765 R_{Fe} + 9730 \times 0.077 R_{Cu} + 7845 \times 0.069 R_{Zn}$
Copper: $659 = 10765 \times 0.017 R_{Fe} + 9730 R_{Cu} + 7845 \times 0.506 R_{Zn}$
Zinc: $680 = 10765 \times 0.003 R_{Fe} + 9730 \times 0.559 R_{Cu} + 7845 R_{Zn}$

Solution by determinants gives $R_{Fe} = 8.8\%$; $R_{Cu} = 4.3\%$; $R_{Zn} = 5.6\%$. Similarly, for precipitate 2 in which the right side of the above equations remains the same, the solution gives $R_{Fe} = 14.3\%$; $R_{Cu} = 5.7\%$; $R_{Zn} = 4.0\%$.

As a check on the unfolding technique, crystal spectrometer data for the three standards and the precipitates were obtained simultaneously with the multichannel analyzer data. Relative X-ray intensities measured directly with the crystal spectrometers are compared with the unfolding data in Table IV. The agreement is excellent, with the relative standard deviation only ±4% of the amount present.

As a second example, Type 317 stainless steel (63%Fe, 18%Cr, 13%Ni, 3.2%Mo, 1.5%Mn) was treated to produce sigma-phase precipitates and examined in the same fashion. Here the three overlapping elements were Cr, Fe, and Ni. Figure 5 shows the standard pulse amplitude distributions and those of the sigma phase and matrix. Relative X-ray intensities by unfolding and by crystal spectrometers are shown in Table V. Again the agreement is excellent.

On a theoretical basis it would be possible to perform the above experiments with a regular single-channel pulse-height analyzer. One would scan first the energy spectrum of each of the standards and then the spectra of the unknowns. The same mathematical treatment would apply. From a

TABLE III

Intensities from Precipitates of
7075 Aluminum Alloy

Ppt.	Fe	Cu	Zn
(Measured intensity* at position indicated)			
1	1015	659	680
2	1610	740	630

*Total counts collected in indicated channel
in 0.4 min.

TABLE IV
Unfolding and Crystal-Spectrometer
Data for Precipitates of Table III

| | (Relative X-ray intensities) | | | |
| | Unfolding, % | | Crystal spectrometer, % | |
Element	Ppt. 1	Ppt. 2	Ppt. 1	Ppt. 2
Fe	8.8	14.3	8.6	14.2
Cu	4.3	5.7	4.2	5.8
Zn	5.6	4.0	4.9	4.0

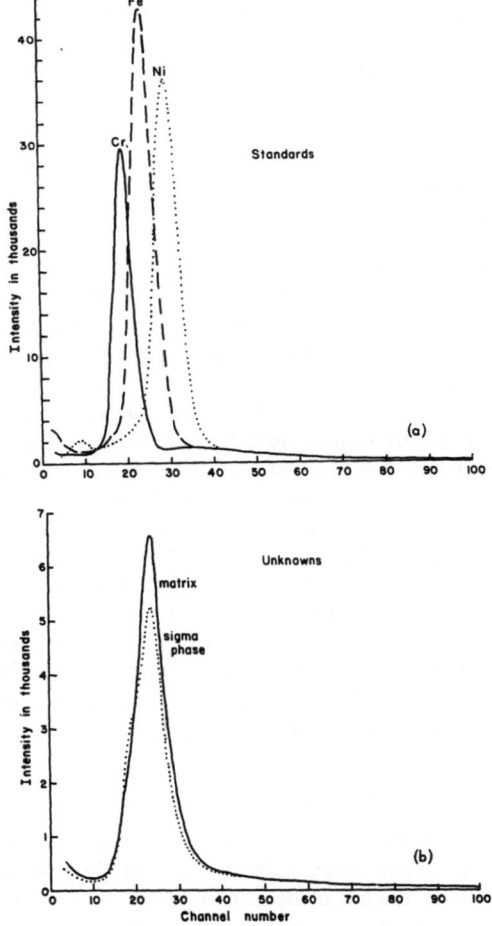

Fig. 5. Quantitative intensity measurements from Cr, Fe, and Ni standards and from sigma-phase and matrix in Type 317 stainless steel (see Table V).

TABLE V
Unfolding and Crystal-Spectrometer
Data for Type 317 Stainless Steel

	(Relative X-ray intensities)			
	Unfolding, %		Crystal spectrometer, %	
Element	Sigma phase	Matrix	Sigma phase	Matrix
Cr	20	13.2	19.5	12.5
Fe	29.4	35.8	29.0	35.0
Ni	4.9	7.0	Not measured	

practical standpoint, however, such a procedure would hardly be feasible, because each spectrum would require at least 10 min (at comparable resolution and statistical precision) even if only twenty one-volt increments were used to cover just the peak of the distributions. This would mean a total of 50 min for three standards and two unknowns, and during that length of time the operating conditions of the electron probe would be likely to shift somewhat. With the multichannel analyzer, the whole process of data collection for the same three standards and two unknowns takes about 3 min.

Unfolding the data from systems of four or more elements is just as straightforward as for three elements, but the number of simultaneous equations increases and is equal to the number of elements. Solution of determinants of more than 3×3 terms is tedious, however, except by computers. From the degree of overlap observed for adjacent elements it can be said that one would usually not need more than five terms in any equation, because the contribution from third neighbors is negligible unless one of the elements is a major constituent and the others are minor constituents (for a major constituent the tails of the distribution might be strong enough to interfere with minor constituents four or five elements away). Actually, electronic circuits available with the multichannel analyzer allow the operator to simulate an unknown spectrum by adding together arbitrary fractions of any number of component spectra. In the summation, a separate background spectrum should be used as one of the components. Preliminary tests with the aluminum and stainless steel specimens gave erratic results because the background was not treated separately. Tests with spectra containing similar peaks but no background intensity gave reproducibility about as good as the mathematical unfolding.

Quantitative Topographic Display

Most present electron probes have provision for automatic electron-beam sweeping and cathode-ray tube display of the specimen in terms of a selected element or in terms of electron current. Figure 6a shows a display of one of the precipitates in the 7075 aluminum alloy of the previous section in terms of Fe K_α radiation. Although the whole precipitate is richer in iron than the matrix, there seems to be a variation in iron content within the precipitate. Because of the small total size of the precipitate and the

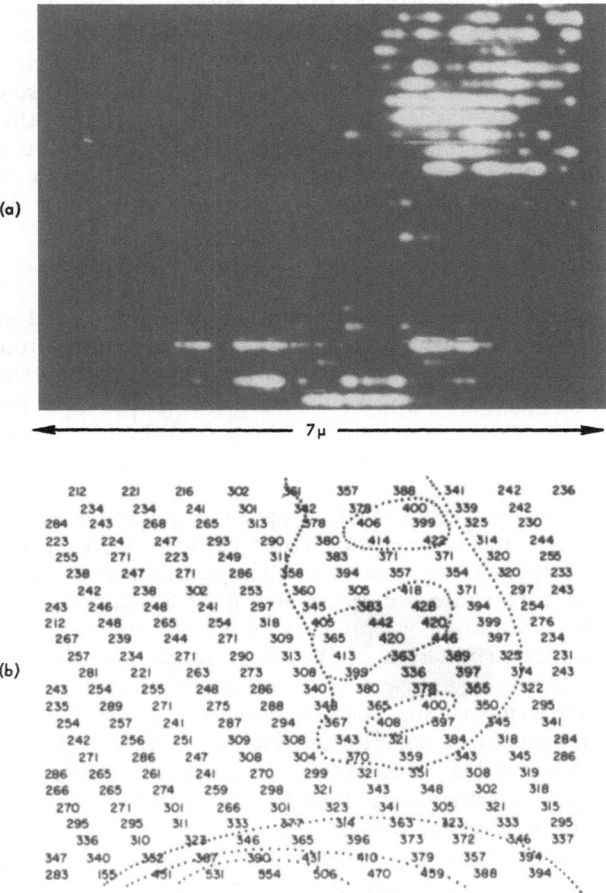

(a)

|← 7μ →|

(b)

```
212   221   216   302   361   357   388   341   242   236
  234   234   241   301   342   378   400   339   242
284   243   268   265   313   378   406   399   325   230
223   224   247   293   290   380   414   422   314   244
255   271   223   249   311   383   371   371   320   255
238   247   271   286   358   394   357   354   320   233
  242   238   302   253   360   305   418   371   297   243
243   246   248   241   297   345   383   428   394   254
212   248   265   254   318   405   442   420   399   276
267   239   244   271   309   365   420   446   397   234
  257   234   271   290   313   413   363   389   325   231
  281   221   263   273   308   399   336   397   374   243
243   254   255   248   286   340   380   378   365   322
235   289   271   275   288   348   365   400   350   295
254   257   241   287   294   367   408   697   345   341
  242   256   251   309   308   343   321   384   318   284
  271   286   247   308   304   370   359   343   345   286
286   265   261   241   270   299   321   351   308   319
266   265   274   259   298   321   343   348   302   318
270   271   301   266   301   323   341   305   321   315
  295   295   311   333   377   314   363   323   333   295
  336   310   323   346   365   396   373   372   346   337
347   340   362   387   390   431   410   379   357   394
283   155   451   531   554   506   470   459   388   394
```

Fig. 6. Topographic display of precipitates in 7075 alloy in terms of
Fe$K\alpha$ radiation: (a) using automatic beam sweep in electron probe;
(b) quantitative topographic display of same area using multichannel
analyzer in multiscaler mode of operation.

low total concentration of iron, it is difficult to reach any quantitative conclusions about the variations.

The multichannel analyzer allows the display to be placed on a more
quantitative basis by use of the multiscaler mode of operation described in
the section on Equipment and Operation. First the input stage was set to
select a small energy increment around the FeK_α peak. Then the analyzer
was set so that each channel would collect all the Fe quanta detected in 1 sec.
The electron beam sweep was arbitrarily set for 30 lines at 10 sec/line,
giving a total of 300 sec to cover the desired area and corresponding to 300
channels in the memory. Readout of the memory is by electric typewriter
set to print 10 columns across corresponding to one scan line. Figure 6b
shows the topographic display with contours drawn in to delineate the Fe-
rich regions. Of course, the contours do not represent sharp concentration
boundaries because of the resolution of the electron probe. However, they

do represent real variations within the precipitate, as may be illustrated by considering the two shaded areas of Fig. 6b drawn to include six readings each and each area corresponding to about 1 square micron. The sum of the readings in one area is 2540 counts, giving a standard deviation of ±50 by usual X-ray statistics. The sum of the readings in the other area is 2218 counts, with a standard deviation of ±47. The difference between the two sums is 322 counts or a difference of more than 6σ, well outside of any possible random fluctuations.

DISCUSSION

One of the shortcomings of both crystal spectrometers and single-channel analyzers is that they measure only a small fraction of the total information available at any one time—that is, they ignore all but one wavelength or one energy increment and effectively waste all of the other radiation. On the other hand, the multichannel analyzer uses all the information reaching the detector. Thus, there is considerable saving in time necessary to collect the same number of data.

From the initial tests run at NRL, it appears that the greatest advantage of the multichannel analyzer comes from using it in conjunction with the regular X-ray optics; that is, one port of the electron probe is used for the multichannel analyzer while the others are used for crystal spectrometers with their better resolution of elements of special interest. Thus one has quick quantitative determination of the most important elements from the spectrometers plus the assurance that no other elements of interest are overlooked. Furthermore, the spectra collected by the multichannel analyzer may be placed on a quantitative basis whenever appropriate by the mathematical unfolding described above. It is difficult at this early stage to evaluate all the possible implications of the quantitative topographic display, but it is certain to be of great value.

For the low-atomic-number elements from Na to C, where intensities are low and suitable crystals are difficult to obtain, the application of the multichannel analyzer and the unfolding technique is probably even more important than for the middle-range elements, but data for light elements are not available as yet for test. The number of light elements is limited, so that few equations are needed for the unfolding process when only light elements are present. If heavier elements are also present, their L or M spectra may overlap the region of interest and complicate the problem of unfolding. The proper detector is, of course, a flow-proportional counter with front and back windows, so that harder radiation will pass through without being detected. Although not tested as yet, it seems likely that a sealed-off xenon proportional counter can be used in back of the flow counter to detect the harder radiation that passes through the flow counter. Signals from both detectors would go to the multichannel analyzer for simultaneous display of the full energy spectrum.

One question always to be considered in adding new equipment is the cost. Although prices vary considerably, the cost of a curved crystal scanning spectrometer and necessary associated electronics seems to be in the range of $8000 to $12,000. The basic equipment necessary with the multichannel analyzer costs about $15,000 and other desirable equipment brings the cost to about \$20,000. For the versatility alone the added expense is well justified. If the time savings are considered, the multichannel analyzer more than pays for itself.

ACKNOWLEDGMENT

We thank J. L. Jamison of Q.E.D. Electronic Sales for numerous helpful comments during our initial adaptation of the multichannel analyzer for soft X-ray measurements.

REFERENCES

1. L. S. Birks, X-Ray Spectrochemical Analysis (Interscience, New York, 1959).
2. R. M. Dolby, Proc. Phys. Soc. 73:81 (1959).
3. H. A. Liebhafsky, H. G. Pfeiffer, E. H. Winslow, and P. D. Zemany, X-Ray Absorption and Emission in Analytical Chemistry (John Wiley and Sons, New York, 1960).

The Development and Use of a Semimicro X-Ray Fluorescence Attachment

W. J. Wittig

Union Carbide Corporation
Stellite Division
Kokomo, Indiana

A simple, inexpensive pinhole collimating device and a movable sample stage were designed for use in X-ray fluorescence equipment. A variety of pinhole sizes permits analyses of minute areas from $1/4$ in. square down to 125 μ square. The construction and operation of the unit are described in detail. Three examples are reported to illustrate the versatility of the unit for qualitative analyses.

The desire to do small-area analyses has always been present, but until the last ten years the facilities and equipment have not been available. With the advent of the electron-probe microanalyzers, analyses have progressed to microscopic areas with excellent results, and micron-size area analyses that were merely dreams a few years ago are now possible. Two problems have grown with this development, however: one is the high cost of equipment, and the other is the inadequate coverage of the gap between gross and micron-size areas. Equipment to analyze this intermediate size is the topic of this paper.

Small areas will be defined as 100 μ up to 2 mm in diameter. Types of samples that fit in this category are minute single crystals, grains in an alloy, inhomogeneities within an alloy, surface inclusions, plated metals on a base material, and a general class consisting of any other sample that can be mounted for metallographic study.

The ideas for a semimicroprobe fluorescence attachment came from two sources. In 1960 General Electric Company advertised the "Heinrich" Miniature Probe [1] for analyses of samples 50 μ, 100 μ, and 1 mm in diameter. Laboratories not having General Electric Company X-ray equipment were at a disadvantage, since the unit would only fit this equipment. A unit similar in design and adaptable to Philips equipment was discussed by Mr. R. D. Sloan at the 1961 Denver Research Institute X-ray conference [2], but his design did not satisfy the equipment versatility and sample motion required at the Union Carbide Stellite laboratory. Therefore, the basic ideas of the two units were modified, and the results will speak for themselves.

Philips Electronics, Inc., equipment (Fig. 1) has been used for both fluorescence and diffraction analyses for nine years in the Technology Department. Permanent modifications for semimicrofluorescence analyses cannot be made to the equipment because of the variety of specimens analyzed on these units. For this reason, a semimicroprobe was developed that could be inserted and removed when necessary.

Fig. 1. X-ray diffraction and X-ray fluorescence equipment used in the Union Carbide Stellite laboratory.

This unit consists of three parts (Fig. 2): a collimator with removable apertures, a collimator chamber, and a movable-stage sample holder. The apertures are the ink wells used in Leroy lettering pens, which are used by many draftsmen and are available at any office supply store. The collimator chamber that holds the apertures (Fig. 3) is wedge-shaped to permit adjustment in the vertical direction, but designed with no tolerance for lateral motion. The chamber is bored on center to permit the apertures to extend out the end and come into close proximity to the sample surface (Fig. 4). The collimator housing (Fig. 5) is the same one used in the standard equipment, except for two minor changes. A groove has been milled on the inside of the bottom plate of the housing for an inset leaf spring used to maintain vertical tension. The vertical adjusting mechanism (Fig. 6) is a screw that presses down on the collimator chamber through a hole in the housing. The sleeved screw is the only permanent change made in the basic X-ray unit, and it can be retracted to eliminate any interference during routine large-area analyses.

The sample stage incorporates a movable sample chamber and a micrometered handle to measure the sample chamber motion (Fig. 7). The stage platform, which is similar to the standard sample stage except for the movable sample chamber which rides on rails within it, will accommodate samples up to $1\frac{1}{4}$ in. long by $1\frac{1}{8}$ in. wide by 1 in. high. One and one-eighth-inch plastic metallographic mounts are used for very small specimens so that they can be conveniently positioned under the X-ray beam. The handle was prepared by cutting the frame from a micrometer, and using the

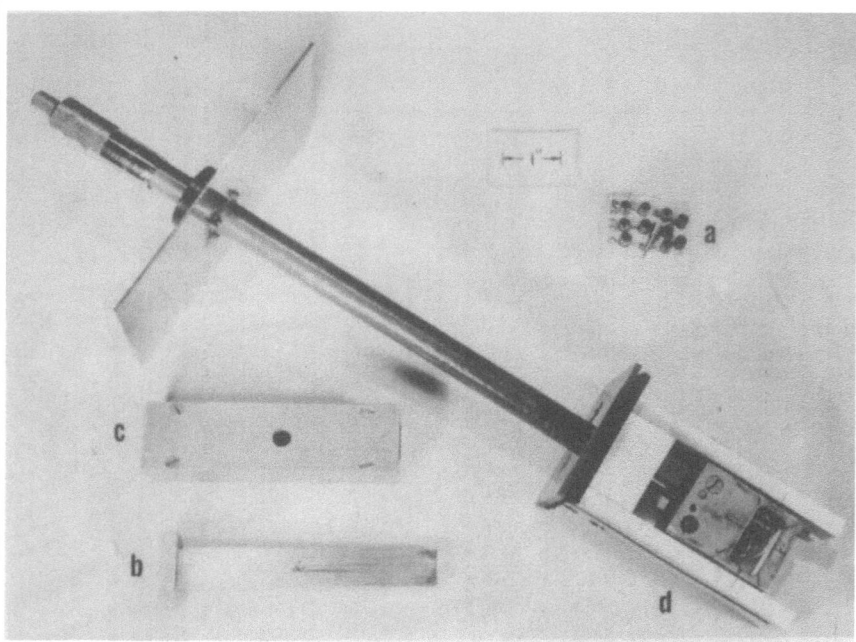

Fig. 2. Components of the removable X-ray fluorescence semimicroprobe: (a) Leroy lettering pen apertures, No. 000 to No. 8; (b) center-bored collimator; (c) collimator housing (formerly the collecting collimator; (d) movable stage sample holder with a micrometer handle.

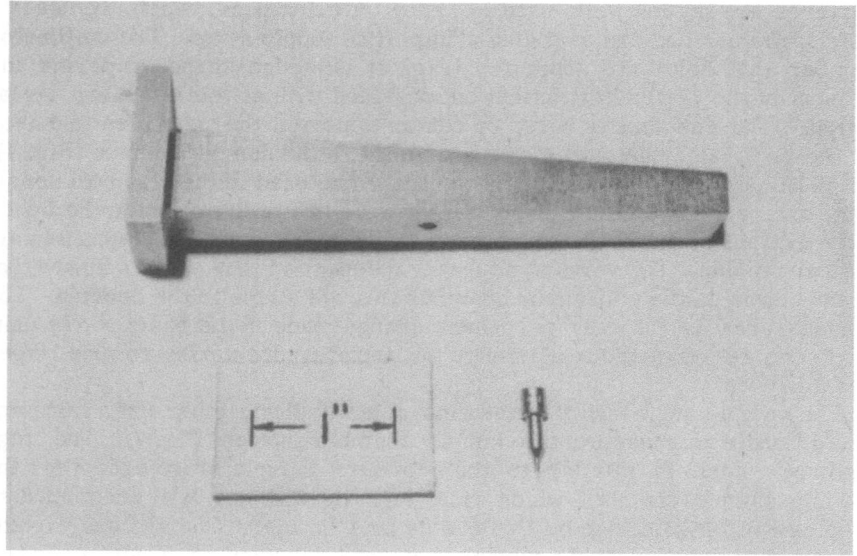

Fig. 3. Collimator with a typical aperture. (Note the wedge shape design.)

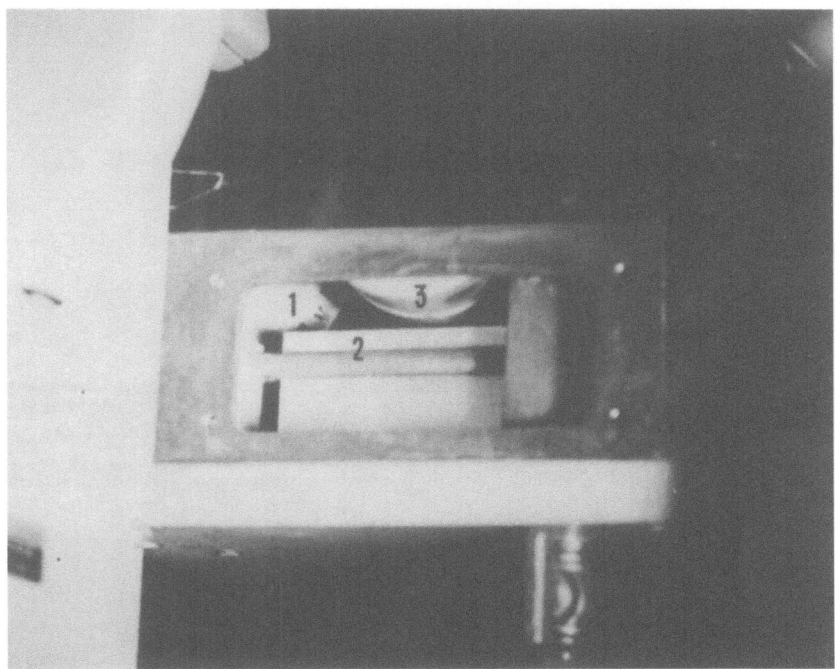

Fig. 4. The X-ray fluorescence semimicroprobe equipment in position, showing the proximity of the aperture (1) to the specimen (2) and X-ray tube (3).

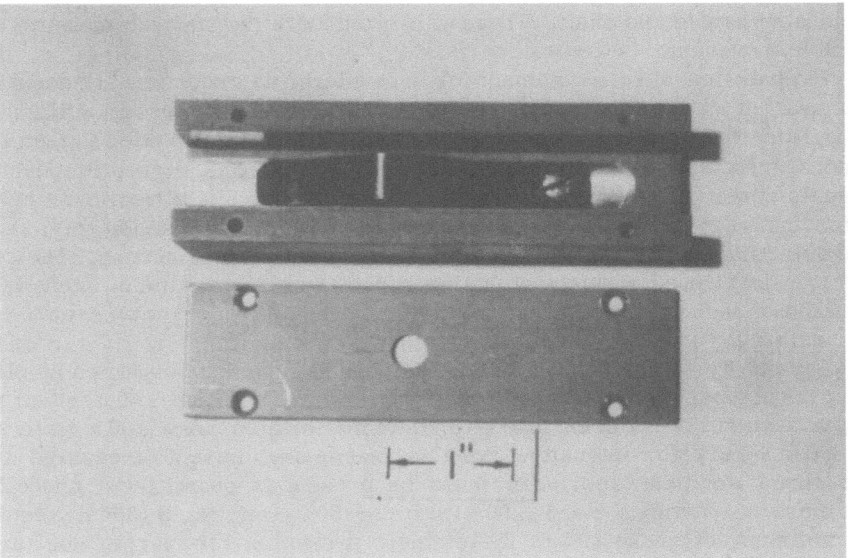

Fig. 5. View of the collimator housing with one side removed. The leaf spring, which is set into a milled groove, exerts continuous pressure on the collimator.

Fig. 6. Adjusting screw mounting. The one permanent change in the equipment is the drilled and tapped hole in the helium chamber. A brass hex-head bolt was drilled and tapped to accommodate the adjusting screw.

graduated handle and shaft. These were attached to the stage to measure the sample movement.

Preparation of the equipment for semimicrofluorescence work requires removal of the crystal holder so that the collimator housing, collimator tube, and the penpoints can be inserted. The crystal holder is then replaced, the unit is turned on, and the equipment is aligned to measure signals at the known 2θ angle of a standard element. Slight movement of the microfocus lever is sometimes required to peak the signal sensitivity, since the detecting crystal was moved from its original precise setting. The collimator is aligned with the adjusting screw to obtain maximum intensity at the known angle. The standard is then removed and an unknown sample can be analyzed.

Sensitivity of the unit using a helium-filled system was studied by plotting counts per second (measured by a flow-proportional counter) vs. the aperture sizes. As seen in Fig. 8, the smaller apertures, 100 μ up to 225 μ, have a very low intensity. For this reason any elements measured with the three smallest apertures must be present in quantities of about 25% or more. Apertures No. 1 (200 μ), No. 2 (250 μ), and No. 3 (300 μ) require a minimum of about 10% of the element present and the larger apertures are relatively less restrictive.

The capabilities of the unit can be described best by three examples taken from the laboratory analyses.

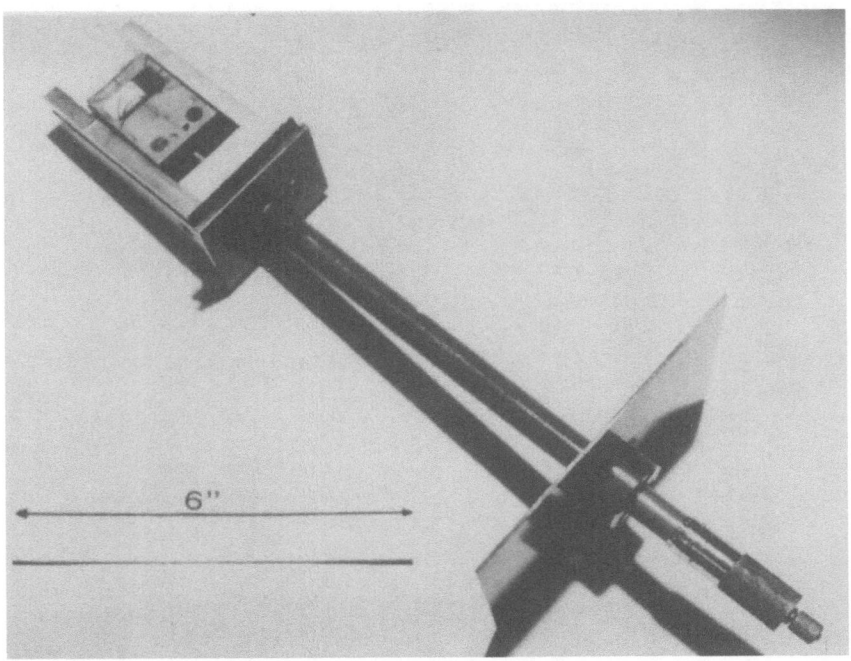

Fig. 7. Sample stage showing the movable chamber and the micrometer handle.

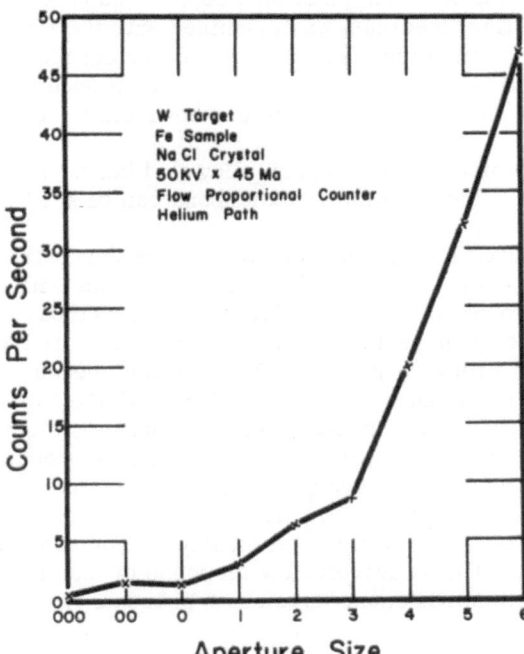

Fig. 8. A relation of aperture size to counting rate for the various apertures of the X-ray semimicroprobe.

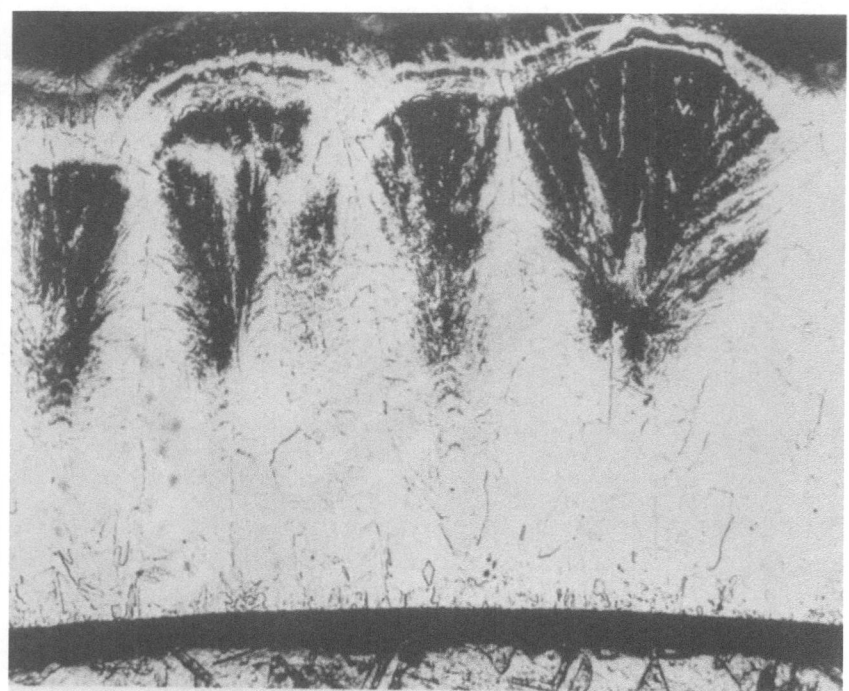

Fig 9. Magnification: 100×. Etchant: Hf + HNO₃+ H₂O (immersion). Fan-shaped areas of discoloration in the $1/_{64}$-in. layer of columbium on copper.

1. Waves of discoloration that could not be explained were visible near the outer edge of a $1/_{64}$-in.-thick columbium plate on a copper rod (Fig. 9). A transverse cross section of this specimen was examined with the 150-μ aperture (No. 0), and three contaminating elements were detected: nickel, iron, and chromium. Examination of the adjacent copper core proved that the elements had come from the plating system and not from the core material.

2. The second study was of a nickel—molybdenum alloy clad between two layers of a second nickel-base alloy. An interlocking bond can be seen in the photomicrograph (Fig. 10) of the clad sheet.

Alloy identification of all three layers and relative element concentrations were determined by this study. Also, the measuring accuracy with a 0.005-in./min stage travel was found to be comparable to optical methods at 250×. Thus a specimen surface can be measured and compared with the X-ray strip chart recording in order to precisely pinpoint any measured variations. A semimicroprobe examination of the clad alloy with the 125-μ aperture (No. 00) showed a gradual change in nickel content caused by slight diffusion at the interface of the two alloys (Fig. 11). Iron and molybdenum were analyzed by scanning at the rate of 0.005 in./min. Iron was detected in the cladding layers (18%), but not in the middle layer (5%). Although the molybdenum content of the two layers differs by nearly 20%, it was not responsive to the X-ray beam. The relatively low response of the argon gas in the counter to the short molybdenum wavelength accounts for this phenomenon.

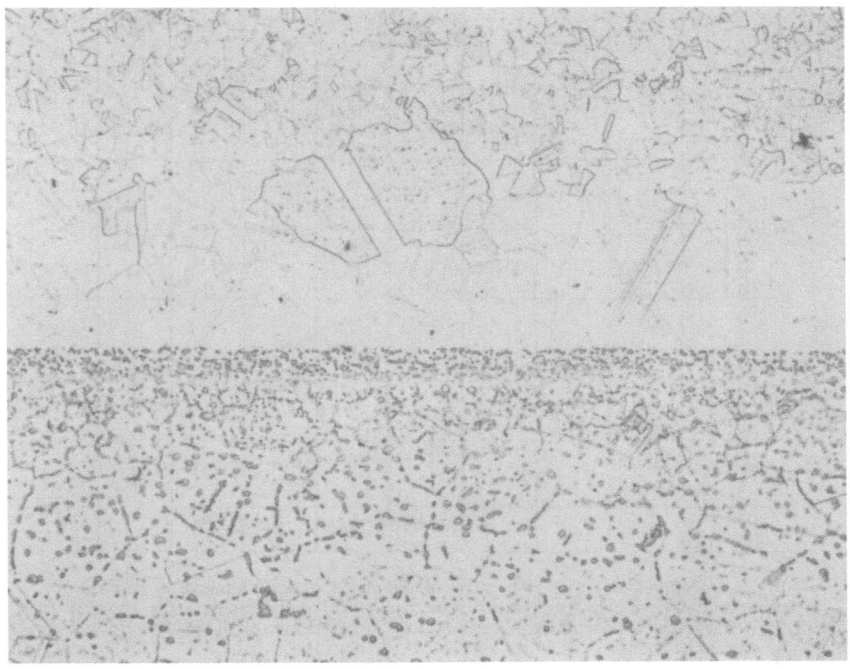

Fig. 10. Magnification: 100×. Etchant: Chrome Regia (3 sec) Hastelloy alloy B. The interface shows a fine dispersion of one alloy in the other, which acts as an interlocking bond. (Hastelloy is a registered trademark of the Union Carbide Corporation.)

3. A thin sheet (0.063 in. thick) of a cobalt-base alloy was submitted for a semimicroprobe analysis. Tearing appeared in circular areas (nearly $1/4$ in. diameter) over a small portion of the sheet. Four elements—nickel, tungsten, chromium, and cobalt—were examined by traversing the specimen under the $200\,\mu$ (No. 1) aperture while keeping the detecting mechanism at the respective 2θ settings. Nickel and tungsten showed no variation across the area, while chromium and cobalt varied inversely to each other. Chromium was low at the point of tearing, while cobalt was higher than normal. At the other side of the torn $1/4$-in. area a cobalt-deficient zone was present. Segregation in the sheet was thus indicated.

Two directions are open to improve the sensitivity of the equipment. Slit mechanisms rather than pinhole collimators have been reported to improve intensity and resolution [3]. The other method of increasing intensity is to use a curved crystal for focusing the fluoresced X-rays on the detector.

The original purposes for designing this unit were to qualitatively analyze compositional differences between small adjacent areas and to relate the analyses to the point being examined by precise measurement of specimen motion. The semimicroprobe can do this when enough of an element is present to be detected through the small apertures. The unit is now a part of the standard equipment in the laboratory.

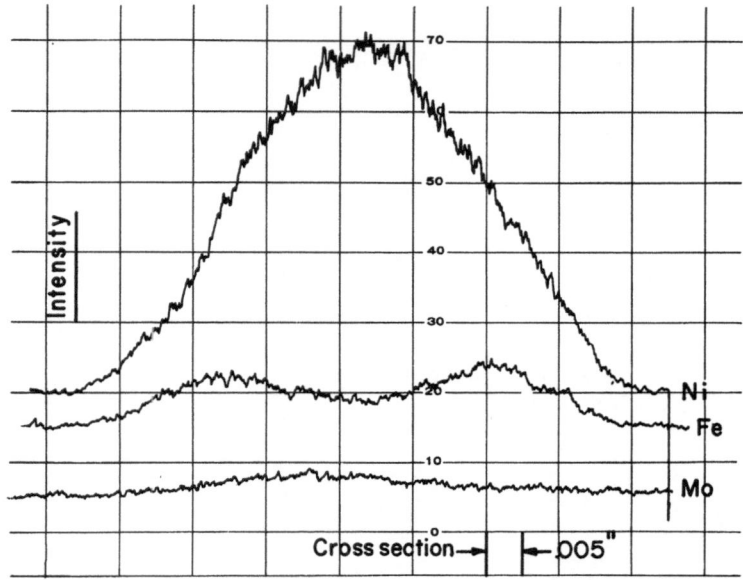

Fig. 11. A 0.005-in./min step scan of the three major elements in a nickel-chromium-iron-molybdenum alloy clad to a nickel-molybdenum alloy showing the relative sensitivity of the elements to the X-ray beam, when they are limited by small apertures.

REFERENCES

1. "New Heinrich Miniature Probe," General Electric Co., X-Ray Dept. Publication 8A-3862.
2. R. D. Sloan, Advances in X-Ray Analysis Vol. 5 (Plenum Press, New York, 1962), p. 512.
3. E. P. Bertin and R. J. Longobucco, Advances in X-Ray Analysis Vol. 5 (Plenum Press, New York, 1962), p. 447.

Process Stream Analysis by X-Ray Spectroscopy

R. H. Munch

Monsanto Chemical Company
St. Louis, Missouri

X-ray fluorescence equipment now available for continuous plant process stream analysis can simultaneously or sequentially analyze for several elements. The type of instrument chosen should depend on process conditions and requirements. Equally as important as the instrument is the sample-handling system. Maintenance of instrument and sampling system are also vital. Finally, optimum use of the data provided by the system must be made. A well-chosen, well-engineered application can have great value.

Usually, when the need for analytical data on in-process material first arises, it is met by sampling the material and taking the samples to the laboratory and making analyses by chemical methods. Often these methods are so time-consuming that the results are history instead of the basis for correcting the process conditions to give optimum results. The next step is likely to be the development of an instrumental analytical method. Such methods usually measure some physical property having a fixed direct relation to sample composition. They cut the time required per analysis but still leave the plant operators with too few data and too much delay between sampling and reporting of the results. When this has been done, the next improvement is to automate the instrument so that it can provide continuous analyses of the flowing process stream with minimum delay between sampling and availability of the data. The ultimate is to use the output of the analyzer for automatic control of the process, eliminating the need for continuous supervision by an operator. In some cases the addition of some form of computer will be required to calculate how much the process variables must be changed to bring the composition of the stream back to the desired one.

What are the factors which lead to the development and use of automatic process stream analyzers? It is a safe bet that the manager who must authorize the expenditure required to install any process stream analyzer will be seeking to increase the profitability of his operation through improvements in one or more of the areas listed in Table I. These boil down to improved product quality, increased production, lower cost, greater safety for personnel and plant, better understanding of the process, and, last but not least, the ability to point with pride to something new and more advanced than the competition has—something that helps make his plant more profitable than those of his competitors.

Naturally, the advantages of the installation must be great enough to outweigh the expenses. Table II lists the expenses involved. People have a habit of thinking only of the purchase price of the instrument. However, in making a realistic estimate of the price paid to achieve the final results desired from the analyzer many other items must be included. Some of these are the price

TABLE I
Advantages of Process Stream
Analyses

1. Improved product quality
2. More uniform product quality
3. Minimum off-quality product
   a. No storage of off-grade product
   b. No blending off
   c. No sale at distressed prices
4. Increased production
5. Reduced costs for
   a. Labor
   b. Raw materials
   c. Maintenance
   d. Capital
   e. Utilities
6. Safety of personnel and plant
7. Better understanding of the process
8. Advertising value

of sample-handling system components, installation costs, research in connection with start-up, operating costs, maintenance and repairs, interest on investment, and depreciation. Installation and research in connection with start-up can easily amount to more than the cost of the equipment.

The characteristics which a good process stream analyzer must have, to as high a degree as possible, are listed in Table III. Basically, these requirements are that the instrument must be sensitive only to the desired component of the sample, accurate, easy to calibrate and maintain, rapid in response, and not too expensive.

Against this background let us evaluate the advantages of X-ray methods as applied to process stream analysis. There are three types of X-ray measurements which are related to composition. The simplest of these is absorption. The law governing the absorption of X rays is Beer's law

$$\log (I_0/I_x) = kx$$

where I_0 is the initial intensity of the X-ray beam, I_x its intensity after passing through a layer of absorbing material of thickness x, and k is a linear absorption coefficient characteristic of the absorber and of the wavelength of

TABLE II
Cost of a Continuous Analysis System

1. Purchase price of instruments
2. Purchase price of sample handling system
3. Installation
4. Research in connection with start-up
5. Operating cost
   a. Utilities
   b. Standardization
6. Maintenance and repairs
7. Interest on investment
8. Depreciation

TABLE III
Desirable Process Stream Analyzer Characteristics

1. A fixed relation between measured variable and concentration of component sought
2. Insensitive to
   a. Temperature changes
   b. Line-voltage changes
   c. Vibration
   d. Dust
   e. Humidity
   f. Aging of instrument components
   g. Other normal sample components
   h. Sample condition
   i. Sample flow rate
3. High signal-to-noise ratio
4. Simple direct method of calibration
5. Rapid response
6. Maintenance simple and infrequent
7. Low first cost

the X rays. In X-ray work this is more often written in terms of the mass absorption coefficient μ, the sample density ρ, and the thickness difference Δx,

$$\log(I_{x_1}/I_{x_2}) = \mu \rho \Delta x$$

Where thickness measurements are being made it may also be written

$$\log(I_{x_1}/I_{x_2}) = \mu_x \Delta x$$

Here μ_x is a linear absorption coefficient. An important fact regarding the mass absorption coefficient of an element is that it is substantially independent of the physical state or state of chemical composition of the element. The absorption coefficient of an element increases smoothly with wavelength until an absorption edge is reached at a wavelength characteristic of the element, then drops sharply to a fraction of its value at the absorption maximum, and rises steadily again until the next absorption edge is reached.

These characteristics of X-ray absorption mean that by measuring this property at well-chosen wavelengths, the composition of a sample or its thickness may be determined. This method has been used for both purposes. An example of the former is the determination of lead tetraethyl in gasoline. This is a fortunate case where a wavelength can be chosen such that the mass absorption coefficient of lead is over one hundred times those of carbon and hydrogen, the major components of the sample. X-ray absorption has also been measured to give a continuous indication of the thickness of sheet metal as it is being rolled. Both these are good applications but, because there are comparatively few applications where the method can be made as specific as in the case of lead tetraethyl, there is at present no manufacturer of X-ray absorption process stream analyzers. Perhaps this method should be re-evaluated in the light of present needs and instrumentation capabilities to see if its capabilities could be extended by making measurements at more than one wavelength and using a simple computer to calculate the desired concentrations. The absorption method is attractive because of its relative simplicity and low cost.

Another possible method is X-ray diffraction. X-ray diffraction methods could, in theory, serve to analyze mixtures for any crystalline components. This theoretical capability is based on two facts. The first is that X-rays are diffracted by the atom planes in the crystal lattice of a substance according to the Bragg diffraction law $n\lambda = 2d \sin\theta$, where n is the diffraction order, λ the wavelength, d the distance between planes, and θ is the angle of diffraction. The second is that the crystal lattice of each substance is unique. In practice there are a number of limitations that make this perhaps the least used X-ray process stream analysis method. To date, it has been used only for laboratory analysis of process streams. One of the serious limitations is that the sample must be finely ground and then packed in a uniform manner with the powder grains in a random orientation. This is not easy to do as a laboratory operation and is extremely hard to accomplish as a continuous operation. Another limitation is that the degree of crystallinity of the sample may be variable. Also, the sensitivity of the method is likely to be low. These factors combine to result in accuracies of the order of ±10% of the component being measured. This is too poor to be useful for most purposes.

The third type of X-ray method suitable for determining the composition of flowing process streams is X-ray fluorescence. This method depends on the fact that when matter is irradiated by X-rays from a source emitting all wavelengths, a line spectrum having wavelengths characteristic of the elements present in the sample will be emitted. Since, as was discovered by Moseley, the X-ray spectrum of each element differs from those of all others, the X-ray spectrum of an element serves as a unique identification. The intensity of the fluorescence spectrum is a function of the concentration of the element emitting it if certain instrument and sample conditions are properly controlled. The method may be used for all elements above magnesium (atomic number 12). It has excellent sensitivity for all these elements except magnesium and the first few elements above it in atomic number. X-ray fluorescence has therefore been widely used as an analytical method, first in the laboratory, and more recently, in plants for continuous automatic analysis of moving process streams.

Figure 1 is a schematic diagram showing the essential components required to excite and measure X-ray fluorescence spectra. A sample is subjected to intense "white" radiation from the target of an X-ray tube in order to excite its characteristic spectrum. The primary collimator selects a parallel beam of the fluorescent radiation from the sample. This beam of radiation is resolved into its component wavelengths by a flat crystal mounted on a goniometer head so that the beam from the primary collimator falls on it. A secondary collimator and a detector are mounted on the goniometer arm, which is geared to rotate at twice the rate that the crystal rotates. The detector is usually a flow-proportional counter for elements of low atomic number or a sealed proportional counter or a scintillation counter for those of higher atomic number. A suitable power supply for the X-ray tube along with a suitable amplifier and scaling circuit complete the apparatus. By using dispersing crystals having different " d spacings" and by rotating the crystal and detector to a suitable angle, the instrument can be adjusted to select and measure the characteristic X-ray emission from a wide range of

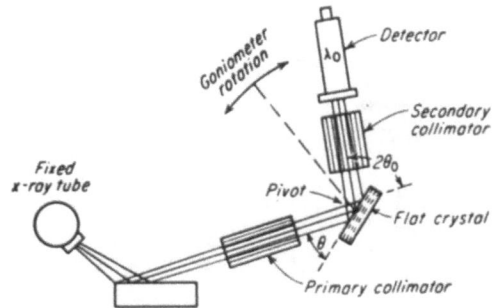

Fig. 1. Essential components for excitation and measurement of X-ray fluorescence spectra.

elements. If an analysis for low-atomic-number elements is to be performed, the X-ray path must be flushed with helium or hydrogen or evacuated.

To maintain a constant relationship between the intensity of the fluorescence radiation and the concentration of the element in the sample it is necessary to maintain the intensity of the exciting radiation constant. Sensitivity of detectors and amplifiers must also be constant. In addition, the sample must be uniform in physical condition. Homogeneous liquid samples present no problems, but slurries or emulsions must be kept uniformly suspended. Solid samples must be ground, preferably finer than 200 mesh, and packed uniformly with a smooth surface. Calibration must be carried out using a set of standard samples which are as like the samples to be analyzed as possible in order to cancel out the effect of the other elements on the one being measured. If elements other than those being measured vary over too wide a range, special measures to avoid the effect of this variation on the element being measured may be required. A method commonly used in laboratory work is fusion in a low melting flux which dissolves and dilutes the sample so that the effect of variations in physical form and chemical composition of the sample is reduced. Fortunately, a single plant process stream usually does not vary so much that such methods are needed; however, separate calibration might be needed for the feed, the concentrates, and the tailings from an ore flotation plant.

If a sample is to be analyzed for several elements, the goniometer must be set successively at the angles required to select the characteristic radiation of each, and an intensity measurement made for each one. This is the usual mode of operation for laboratory use, particularly where the results are not required for process control.

Another possible mode of operation is to use separate collimators, dispersing crystal, and detector for each element to be determined. This has two advantages. The first is that all the elements may be determined simultaneously. The second and more important one is that the parts for each spectrometer may be securely locked to a rigid base plate. This type of construction is less expensive than an adjustable goniometer and, even more important, it is more rugged, less subject to wear, and less likely to get out of adjustment from vibration.

For some applications, a rather different spectrometer arrangement is desirable. This type, shown in Fig. 2, uses a curved crystal mounted along with entrance and exit slits on a Rowland circle in such a way that the desired characteristic radiation is focused onto the exit slit and passes through

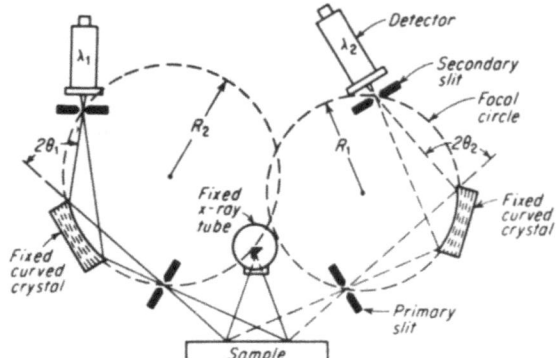

Fig. 2. Curved-crystal
spectrometer.

it to the detector. The advantage of the curved-crystal spectrometer over
one with a flat crystal is that it has a greater angular aperture and gathers
a greater fraction of the fluorescent characteristic radiation from the sample.
This is important when it becomes necessary to obtain high sensitivity.

As is shown in Fig. 2, a number of spectrometers each set to the characteristic wavelength of a particular element may be arranged symmetrically
around the sample so that a corresponding number of elements can be determined simultaneously. Of course, this can also be done with fixed-wavelength flat crystal spectrometers.

Figure 3 shows an interior view of the analyzer head of a Philips process
stream analyzer with two fixed-wavelength spectrometers mounted on it. As
many as six can be accommodated. Note the simple rugged construction.
With its cover on, the unit is well protected from dust and dirt, chemical
fumes, and moisture.

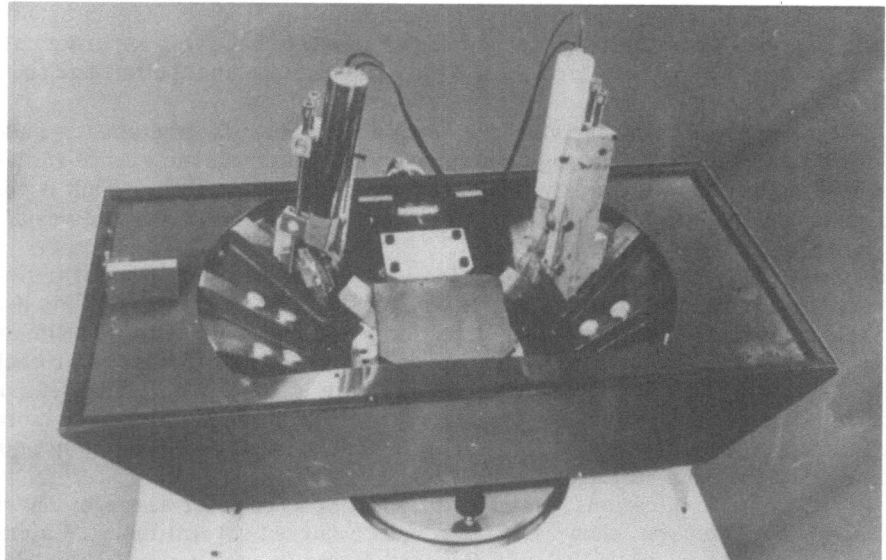

Fig. 3. Process stream analyzer head with two mounted fixed-wavelength spectrometers.

It should be noted here that manufacturers of commercial equipment of this type have gone to great lengths to ensure that their equipment cannot harm those who use it or work near it. Great care is taken to shield the unit so that harmful radiation cannot escape, and interlocks are provided so that no one can get any part of his body into the X-ray beam without shutting off power to the unit. The high-voltage system is well insulated and protected by cabinets provided with interlocks which shut off the power if the cabinet is opened.

As was mentioned earlier, instrument conditions, particularly intensity of the exciting radiation and sensitivity of detectors and amplifiers, must be maintained constant to keep a constant calibration. This is somewhat difficult to do with sufficient precision. These requirements may be made somewhat less stringent, or improved calibration stability may be achieved, by measuring the ratio of the fluorescent radiation to that of the exciting radiation. This automatically cancels out the effect of fluctuations in tube current and progressive loss of tube efficiency, and will also help minimize the effect of line-voltage variations and aging on the detector system. Another method for maintaining continuous calibration accuracy is automatic standardization at frequent intervals. One or both of these techniques is almost essential in a process stream analyzer if it is to give continuously reliable results.

When it is necessary to determine the percentage of some element in the solid portion of a slurry, the instrument calibration will depend on the ratio of solid to liquid. This ratio must therefore be kept constant, or else some method of compensating for variations in it must be developed. This could be done by measuring the density of the slurry discharged from the process

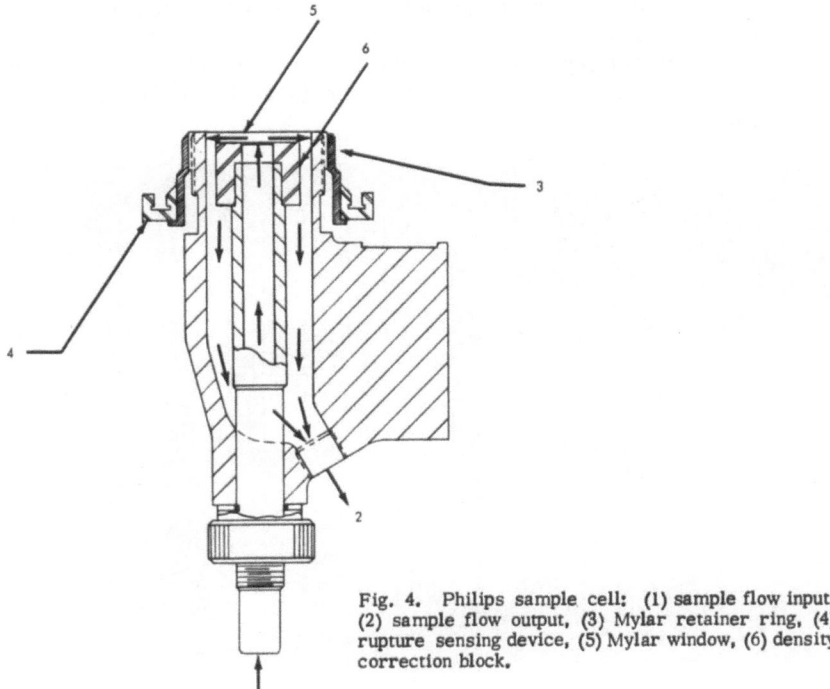

Fig. 4. Philips sample cell: (1) sample flow input, (2) sample flow output, (3) Mylar retainer ring, (4) rupture sensing device, (5) Mylar window, (6) density correction block.

stream analyzer and making an appropriate correction. The newer X-ray fluorescence process stream analyzers make a suitable correction automatically.

One method by which this can be done is to incorporate a block of a chemically inert element which is not a sample constituent into the sample cell in such a way that one analyzer channel measures the fluorescent radiation from it through the slurry. The intensity of the radiation from this element is then a function of the slurry density and provides a signal which can be used to automatically compensate for the effect of slurry density on the calibration curves for the elements being measured. The Philips Electronic Instruments sample cell shown in Fig. 4 incorporates this feature. In this cell, sample flows in at 1 up the center tube and impinges on the Mylar window 5 then down the outer channel and out at 2. The density correction sample is an annular ring of metal shown at 6.

This cell incorporates two other features worth noting. Its design is such that there are no dead spaces where sample solids may settle and collect, and it is provided with a device for sensing breakage of the Mylar window. Of course, it can be used equally well for clear liquids as for slurries. A flow rate of about 1 gal/min is normally used.

Figure 5 shows a complete X-ray fluorescence process stream analyzer set up to analyze four continuous samples. From left to right the units are:

1. The power supply for the X-ray tube.
2. A cabinet containing the amplifiers, counter circuits, data processing circuits, and programmer.
3. The analyzer head mounted on rails which permit it to be successively positioned over a standard sample and up to 15 continuous flow type sample chambers.
4. A rack containing four strip-chart recorders to give a continuous record of concentration of the elements to be analyzed for in each sample.

Fig. 5. Complete X-ray fluorescence process stream analyzer.

This analyzer made by Philips Electronic Instruments is used as an example. Other makers of X-ray process stream analyzers are General Electric Company, X-ray Department, and Applied Research Laboratories. The process stream analyzers made by these companies operate on similar basic principles.

In addition to liquids and slurries, solid samples are often of interest. In some cases, as in the control of the thickness of tin electroplated onto sheet steel, the sample is available in continuous sheet form. This presents a relatively simple sampling problem since the analyzer head of the instrument need only be mounted above the continuously moving sheet. In this case if a measure of the thickness of the tin on both sides of the sheet is required, two analyzer heads, one above and one below the sheet are required.

Finely divided powdered solids can also be handled without too much difficulty. To do this the sample is conveyed onto a horizontal disc rotating about a vertical axis. As the disc turns the sample is carried under a smooth metal shoe to remove excess sample and compact and smooth the surface, then under the analyzer head, and finally to a scraper which removes the sample from the wheel after it has been analyzed.

If the sample is in the form of chunks or granules it will be necessary to provide continuous grinding equipment to powder the sample so that it can be analyzed.

Pasty materials, such as white lead or tooth paste, could probably be spread on a disc or continuous belt and passed under the analyzer head. A moist material such as raw phosphate rock, which looks like moist clay with sand and gravel in it, would be extremely difficult to prepare by a continuous process. It might perhaps be done by continuous drying followed by sample grinding, or by adding more water and wet-milling to a slurry which could be analyzed in that form.

If the results of any analysis are to be meaningful, the sample on which the analysis was performed must be representative. This applies just as much to analyses from continuous process stream analyzers as to laboratory analyses. Unless the entire stream can be passed through the analyzer, considerable care must be given to the selection of the part which will be fed to the analyzer as sample.

One fundamental method for selecting the sample to be analyzed from the main process stream is to pass a sample cutter back and forth through a free falling stream. A Denver Equipment Co. automatic sampler designed for this purpose is shown in Fig. 6. It consists of a motor-driven traversing mechanism, shown with its cover removed and resting behind it, and the cutter head shown in the lower left of the figure. The material to be sampled may be liquid, slurry, dry powder, granular, or lump. However, the opening in the top of the cutter head must be sized with due consideration to the particle size and the particle-size distribution. In some cases sampling must be done in more than one stage to cut a small enough sample from a large stream.

Sometimes where there are a number of process streams to be analyzed and process conditions do not change too rapidly, it is more economical to use one analyzer to analyze all the streams sequentially. Figure 7 shows a sampling system of this type used by Anaconda Copper Company in an ore flotation mill to analyze 13 streams. The complete sequence requires 26 min. The main process stream is cut twice to reduce the volume sufficiently to give a satisfactory sample, and pumped to the laboratory, where it is re-

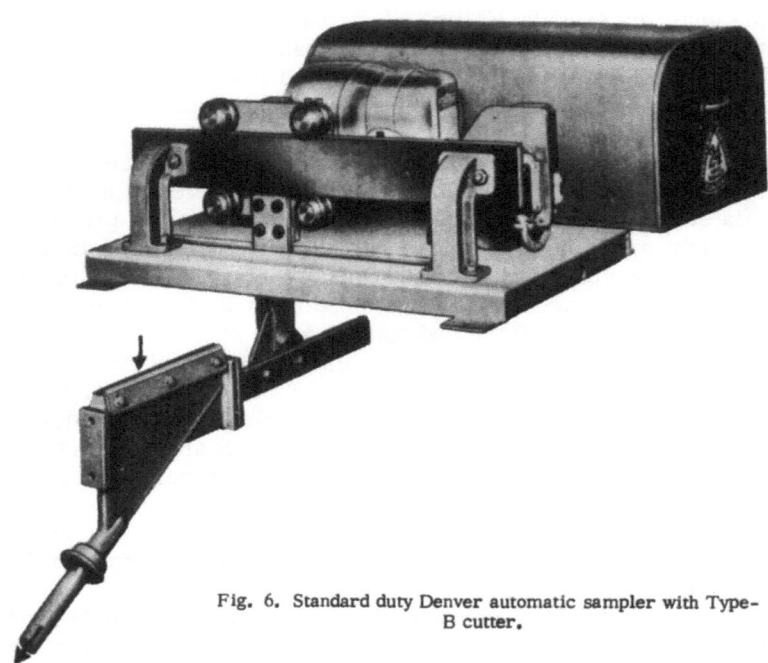

Fig. 6. Standard duty Denver automatic sampler with Type-B cutter.

duced in volume again, screened on a vibrating screen to remove wood chips, and fed to an agitated feed tank. From this it flows to the analyzer sample chamber by gravity. The analyzer chamber is emptied and flushed before the next sample is analyzed. The figure also includes equipment for securing suitable samples of the same streams for periodic chemical analysis.

Figure 8 shows a typical sample processing setup for coarse granular dry materials. Feed falling from the end of a belt conveyor is cut by a continuous sampler which discharges onto a belt conveyor which carries the

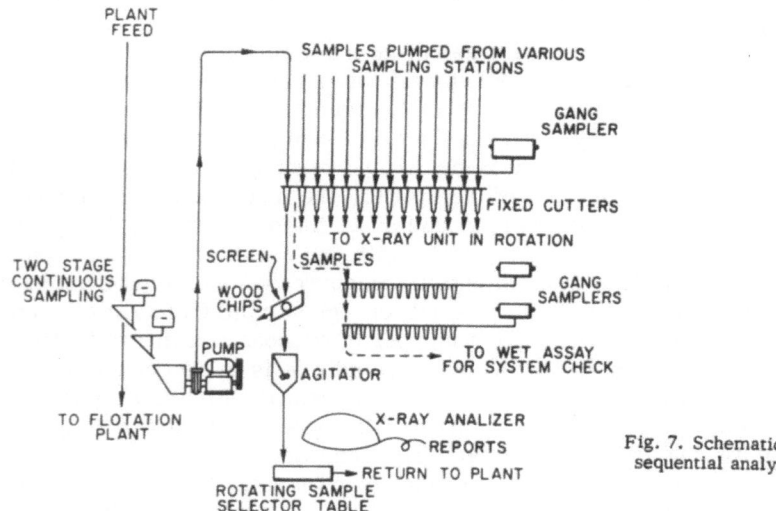

Fig. 7. Schematic of a sequential analyzer.

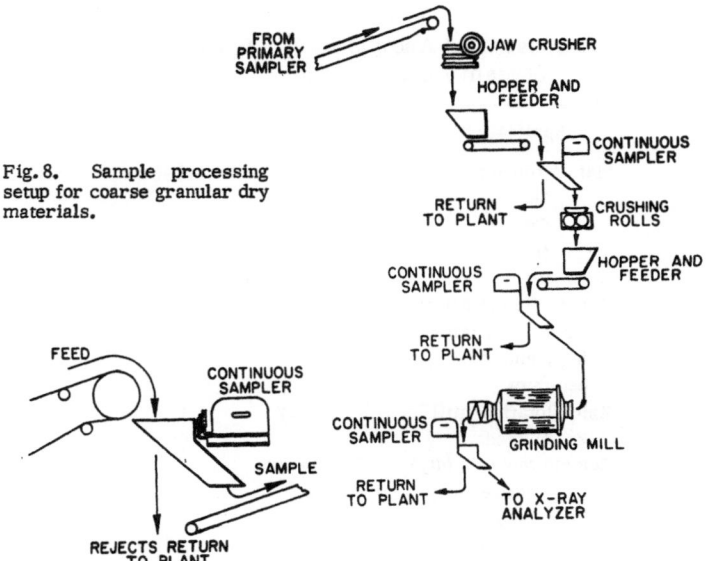

Fig. 8.   Sample processing setup for coarse granular dry materials.

primary sample to a jaw crusher.   The crushed material is cut again and further reduced in size by a roll crusher, cut again, and fed to a mill which reduces it to a particle size which can be analyzed by the X-ray emission process stream analyzer. It must be recognized that considerable time may be required for the sample to pass through a system such as this. In addition, considerable back mixing will take place tending to reduce the high concentrations and raise the low concentrations present in the original sample. In extreme cases, the time lag and the averaging effect of a complex sample preparation system may be so great as to make the whole system worthless. Demonstrated applications of automatic X-ray process stream analyzers are listed in Table IV.

One might predict that where there is sufficient economic incentive, process stream analyses now accomplished manually using laboratory X-ray fluorescence methods will be converted to automatic continuous analyzers or to automated intermittent analyzers as soon as sample-handling problems involved can be solved.

It is difficult to make statements about the accuracy of continuous automatic process stream analyzer results. The analyzers themselves are capable of essentially the same accuracy as laboratory X-ray fluorescence methods in a given application. However, it should be remembered that the overall accuracy is no better than the sample supplied to the instrument.

Perhaps it would be worthwhile to compare some of the characteristics of process stream analyzers based on X-ray fluorescence with those of several other types.   The X-ray type is almost unique in that it measures the concentration of one or more elements in the sample without regard to the state of chemical combination.   Most others, including infrared and ultraviolet absorption analyzers, as well as those based on refractive index and vapor-phase chromatography, to name but a few, measure the concentration of a particular molecule of interest. The X-ray type is highly specific as opposed to those based on refractive index, density, and thermal conductivity.

TABLE IV
Demonstrated Automatic X-Ray Process
Stream Analyzer Applications

Application	Sample form
Plating thickness	Continuous sheet steel
Sn	
Cu	
Ni	
Zn	
Copper flotation mill	Slurry
Cu in feed, tail-	
ings, and con-	
centrates	
Zinc flotation mill	Slurry
Zn in feed	
Cement raw mix (dry)	Powder
Fe_2O_3	
SiO_2	
$CaCO_3$	
Al_2CO_3	

These can be used only on two-component sample streams or on pseudo two-component streams. X-ray process stream analyzers are unique in their ability to analyze samples which are slurries or powdered solids. Most other analyzers will tolerate only gaseous or liquid samples. We should note that X-ray analyzers are in the general price range of $20,000 to $50,000, while process analyzers based on infrared or ultraviolet absorption or gas chromatography are in the $3,000 to $7,000 price bracket. Others, such as those depending on thermal conductivity or density or electrical conductivity, can range from $500 to $2,000. The price of the X-ray analyzer means that if another type will do the job adequately the X-ray instrument will not be considered.

There are two basic kinds of use for continuous automatic X-ray fluorescence analyzers. The first is for accounting purposes, to determine the amount of a valuable component being delivered by a seller to a buyer. Here it is necessary to have an integrated value of the product of the rate of delivery and the percentage of the desired component. The second is for control of a process. The analysis of feed, concentrates, and tailings from a flotation mill is an example. For this kind of application it is necessary that the speed of response of the analyzer, including the sampling system, be about ten times faster than that of the process it is to be used to control. At present X-ray process stream analyzers are being used to furnish data to operators who manually adjust appropriate process conditions to give the desired performance. Automatic, closed-loop control is difficult to justify because most of the processes respond slowly and because the economic gain from completely automatic control is not sufficient to justify the cost.

An important benefit from a process analyzer installation is that the data it will make available will almost surely lead to a better fundamental understanding of the interrelationships among the process variables. Usually what happens is that the analyzer shows unexpected fluctuations. The plant operators will probably tell the people responsible for its installation that it is not operating properly. After suitable checks are made it will most likely be shown that the analyzer is operating properly and that the fluctuations are real, necessitating a search for their cause. This will bring to light the effects of equipment poorly designed for its purpose or poorly maintained, or poor control of some process variable such as the temperature of a reactor, or poor operating procedures such as "slugging" a continuous process with some recycle stream. Realization of the existence of these factors makes it possible to correct them, frequently leading to important operating economies or product improvement.

In summary, it can be said that management expects process stream analyzers to improve product quality, increase production, lower costs, and promote safety of personnel and plant. Overall costs, including equipment, installation, research in connection with start-up, operation, maintenance, interest on investment, and depreciation must be small enough compared to the benefits gained to show a reasonable profit. In well-chosen applications, the profit can be very attractive. One installation has been said to result in annual savings of $835,000, an extremely attractive return on the investment required for an analyzer installation. The characteristics which a process stream analyzer must have in as high a degree as possible are that it must be sensitive only to the desired component of the sample, accurate, easy to calibrate and maintain, rapid in response, and not too expensive. X-ray fluorescence spectroscopy provides the basis for process stream analyzers which are highly developed and satisfy these requirements for many applications where it is desirable to have information on the concentrations of one or more elements in a flowing process stream with a minimum of time lag and expense. By taking advantage of the experience and engineering know-how of the user, the manufacturer of the analyzer, and the manufacturer of the sample-handling equipment, each application proposed can be evaluated, and only those which will promote management's aim of increased profitability chosen.

REFERENCES

1. M. L. Fuller and P. E. McNarry, Engineering and Mining Journal (April, 1962).
2. F. L. Holderreed and W. Lucy, Mining Congress Journal (July, 1960).
3. William S. Hoffman and J. W. Pastorius, "X-Ray Analysis and Control of a Cement Plant Raw Mix," presented at A.I.E.E. Cement Industry Conference, April 3-5, 1962, St. Louis, Mo.
4. P. E. Cook, Nonmetallic Minerals Processing (March, 1962).
5. J. W. Kemp, "Instruments for Production and Process Control," in Handbook of X-Rays, Emmett Kaelble (editor) (in preparation).
6. W. R. Kiley, "Application of On-Stream Analysis to Various Mineral Beneficiation Processes," Norelco Reporter Vol. X, No. 2 (1963), p. 66.
7. F. L. Abboud, F. Bernstein, A. D. Furbee, and C. H. Hailey, "The X-Ray Emission Gauge," A.I.E.E. paper 62-656, Great Lakes District Meeting, Ft. Wayne, Indiana, April 25-27, 1962.
8. A. D. Furbee, "On-Line Type X-Ray Emission Analyzer Systems," Tenth Annual Conference on Applications of X-Ray Analysis, August 7-9, 1961.
9. W. F. Loranger, "On-Line Process Analysis by X-Ray Emission Technique," 12th Annual Symposium on Spectroscopy, Chicago, Illinois, May 16, 1961.
10. W. F. Loranger and R. G. Edholm, Mining Congress Journal, p. 51 (October, 1961).
11. W. S. Hoffman and J. W. Pastorius "Whitehall's Experience with the X-Ray Emission Gauge," GTC 1962 Fall Meeting, September 10-13, 1962, Baltimore, Maryland.
12. "Automatic Control of Chemical and Petroleum Processes," Gulf Publishing Co., Houston, Texas, 1961.
13. P. S. Goodwin, "The Quantrol as Tin Coating Weight Gauge," Analytical Systems Division, Application Report, Applied Research Laboratories.
14. P. S. Goodwin and R. H. Prowett, "Plating Thickness Measurements of Electrolytic Brass, Copper, Nickel and Zinc on Mild Steel Using the A.R.L. Quantrol," ibid.
15. P. S. Goodwin, Control Engineering, 5:94-99 (August, 1958).
16. A. E. Bernhard, "Some On-Line Applications of X-Ray Fluorescence Spectrometry," I.S.A. Winter Automation Conference, St. Louis, Mo., January 17-19, 1961, Preprint No. 18SL61.

The Use of Digital Techniques to Aid in the Phase Analysis of Multicomponent Mixtures by X-Ray Diffraction

G. J. C. Frohnsdorff and P. H. Harris

American Cement Corporation
Riverside, California

X-ray diffraction is the most powerful tool available for the determination of the phase composition of multicomponent crystalline mixtures. In order to take advantage of the great potential of digital computers for aiding quantitative X-ray diffraction analysis (QXDA), equipment has been set up to obtain X-ray diffraction data directly in digital form on punched paper tape. The method of using a computer to produce a quantitative analysis from the recorded data is described and results obtained for the analysis of Portland cements are used as an illustration.

INTRODUCTION

X-ray diffraction is the most powerful tool available for identification of the components of mixtures of crystalline materials. Although it can also be applied to the quantitative analysis of mixtures, X-ray diffraction has not been used very extensively for this purpose. However, its potential value to the cement industry is now widely appreciated, and papers describing the application of quantitative X-ray diffraction analysis (QXDA) to the analysis of cements have recently been published in the United States [1], England [2], France [3], Germany [4], and Japan [5]. The purpose of the present paper is to outline a simple method of using a digital computer to analyze X-ray diffraction patterns which have been recorded in digital form. The method should bring substantial savings in time and human effort, and an increase in precision to the routine quantitative analysis of mixtures with known components. As an example, the application of the method to the analysis of Portland cements is described, but it is quite general for the analysis of multicomponent mixtures of crystalline compounds for which reproducible X-ray diffraction patterns, free from the effects of preferred orientation, can be obtained. The conditions for obtaining such reproducible diffraction patterns have been discussed by Klug and Alexander [6].

THE ACQUISITION OF X-RAY DIFFRACTION DATA IN DIGITAL FORM

A conventional X-ray diffraction powder pattern obtained using a Norelco diffractometer is shown in Fig. 1. It is the result of scanning the radiation diffracted from a finely powdered sample at a rate of $1/8°$ 2θ per min. In this case, the output from the X-ray counting tube was fed through the rate meter

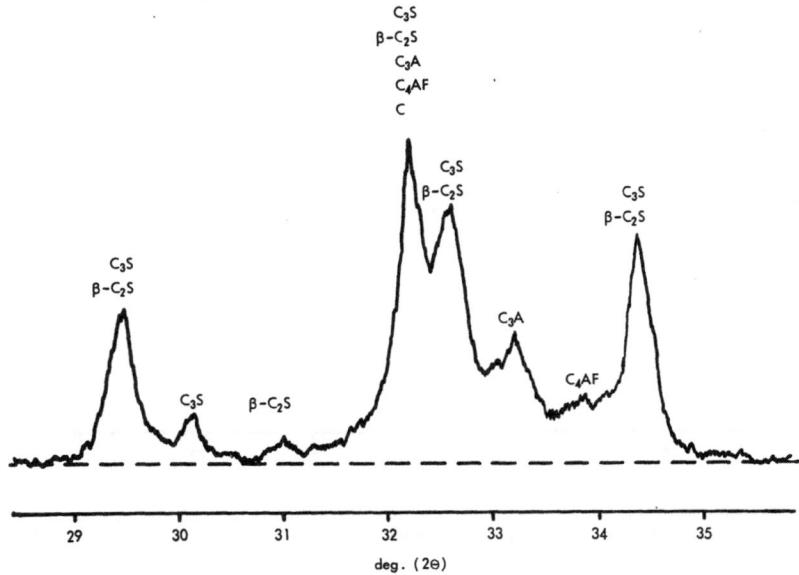

Fig. 1. Portion of X-ray diffraction chart for P.C.A. cement LTS-12. Radiation: CuKa, nickel filter; scanning rate $\frac{1}{8}$° 2θ per min. Phases contributing to the main peaks are indicated in cement industry nomenclature: C = CaO, S = SiO$_2$, A = Al$_2$O$_3$, F = Fe$_2$O$_3$. The broken line represents the background radiation intensity.

into a strip-chart recorder. For the purpose of low-cost digital recording in a form suitable for processing with a computer, the output pulses from the counting tube are fed into a CMC (Computer Measurements Corporation) electronic counter (Model No. 2707A) connected, through a CMC coupler (Model No. 403AD), to a Tally paper tape punch (Model No. 420PR-24-6). The counter totalizes the number of pulses recorded while scanning at a constant rate over consecutive narrow 2θ ranges of predetermined width (e.g., $\frac{1}{16}$°), the total for each range being recorded on the punched tape in response to pulses from a switch attached to the goniometer gears. The numbers x_i recorded on the punched tape represent the areas within consecutive vertical strips under a continuous curve, such as that of Fig. 1.

Figure 2 gives a graphical representation of the digital record corresponding to Fig. 1. Clearly, for a given rate of scanning, the digital record can be made to define the continuous X-ray diffraction pattern as precisely as necessary by adjusting the width of the ranges for which the counts are totalized. A great advantage of digital, as opposed to strip-chart, recording is that errors which arise from slowness of pen movement are avoided.

THE ANALYSIS OF X-RAY PATTERNS OBTAINED IN DIGITAL FORM

The series of numbers in the digital record may be considered as a vector x where

$$x = [x_1, \ldots, x_n] \tag{1}$$

and $x_1, \ldots, x_n$ are the individual elements; x may be called the "X-ray intensi-

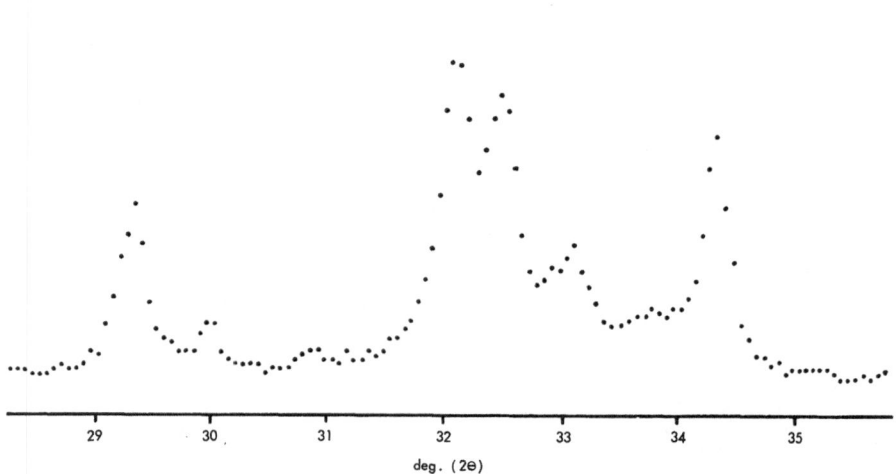

Fig. 2.   Plot of digital record representing the X-ray diffraction pattern of P.C.A. cement LTS-12.
Each dot represents the area under a $1/16$° 2θ-wide strip of a smooth curve such as that of Fig. 1.

ty distribution vector" for a sample scanned under the defined conditions.
(For convenience in the rest of this discussion, the single word vector will
be used to imply the whole of the above expression in quotation marks, and it
will be assumed that all vectors cover the same range of values of 2θ in the
same number n of equal steps, and that they apply to the same quality of primary radiation.)   In general, x will contain contributions from: (a) background radiation ($K.b$); (b) radiation diffracted by each of the crystalline compounds in the mixture; and (c) counting noise ($K.v$).   These contributions are
additive and, in the case of the diffraction contribution from suitably prepared
samples, are proportional to the quantities of the compounds present [6].
Thus,

$$x = K.(b + \sum_{i=1}^{w} p_i\, a_i + v) \tag{2}$$

where K is a factor which is proportional to I (the intensity of the primary
X-ray beam) and inversely proportional to the rate of scanning and to μ (the
mass absorption coefficient of the sample for radiation of the predominant
wavelength) and which allows variations of I and μ to be taken into account; b
is a background vector; the vectors a_i are "standard vectors" representing
the radiation which would be diffracted by the individual compounds under
standard conditions characterized by one particular value of K; the p_i's are
the weight fractions of the w individual compounds in the mixture; and v is a
counting error vector with elements which are randomly distributed and have
a mean value close to zero. For the purpose of ZXDA, the values of b (or $K.b$),
v (or $K.v$), and K must be known before the p_i's can be determined from x.
       In the case of strip-chart records, the background, which is the counter-
part of $K.b$ in the present treatment, is usually determined by drawing a line
tangent to the diffraction curve at two widely separated minima, allowance
being made for the counting "noise": as an example, the background level in

Fig. 1 is represented by the broken line. The corresponding procedure can be used to determined $K.b$ from x by making the elements of $K.b$ form an arithmetic progression which, within the limits of experimental error, makes at least two widely spaced elements of $(x - K.b)$ equal to zero without making any elements less than zero. The vector $(x - K.b)$, which includes the counting errors, will be called the diffracted intensity distribution vector d; thus

$$d = (x - K.b) = K. \sum_{i=1}^{w} p_i.a_i + v \tag{3}$$

The next step in the analysis is to decide which compounds may be present in the mixture and to select standard vectors for suitable forms of these compounds; (methods of obtaining the standard vectors for the individual compounds are discussed in a later section). Forms of a compound which could be included in the analysis are different crystalline modifications or samples of the same crystalline modification but with different degrees of crystal perfection. The set of selected standard vectors will be written $a_1', \ldots, a_y'$, where $y \geq w$ and, for brevity, it will be assumed that they are arranged in an order such that a_i' is the member which correlates most closely with a_i.

Analysis in the Ideal Case, Where $a_i' = a_i$

The technique of analysis is to form the linear combination

$$\sum_{i=1}^{y} k_i.a_i'$$

of the vectors a_i' which correlates most closely with the diffracted intensity vector d for the unknown mixture. The required set of values of the constants k_i is obtained by minimizing the sum of the squares,

$$\sum_{i=1}^{n} z_i^2$$

of the elements of z, where z is defined by

$$z = d - \sum_{i=1}^{y} k_i.a_i' \tag{4}$$

The constants k_i must satisfy the partial differential equations

$$\left[\frac{\partial \sum_{i=1}^{n} z_i^2}{\partial k_j} \right]_{k_1, \ldots, k_{j-1}, k_{j+1}, \ldots, k_y} = 0 \tag{5}$$

They are found most efficiently by setting $z = 0$ and applying multiple linear regression [7] to the n simultaneous equations formed by its elements:

$$d_j - \sum_{i=1}^{y} k_i.a_{i'j}' = 0 \qquad \text{for } j = 1, \ldots, n \tag{6}$$

This can only be done unambiguously if $n \geq y$, which will usually be true. If n is larger than about 5, it would not be practical to carry out the calculations routinely without a computer. However, with a computer, values of n which are larger than 100 can be handled without difficulty.

In the ideal case where, for $i \leq w$, $a'_i = a_i$, equation (4) may be written

$$z = \sum_{i=1}^{w} (Kp_i - k_i).a_i - \sum_{i=w+1}^{y} k_i.a'_i + K.v \tag{7}$$

and it follows that, apart from the effect of random errors represented by $K.v$, Σz_i^2 will be a minimum when for $i \leq w$,

$$k_i = Kp_i \tag{8a}$$

and for $i > w$,

$$k_i = 0 \tag{8b}$$

The minimum value, which is $K.\Sigma v_i^2$, provides an indication of the magnitude of the error. It can now be seen that the p_i's, which are the desired end results of the analysis, may be determined from equation (8a) if K can be evaluated. This can be done if the sum $(p_q + \cdots + p_r)$ of any group of one or more of the p_i's is known, since it follows from equation (8a) that

$$K = \frac{k_q + \cdots + k_r}{p_q + \cdots + p_r} \tag{9}$$

The use of an internal standard added to the sample in known quantity ensures that at least one p_i is known so that K can be determined from

$$K = k_s/p_s \tag{10}$$

where the subscript s denotes that the values are for the internal standard.

Another useful special case is that in which it is known that all compounds present in the mixture contribute to d, for then

$$\sum_{i=1}^{w} p_i = 1$$

and

$$K = \sum_{i=1}^{w} k_i \tag{11}$$

Analysis in the General Case, Where $a'_i = a_i + e_i$

In practice, the condition $a'_i = a_i$ for all $i \leq w$ will not hold exactly. Instead, it will be generally true that $a'_i = a_i + e_i$, where e_i is an error vector representing the difference between the actual standard vector a_i for compound i and the assumed standard vector a'_i. If the elements of each e_i are randomly distributed and have a mean value of zero, then the minimum value of Σz_i^2 will be obtained when

$$z = K.v + \sum_{i=1}^{w} k_i.e_i$$

and this will have the effect of increasing the minimum value of Σz_i^2 while leaving its elements randomly distributed about zero. This should not cause large errors in the determination of the p_i's by the procedure outlined in the previous paragraph provided Σz_i^2 is much smaller than Σd_i^2. As before, the magnitude of Σz_i^2 can be used as an indication of the degree of success in fitting $\Sigma k_{j}.a_i''$ to d; for a perfect fit, Σz_i^2 would be zero.

A vector e_i is only likely to cause serious analytical errors when its elements are not randomly distributed about zero, but show a correlation with the corresponding elements of one or more of the vectors a_j which contribute to d. This could result from errors in the determination of the vector a_i' by the procedure outlined in the following section, if the p_i's used for the "known" mixtures were incorrect. It could also result from the a_i' chosen to represent a_i in the analysis being the standard vector for a form of compound i which has a slightly different X-ray pattern, e.g., broader peaks due to less perfect crystallinity; the effect of peak width is discussed in the Appendix.

Evidence about the adequacy of the chosen set of vectors a_i' can be obtained from a plot of the elements of z. For example, if the plot shows z to have positive elements where the elements of a_i' are a maximum, it is likely that the form of compound i which was used to obtain a_i' has broader and less intense peaks than the form of the compound in the mixture being analyzed. To facilitate this aspect of the analysis, it is desirable to arrange for the computer output to include the elements of z in a form suitable for mechanical plotting.

THE DETERMINATION OF STANDARD INTENSITY DISTRIBUTION VECTORS FOR INDIVIDUAL COMPOUNDS

It will have been noted that the preceding treatment depends upon the standard vectors for the individual compounds in the sample being known. These standard vectors can be obtained from vectors for individual compounds mixed with an internal standard, or from the vectors for a sufficiently large set of known mixtures of the compounds, each known mixture containing a known quantity of an internal standard. The internal standard is an aid in establishing the K factors for the conditions under which the experimental data are obtained. The following relationships will hold for a set of mixtures:

$$d_q/K_q = \sum_{i=1}^{w} p_{qij}.a_i + v_q \text{ for } q = 1, \ldots, t \qquad (12)$$

If the number of mixtures, t, is greater than or equal to w, and if all the p_{ij}'s are known, then setting $v_q = 0$, the set of t equations for each of the n integral values of j such as

$$d_{qij}/K_q - \sum_{i=1}^{w} p_{qij}.a_{ij} = 0 \qquad (13)$$

can be solved by multiple linear regression to obtain the values a_{ij} of the elements of the standard vectors for the individual compounds. This would be difficult to do by hand calculation,but it is a task which is readily performed with the digital computer. The technique is particularly valuable when the peaks for the individual compounds overlap and would be difficult to resolve in any other way. It can be seen from equation (3) that d should be close to zero when $p_i = 0$, for all i; this can be made a condition of the solution of equations (13) by doubling the number of equations in the set by including ones formed by multiplying both sides of the existing equations by -1.]

RESULTS OBTAINED FROM THE APPLICATION OF THE PROPOSED TECHNIQUES TO THE ANALYSIS OF PORTLAND CEMENTS

The techniques outlined above have been tested using twelve Portland cements of known composition. These cements were kindly provided by the Portland Cement Association. From the known compositions [1, 8, 9] and from experimentally determined intensity distribution vectors representing the X-ray diffraction patterns of mixtures of the cements with 10% of rutile as an internal standard, a standard vector for each of the four major phases was determined by the procedure outlined above; an IBM 1620 computer with a 20,000-decimal-digit memory was used. The four phases, which account for about 90% of a typical Portland cement, are tricalcium silicate, β-dicalcium silicate, tricalcium aluminate, and the aluminoferrite phase commonly referred to as tetracalcium aluminoferrite. The results are plotted in Fig. 3. The plots for the silicates are similar to diffraction patterns which have been obtained for the pure compounds (e.g., see [4]), but the aluminate and aluminoferrite plots seem to have been imperfectly resolved from the tricalcium silicate. This may be due to errors in the accepted compositions of the cements or to some, as yet undiscovered, source of error in the outlined procedure; this point will be checked as more information becomes available.

Subsequent to the establishment of standard vectors for the individual compounds, the analytical method described in the earlier section on analysis of X-ray patterns was applied to the determination of the compound compositions of the twelve cements; with the IBM 1620, the computer time required for each analysis need be no more than 5 min, though actual time was considerably longer due to the use of an unnecessarily large number of steps in the program.

The results for the silicates were quite satisfactory (see Table I), but the results for the aluminate and aluminoferrite phases were not so good and are not shown. The cause of poor agreement with the previously established values is believed to be variation in the extent of the overlap of the silicate peaks into the important aluminate and ferrite diffraction region between 33 and 34° 2θ (see Fig. 1). This appears to be confirmed by the fact that, when an empirical correction for the overlap was made by the numerical equivalent of the geometrical construction used by Copeland et al. [1], much improved results were obtained [10]; the computer was used for these calculations. It may be noted that the precision of the silicate determinations, which is expressed as the standard deviation of the standard deviations for duplicate determinations listed in Table I, appears to be superior to that obtained by previous workers (see Table II); however, it must be pointed out that it is difficult to be sure that the published figures are exactly comparable. That the precision should be superior is to be expected since the

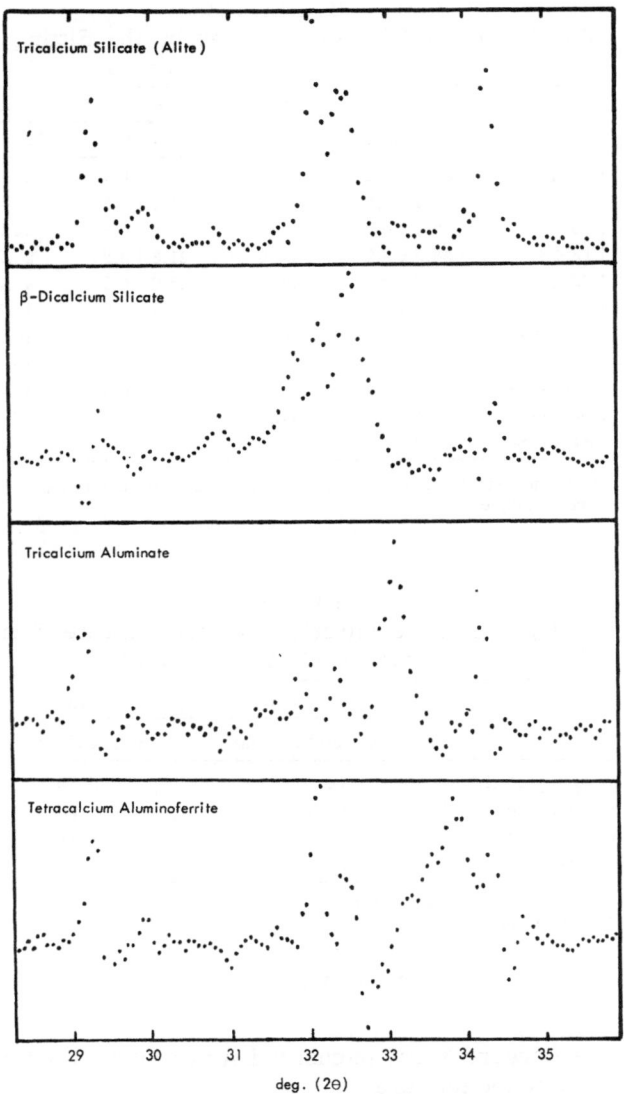

Fig. 3. Plots of calculated values of the elements of the X-ray intensity distribution vectors for the four major phases in Portland cement.

proposed method of analysis permits the whole of the measured X-ray pattern, including the intense overlapping silicate peaks, at about 32° 2θ (see Fig. 1) to be used, whereas previous methods have relied on the planimetering of isolated peaks of relatively low intensity.

In conclusion, it may be stated that the proposed method appears to provide a practical basis for the quantitative X-ray diffraction analysis of multicomponent mixtures though refinements such as the "weighting" of certain portions of the diffraction pattern may prove to be desirable in some cases.

TABLE I
The Results of Silicate Analyses by the Fitting of Computer-Synthesized Curves

Sample	% tricalcium silicate		%β-dicalcium silicate	
	Computer	P.C.A.	Computer	P.C.A.
LTS-12	51.5 ± 2.0*	51.0	25.8 ± 0.7	27.4
LTS-15	62.4 ± 1.0	60.9	11.4 ± 0.1	12.5
LTS-17	55.9 ± 0.2	58.6	21.4 ± 0.6	20.2
LTS-23	55.8 ± 3.2	55.6	21.3 ± 3.2	23.2
LTS-25	37.0 ± 0.6	38.7	45.9 ± 0.5	43.0
LTS-31	60.6 ± 0.8	57.7	18.8 ± 0.5	17.1
SBR-15497	59.2 ± 4.4	55.6	14.3 ± 3.2	16.2
SBR-15622	44.3 ± 1.3	52.6	36.6 ± 0.4	30.6
SBR-15669	31.5 ± 0.1	29.7	55.2 ± 0.5	57.8
SBR-15698	60.1 ± 0.5	65.0	17.4 ± 0.8	16.4
SBR-15754	49.2	52.0	29.9	25.2

*Indicated standard deviations are calculated from results for two different samples.

TABLE II
Precision of Determinations of the Silicate Phases of Portland Cement by QXDA

	Reproducibility (% absolute)	
	Tricalcium silicate	β-dicalcium silicate
Midgeley et al.*	±4	±5
Smolczyk [4]	±2.5	±3
Von Euw [3]†	±2 to 3	±6 to 9
This work	±2.0	±1.6

*Figures quoted by Nurse [11].
†Used an X-ray camera.

NOTATION

Vectors

a_1 The standard X-ray intensity distribution vector for the ith compound in the mixture

a_i' The standard vector similar or identical to a_i which takes the place of a_i in the analysis

b The vector in terms of which the background vector $K.b$ is defined

d The diffracted intensity vector: $d = x - K.b$

d_q The vector d for the qth mixture

e_i The error vector: $e_i = a_i' - a_i$

v The vector in terms of which the counting error $K.v$ is defined

v_i The vector v for the ith mixture

x The X-ray intensity distribution vector for the mixture of unknown composition

z The vector which represents the difference between d and the value of d which would be predicted from the analysis

Scalars

a_{ij}	The jth element of vector a_i
a'_{ij}	The jth element of vector a'_i
d_{qj}	The jth element of vector d_q
I	The intensity of the primary X-ray beam
K	A factor which is proportional to I/μ and to the scanning rate
K_q	The factor K for the qth mixture
k_i	The constant by which the vector a'_i is multiplied in forming a vector similar to d
p_i	The weight fraction of the ith component of the mixture
p_{ij}	The weight fraction of component j in the ith mixture
v_i	The ith element of vector v
v_{qj}	The jth element of vector v_q
x_i	The ith element of vector x
z_i	The ith element of vector z
s	Subscript denoting values for the internal standard
$i, j, n, q,$ r, t, w, y	Integers used as subscripts
μ	The mass absorption coefficient of the mixture

ACKNOWLEDGMENTS

The authors wish to thank Mr. J. P. Giles, President of the American Cement Corporation, for granting permission to publish the paper. They also wish to acknowledge their gratitude to Dr. S. Brunauer, Dr. L. Copeland, and Dr. D. Kantro of the Portland Cement Association, and to Dr. P. Halstead and Miss A. E. Moore of the Cement and Concrete Association for comments which have materially helped the progress of this research. Finally, the programming and experimental assistance of Mr. S. T. Suess is gratefully acknowledged.

APPENDIX

The Effect of Differences in Peak Width Between the Unknown and Standard Samples on the Calculated Composition

While it is not possible to draw any generally applicable, quantitative conclusions about the effects of differences in peak width between the form of a component in the unknown mixture and of the form of the same material used as a standard for the analysis, it is informative to make calculations for hypothetical peak shapes. As an example, calculations were made for two different shapes of peaks, (1) triangular and (2) Gaussian, to see what errors might be caused in an analysis due to variations in the width, but not area, of the peak for the standard.

For the purpose of the calculations, it was assumed that a single peak of an X-ray pattern was to be fitted by a standard peak of the same shape but different width-at-half-height, and having the same base-line. The Gaussian peaks were cut off to zero at a distance of one width-at-half-height on each side of the maximum to facilitate the calculations. In each case, the peak

which was to be considered as the "unknown" was represented by a vector
with 16 nonzero elements, and peaks which were taken as standards had pro-
portionally more or less nonzero elements depending on their width. In the
following table the ratio of the peak area for the unknown and the standard
fitted to it is listed beside the corresponding peak width ratio. (Since both
Gaussian and triangular peaks gave the same peak area ratios, only one
column has been used for both.)

Peak width ratio, standard/unknown	Peak area ratio, standard/unknown
0.75	1.13
1.00	1.00
1.25	0.88
1.50	0.78

These values suggest that, if the peak width for the standard is less than
that for the unknown, a high analytical result will be obtained, whereas for a
broader standard peak, a low result would occur.

REFERENCES

1. Copeland, Brunauer, Kantro, Schulz, and Weise, Anal. Chem. 31:1521 (1959).
2. Midgley, Rosaman, and Fletcher, Proc. Fourth International Symposium on the Chemistry of Cement, NBS Monograph No. 43, Vol. 1, p. 69, NBS, Washington (1962).
3. Von Euw, Silicates inds. 23:643 (1958).
4. Smolczyk, Zement-Kalk-Gips 14:558 (1961).
5. Yamaguchi, Tanaka, and Kajii, Rev. of 13th General Meeting, Japan Cement Engineering Association, p. 2 (1959).
6. Klug and Alexander, X-ray Diffraction Procedures (John Wiley and Sons, New York, 1954).
7. Crow, Davis, and Maxfield, Statistics Manual (Dover, New York, 1960).
8. Lerch and Ford, Proc. Am. Concrete Inst. 44:745 (1948).
9. Powers and Brownyard, Proc. Am. Concrete Inst. 43:306 (1947).
10. Frohnsdorff and Harris, Proc. Symposium on the Analysis of Calcareous Materials (London, 1963), to be published.
11. Nurse, Proc. Fourth International Symposium on the Chemistry of Cement, NBS Monograph No. 43, Vol. 1, p. 9, NBS, Washington (1962).

Practical Methods of Solving Absorption and Enhancement Problems in X-Ray Emission Spectrography

E. L. Gunn

Research and Development
Humble Oil & Refining Company
Baytown, Texas

A description of the manner in which absorption and enhancement effects produce nonlinear relationships between intensity and concentration in X-ray fluorescence spectrography is given. Under certain circumstances enhancement may pose an asset in providing a means of increasing the detectability of an element in low or trace concentrations. Since absorption and enhancement are fundamental properties of matter, the practical approach in X-ray spectrographic analysis is to employ methods or techniques which surmount or circumvent their effects. Those used by various workers to accomplish this may be classified into three categories: (1) minimize or reduce these effects to negligible levels by attenuation or by sample dilution; (2) compensate for them by comparison or by internal standardization; (3) apply mathematical corrections to the observed measurements based either on empirical knowledge of the magnitude of these effects or on theoretical considerations. A discussion of examples from the methods of various workers which illustrate these general approaches is given.

Absorption and enhancement effects that produce nonlinear correlations between intensity and concentration probably constitute the primary problem encountered in X-ray spectrographic analysis. Von Hevesy [21] recognized and measured these effects early in the development of the X-ray fluorescent technique. The title of a current well-known book, "X-ray Absorption and Emission in Analytical Chemistry," presumably was chosen by Liebhafsky et al. [13] because its content deals with X-ray absorptiometry as well as with X-ray emission spectrography. Considering emission spectrography alone, one could infer a didactic implication in the title which these authors did not explicitly intend, namely, that having to do with the essential mutuality existing between absorption and emission in X-ray fluorescence. We are thinking here not just of the basic principle of generating fluorescent X-rays in a sample by causing more energetic X-rays to impinge upon it, but beyond this to the manner in which the impinging beam is modified as it penetrates the sample, the fate of the generated X-rays as they emerge to the sample surface, and of the possible influence these secondary X-rays may have on the emission of some other coexisting element in the sample. These interactions account for the nonlinear correlation between fluorescent intensity and concentration.

It is not absorption and enhancement effects per se which account for nonlinearity; rather, it is changes in these effects concurrent with change in composition which produce nonlinearity. The well-known curves of Fig. 1 display in a typical fashion the effects of varying composition, and in turn absorption, on intensity. The extreme bow of curve 1 is observed when a

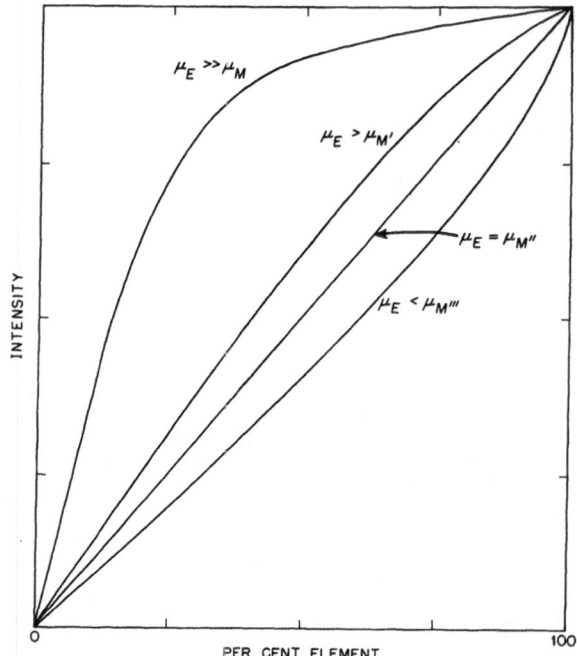

Fig. 1. The relationship of intensity to composition for an element in various matrices.

high-atomic-number element is measured in a low-atomic-number matrix or carrier medium. The absorption constant of the element for its own characteristic fluorescence X-rays is much greater than is that of the matrix for the same X-rays.

As an example, in a binary mixture of elemental carbon (A. N. 6) and iron (A. N. 26) the mass absorption coefficient for iron K_α X-rays increases 72-fold as the composition ranges from that of a pure carbon matrix to that of pure iron.

Curve 2 applies where the element is nearer the matrix but still is greater in absorption; and curve 3, perhaps hypothetical, where the element and the matrix have the same absorption constant. Curve 4 is typical of curves obtained where the absorption constant of the element for its fluorescent X-rays is less than that of the matrix. For example, in the measurement of titanium (A. N. 22) in mixture with zinc (A. N. 30), the absorption coefficient for Ti K_α X-rays decreases by a factor of 0.4 in a matrix ranging from that of pure zinc to pure titanium.

There is an effect on intensity over and above that of absorption which is produced under certain circumstances by element–matrix interaction as suggested in the title of our present discussion, namely, that of enhancement. This effect, which further complicates the nonlinearity problem, occurs when the characteristic radiation of the higher-atomic-number element occurs on the short-wavelength side but near the absorption edge of a coexisting matrix element of lower atomic number. Sherman [19] has shown that under certain conditions mutual enhancement of the lower atomic number element actually may be several times greater than the intensity generated by the primary beam itself. The basic rationale for the derivation of fluorescent intensity from theoretical considerations has been developed by Sherman. He points out that the derivation is complicated by the fact that the fluorescent radiations from elements of higher atomic number do excite additional radiation in those of lower atomic number. In his fundamental approach to the correlation of intensity with composition Sherman also points out that the polychromatic character of the conventional excitation beam imposes additional complexity to the theoretical elucidation of the relationship of intensity to composition.

Sherman's work [19] on the theoretical derivation of expressions for intensities of elements in mixtures followed certain steps. By assuming a monochromatic incident beam and single-line fluorescent spectrum, narrow incident and exit beams, and separation of atomic numbers of a three-element system so that emission lines fall on the short-wavelength side of absorption edges of lower atomic number elements, complex integral expressions were developed for intensity. Formulas for the evaluation of the integrals also were computed. Assuming an incident wavelength of 1.0 to 1.2 A, Sherman applied theoretical calculations to binary mixtures of zinc and iron and to ternary and quaternary systems containing zinc, nickel, iron, and chromium. Computed and experimentally observed intensity ratios of the elements were in good agreement. Sherman's work provides clarification of the fundamental nature of the sample absorption fluorescent X-rays and at the same time gives others working in this field an appreciation of the complex nature of the interactions between X-rays and matter which are involved in the process.

Although enhancement generally introduces further complexity in the absorption—enhancement problem, it can under selected conditions be turned into an asset. Salmon [18] found that silver used as a substrate behind a thin layer of columbium oxide enhanced the intensity of columbium K_α by 10 to 25% over that observed in the absence of substrate. The author's laboratory employs X-ray fluorescence routinely for the determination of nickel in oils and residua at the level of a few parts per million. Means of improving the ratio of the nickel line to background were sought so as to improve the sensitivity of detection. A study was made in which thin filters at the X-ray tube window were used to obtain selective enhancement of the measured nickel line.

An inspection of mass absorption coefficient data for nickel indicates the wavelengths of X-rays which are effective in exciting Ni K_α fluorescence; these are shorter than the K-absorption edge of nickel at 1.488 A. Tungsten radiation was used as a means of excitation. To provide a concept of the excitation situation, a profile of the energy distribution of tungsten is shown in Fig. 2. This profile was obtained by measuring the X-ray scattering properties of white oil. The peaks produced by Comptom scattering, which is typical of low atomic number substances, may be noted on the long-wavelength side of the characteristic WL peaks. Other low-atomic-number substances of high purity such as carbon, starch, lithium carbonate, and polymers are also satisfactory for this purpose. A scintillation counter was used as the detector. Absorption edges and lines for several elements are indicated in Fig. 2 so that the distribution of energy in the primary radiation which is most effective in exciting them can be qualitatively assessed.

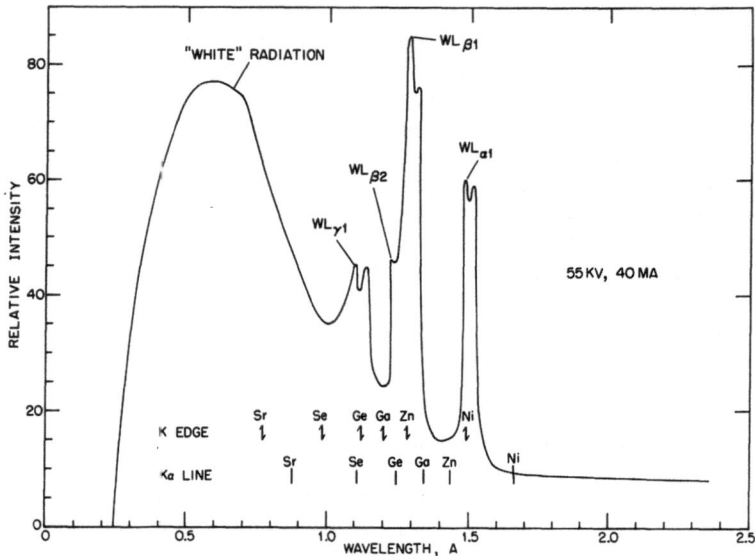

Fig. 2. Energy distribution of tungsten X-ray tube emission scattered by white oil.

Thin filters composed of zinc, gallium, germanium, selenium, and strontium, respectively, were investigated. These filters were prepared by depositing a fine powder of an element or its compound onto a 1-mil foil of aluminum metal. The number of filter thicknesses covering the tube window was selected by trial to provide the level of intensity desired.

The intensity ratio of the Ni K_α line at 1.659 A (48.7° 2θ) to that of background at 1.687 A (49.5° 2θ) is used as the parameter for measuring nickel concentration in an oil. Reference oil standards were measured for each type of filter. The character of the spectral profile for this region of the spectrum using a strontium filter is exemplified in Fig. 3.

A plot showing the relation between increase in intensity ratio and intensity for the reference standards is given in Fig. 4. The increase in intensity ratio is relative to that in which no filter was used. For example, by using germanium, the intensity ratio has been increased by a factor of 1.6 over that in which no filter was used. However, along with this enhancement in intensity ratio produced by the filter, a loss in intensity to only 17% of that obtained without a filter was sustained. Thus the achievement in filter enhancement is somewhat offset by a serious loss in intensity. Presumably an additional increase in filter thickness would give a larger value for the intensity ratio, but the penalty in intensity loss would become more severe. Since the intensity ratio is obtained as the quotient of two count numbers, these numbers must be relatively large to provide adequate statistical significance. Therefore, there must be a judicious choice of balance between the enhancement desired and the number of counts to be taken.

The zinc filter significantly reduced the intensity without observable enhancement of the nickel peak-to-background ratio. This effect was attributed to the high absorption of the effective WL lines in the primary

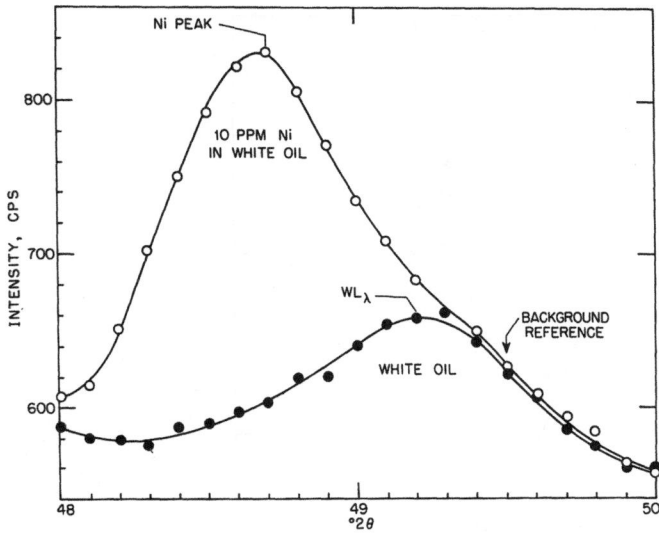

Fig. 3. Fluorescence profile of oils with strontium filter at tube window (LiF crystal).

spectrum, especially of WL_{β_1} and WL_{β_2}, while at the same time contributing insufficient characteristic zinc radiation to offset this absorption loss through a significant enhancement of the nickel radiation. An inspection of Fig. 2 indicates the location of the zinc absorption edge at a spectral position in which the WL lines are strongly absorbed.

The curve of Fig. 4 indicates a much higher correlation of enhancement with intensity than with the type of filter; the filters do not differ markedly in their enhancement effects. Strontium appeared to provide the most favorable enhancement–intensity property, however, and it was selected for use in measuring nickel in oil.

An enhancement factor of 1.5 is obtained using strontium, with a simultaneous reduction in intensity to about 40% of that obtained without a filter at the window. When considered in terms of the number of counts required to achieve a given precision and the time required for counting, this level of strontium filtration appeared to be about optimum. This filter has been used routinely in measuring nickel in oil in concentrations of a few parts per million.

Since absorption–enhancement problems originate in the fundamental properties of matter, their presence is unavoidable. The analyst therefore must devise techniques which either surmount or circumvent their effects. Practical methods of doing this—which is the subject of the present discussion—fall into three general categories. This classification, although not rigorous, is as follows: (1) minimize or reduce these effects to negligible levels by attenuation or sample dilution; (2) compensate for them by comparison or by internal standardization; (3) apply mathematical corrections to the observed measurements based either on empirical knowledge of the magnitude of these effects or on theoretical considerations. The purpose in all cases is to interpret intensity measurements in a manner which will provide accurate results in the estimation of true composition. The follow-

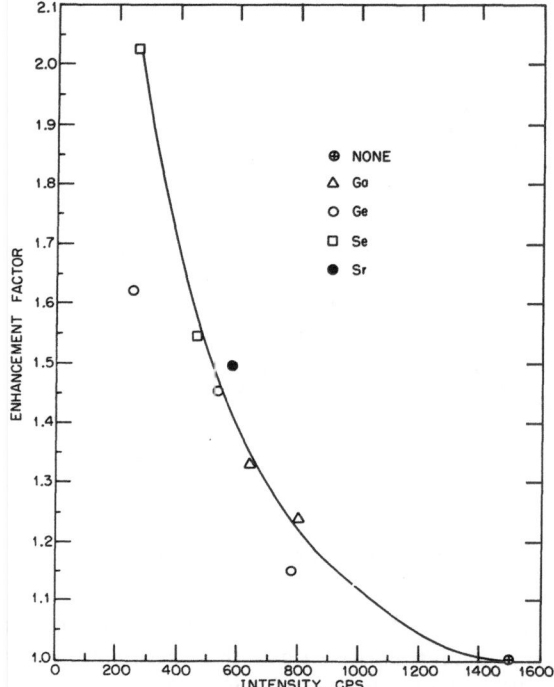

Fig. 4. Relationship of enhancement to intensity for different
filters.

ing discussion will consider these three categories in detail, and will include examples of successful and practical uses of each of the techniques.

ATTENUATION AND SAMPLE DILUTION

The effects of absorption and enhancement often are minimized or reduced to a level at which their effects become insignificant by (1) attenua-
ting the specimen as a thin film, or (2) by diluting the sample to a high
degree in a carrier material. The expressions "minimize" and "reduce to
an insignificant level" imply that these effects are never totally eliminated
in the system examined but only that their effectiveness has been decreased
to a level at which they are comparable in magnitude with the inherent
detection limit of the measuring device itself.

Use of Thin Film

Basic Considerations

The analysis of a minute amount of sample supported by thin film
probably is the simplest system one encounters in X-ray analysis because,
within a limit, it is essentially free of the interaction effects produced by
absorption and enhancement in specimens of normal depth [13]. Furthermore, it provides a means of obtaining the greatest possible degree of
sensitivity using conventional apparatus. Thin film is an appropriate first
choice for discussion among the various quantitative techniques in which

solution of the absorption and enhancement problem is considered. The simplicity and high sensitivity afforded by thin-film analysis can be understood in terms of certain basic principles of absorption and emission.

It is assumed that a thin, uniform film deposit is irradiated by an X-ray beam, exciting a characteristic fluorescent line of an element. The intensity I of the fluorescent line from a minute unit of volume in the deposit having a depth dx is given by the following expression:

$$dI = C \csc \phi_1 I_0 \exp [-(\mu_1 \csc \phi_1 + \mu_2 \csc \phi_2)\rho x]\, dx \qquad (1)$$

where C is a proportionality constant, I_0 is the incident intensity, ϕ_1 is the angle between the incident beam and the sample surface, ϕ_2 is the angle between the emergent beam and the sample surface, μ_1 is the effective absorption coefficient of the incident beam, μ_2 is the effective absorption coefficient of the emergent beam, ρ is the sample density, and x is the sample thickness.

If the sample deposit is kept very thin, its absorption of X-ray will be very small or practically zero. This signifies that the exponential term of equation (1) approximates a value of unity, or

$$dI = C \csc \phi_1 I_0\, dx \qquad (2)$$

If equation (2) is expressed in terms of increments, it becomes

$$\Delta I = C \csc \phi_1 I_0 \Delta x \qquad (3)$$

For a constant deposit area,

$$\Delta N \sim \Delta x$$

where N is the number of atoms of the element in the deposit which emit fluorescence. Hence

$$\Delta I = C' \csc \phi_1 I_0 \Delta N \qquad (4)$$

Thus, in a region of limited thickness, the physical significance of the last expression is that a direct linear proportionality will be observed between concentration and intensity. This proportionality will be maintained as long as sample absorption is negligible. Illustrations of thin-film applications follow.

Film Measurement of Surface Oxidation

Rhodin [17] has employed thin film as a support in measuring the surface effects of passivity and oxidation on steels. In this application ultrathin deposits on the film were made both by evaporation (10^{-5} mm Hg) and by chemical action on the metal surface. The effect of film thickness on the intensity of metal film is given in Table I. Different amounts of chromium, iron, and nickel were deposited. The constancy of the specific intensity for each metal from sample-to-sample is evidence of the proportionality existing between intensity and concentration for these thin films. The agreement of X-ray with microchemical analysis for the metals on the same films was found to be excellent.

TABLE I
Effect of Film Thickness of Pure Metals
on Specific Radiation Intensity *

Metal	Surface conc. γ/cm^2 †	Specific intensity (counts/sec)/γ-cm^2
Iron	7.7	16.6
Nickel	6.3	26.9
Chromium	12.0	8.5
Iron	12.1	16.5
Nickel	11.1	26.8
Chromium	20.5	8.5
Iron	19.0	16.5
Nickel	17.8	26.7

*From Rhodin, Anal. Chem. 27:1857 (1955).
†Obtained by dividing weight of film by its area.

Film Depth and Linearity

The amount of sample deposit which can be placed on thin films without absorption loss is significant because an excessive depth of sample causing such loss will, as has been pointed out, affect the linear relation between intensity and concentration.   Gunn [8] has made a study of depth and absorption on several inorganic salts deposited from solution onto Mylar film. Although uniform deposition onto the film is difficult, the results do give a reasonable estimate of the greatest amount of deposit for which linearity will be maintained.   Figures 5 and 6 illustrate the influence of absorption for both the X-ray emission and transmittance of a selected wavelength through thin deposits of iron.   The experimental points are somewhat scattered at higher concentrations because of the nonuniformity of the deposits, but departure from linearity is shown at about 150 γ in both figures.   The transmittance properties shown in Fig. 6 are for wavelengths on either side of that for the K-absorption edge of iron (1.743 A).   Although the scatter of experimental points makes the fit of the curves only an uncertain approximation, it appears that the absorption of 1.54 A is greater than that of 1.94 A, which should be expected because of the higher absorption coefficient of iron for 1.54-A X-rays.

Such measurements made on aqueous deposits of various elements indicate the manner in which the intensity of each is affected by deposit depth. The linear sensitivity, upper concentration limit of linearity, and concentration for maximum change in sensitivity shown in Table II were obtained by these measurements.

The sensitivity of the detector, absorption by the nonmetallic components in the deposit, optical geometry, fluorescent yield, and distribution of energies in the primary excitation spectrum are factors which account for the marked differences between elements shown by these data in Table II. The upper limit of sensitivity and point of maximum change tend to increase with the atomic number of the element.

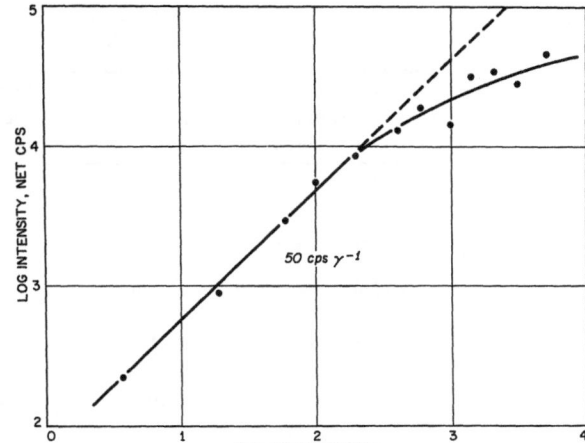

Fig. 5. Relation of fluorescent intensity of Fe K_α to concentration of iron (ferric oxalate).

Large Particles Supported on Film

The foregoing properties of thin film suggest a question as to the fluorescent behavior of relatively large uniform particles which are widely dispersed or separated on the film. Under ideal conditions each particle would be bathed by the X-ray excitation beam but be independent of the influence of adjacent particles. If each particle emitted independently, the observed intensity would be directly proportional to the number of particles. Thus by avoiding interaction or shielding effects between particles, linearity might be achieved, which would indicate a practical technique in analysis through the use of film-supported particles of relatively large size. The

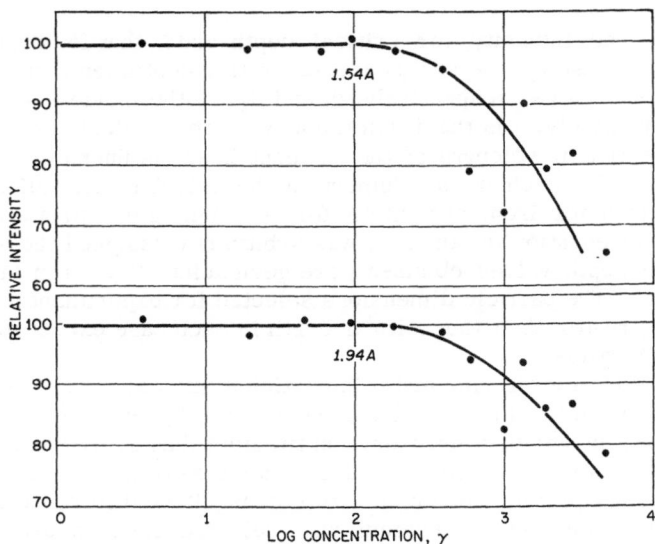

Fig. 6. X-ray transmittance of ferric oxalate. [From Gunn, Anal. Chem. 33:921 (1961).]

TABLE II
The Effect of Deposit Depth on Intensity*

	Sensitivity, cps/γ	Linearity limit, γ	Maximum sensitivity change, cps/γ^2
Ca (20)	0.7	400	–
Ti (22)	5.9	100	200
Fe (26)	60	100	250
Cu (29)	63	100	300
Sr (38)	4.9	1000	2500
Mo (42)	1.4	> 20,000	–

*From Gunn, Anal. Chem. 33:921 (1961).

measurement of uniform-size particles of an element or those of several elements of different atomic number could be used as an experimental test of the X-ray fluorescent behavior of film-supported particles. In a study carried out in the author's laboratory, germanium, iron, and titanium metals were selected for an experiment to determine the relationship of fluorescent intensity of particles supported on film to the quantity by weight of those particles.

Particle dimensions were considered in reference to critical absorption depths for germanium, iron, and titanium. The basic relation is [13]

$$I_c/I_\infty = 1 - \exp(-a\rho x_c) \tag{5}$$

where I_c is the intensity for critical depth and I_∞ that for infinite depth, $a = (\mu_1 \csc \theta_1 + \mu_2 \csc \theta_2)$, μ_1 and μ_2 are absorption coefficients of the incident and emergent beams, respectively, θ_1 and θ_2 are the incident and emergent angles, respectively, ρ is the density, and x_c is the critical depth.

The effective wavelength of the incident beam is unknown, and a value was assumed for each metal element in the calculations: 1.05 A for germanium, 1.60 A for iron, and 2.28 A for titanium. Appropriate units were substituted in equation (5) and I_c/I_∞ was arbitrarily assigned the value 0.99. The critical depth values obtained were germanium 43 μ, iron 11 μ, and titanium 12 μ. The particle dimensions selected for experimental measurement were greater than the critical depth in each case except for the 20 μ germanium particles.

A method of segregating metallic particles into selected size ranges by micro and conventional sieving has been described previously [9]. Selected particle sizes of germanium, iron, and titanium thus segregated were used in the present study. Since their different-size-particle segregates were available from a previous study, an order of magnitude difference in particle size of both iron and germanium was selected for measurement in the present study. The film used for supporting the metal particles was 0.00025-in. Mylar, which is essentially transparent to X-rays. The addi-

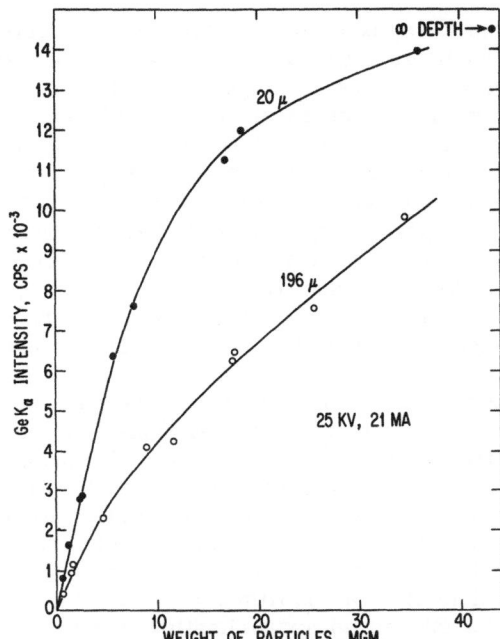

Fig. 7. Fluorescent intensity of germanium metal particles on film.

tional contribution of the film preparation to air and instrumental X-ray scatter was found to be practically negligible, but correction for the intensity contributed by the latter two sources was applied to all experimental measurements to obtain a net value for the particles.

Mylar-film circles (2.8 cm²) were used in the preparation as support for the metal particles by applying a thin coating of petroleum grease to a circle surface so that the particles would be held in position on the film. The particles were deposited onto the film through an appropriately fine sieve from a distance of 12 in. or more above the Mylar circle. The circle was weighed to five significant places with a microbalance before and after the deposit of particles to obtain a tare weight. A second circle, also coated with grease, was applied so that the metal particles would be sandwiched between and held firmly by the layers.

The cell containing the Mylar-deposit preparation was rotated at a rate of 30 rpm during the fluorescence measurement to minimize the effect of the nonuniformity of deposit on the film. Three or more replicate measurements were then made in which 25,600 or more counts, depending on the amount of deposit, were taken. Tungsten radiation was used for exciting each deposit. A scintillation counter was used as the detector, with pulse-height discrimination of the resolved spectrum from a LiF crystal.

Figure 7 shows graphically how the fluorescent intensity of germanium varies with the weight of the particles deposited on the film. Plots are given for 20- and 196-μ particles. The correlation points of intensity vs. weight show a greater scatter from the curve for the large particles than for the small ones. It may be noted that the intensity for a weight of 36 mg of 20 μ particles was 96% that of an "infinite depth" (several grams on the same area) of these particles, i.e., for a depth exceeding the depth of penetration of the X-ray beam into the sample. The curves—especially that for 20μ —are approximately linear for only a few milligrams of deposit. The tendency toward decrease in slope with weight of particles deposited on the film is shown in both curves, the decrease being more marked for the smaller size particles. Further interpretation of particle fluorescence can be derived from these curves in terms of change in sensitivity with the quantity of particles deposited and with the fraction of the total film area covered by them.

The sensitivity (change in intensity with unit change of weight) is shown graphically as a smooth curve for each particle size in Fig. 8. For the $20-\mu$ particles the sensitivity for a 36-mg deposit is only 8% of that for 1 mg. The corresponding decrease for the $196-\mu$ deposit is 28%, i.e., the rate of change in sensitivity over the same range is much greater for the small particle size of germanium than for the large. The sensitivity for a 1-mg deposit of $20-\mu$ particles is almost twice that for $196-\mu$ particles.

An examination of the relationship of sensitivity to the film area covered by the particles is of interest. The shape of each particle was, of course, not that of a uniformly symmetrical solid, but varied somewhat from one particle to another. Furthermore, the particles were deposited in a random manner onto the surface and therefore they were not spatially separated in a uniform or regular manner, either in lateral distance or in depth. Hence a calculation of the film area covered by a given weight of particles can be made only as an approximation, but this approximation is probably a fair estimate of the correct magnitude for the area covered. The fraction of the area covered was calculated from the known area of the film and density of the metal, assuming that the particles were uniform cubes of the respective nominal sizes.

Figure 9 shows the relationship of sensitivity to calculated area. The curves, of course, have the same general symmetry of those of Fig. 8, but the abscissa (percent area) scales are different. Separate area scales were used to plot the relationships for the small and large particles.

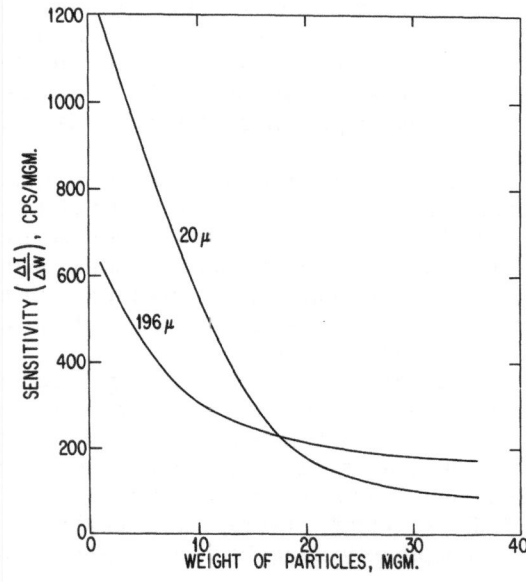

Fig. 8. Relationship of sensitivity to weight of germanium particles.

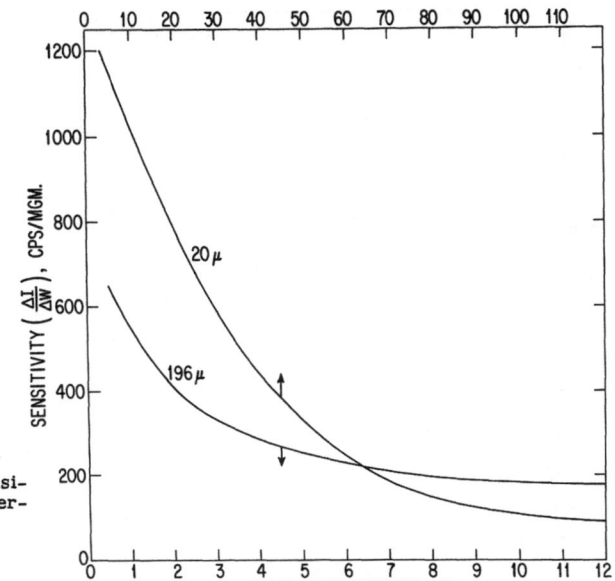

Fig. 9. Relationship of sensitivity to area covered by germanium particles.

The sensitivity for the 196-μ particles decreased as the area covered increased, although the maximum area covered was only 12% of the total film area. This result indicates that extensive linearity was not achieved by the separation of the deposited particles. Although inhomogeneity of the deposit may produce clumps or aggregates which affect absorption, and in turn linearity, the appearance of the deposits in the film preparations did not indicate that this occurred to a significant extent. It is concluded, therefore, that sensitivity is lessened by increase of the effective absorption of the deposit, even with a particle attenuation in which only 12% or less of the surface is covered. The effective absorption appears to be further increased by particle interaction.

The above results indicate not only that the linear range in amount of deposit is rather narrow, but that the particle size and distribution of the deposit are highly significant in its effect on the emitted intensity.

The general symmetries of the curves for iron particles in Fig. 10 are similar to those of Fig. 7 for germanium. A deposit of 33 mg of 20-μ particles on the 2.81 cm^2 film area gave an intensity which was 93% of that for "infinite depth" of the particles. The curves exhibit approximate linearity for small amounts of deposit—up to about 3 mg of 20 μ and 18 mg of 180 μ — then decrease in slope as the amount of deposit is further increased.

The sensitivity decreased 40-fold as the quantity of 20-μ particles is increased from 1 to 35 mg. The same increase in quantity of 180-μ particles decreased the sensitivity only about threefold.

The plots of sensitivity vs. area covered by the iron particles demonstrated, as with the germanium particles, that extensive particle separation does not provide a constancy in sensitivity, even for the 180-μ particles.

Figure 11 exhibits approximate linearity for 20-μ titanium particles up to about 5 mg of deposit. A deposit of 36 mg emitted an intensity which was 81% of that for "infinite depth."

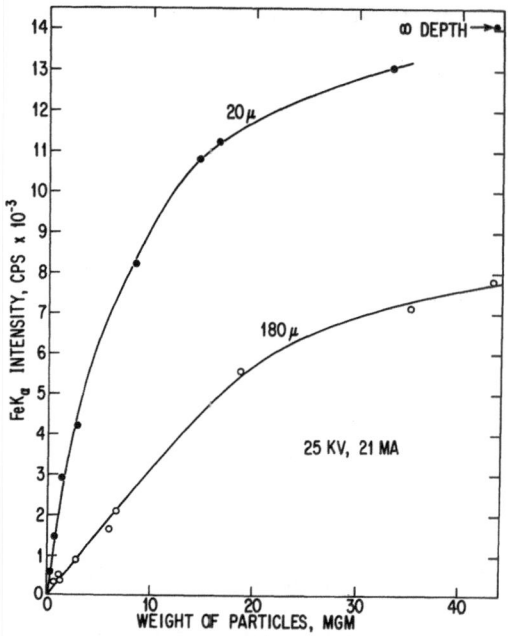

Fig. 10. Fluorescent intensity of iron metal particles on film.

Sample Dilution

The second method of reducing absorption–enhancement effects to negligible levels is through extensive dilution of the sample in a carrier material—often a material of low atomic number · The analyst may sometimes have the good fortune to receive a sample type already diluted, "ready-made" for his inspection. The carrier not only minimizes interaction between the sample components themselves but also provides a specimen type with an almost constant absorptive property. This may be accomplished by liquid solution, a fusion melt, or dilution of a powdered sample in another powder. In either a liquid solution or fusion melt an

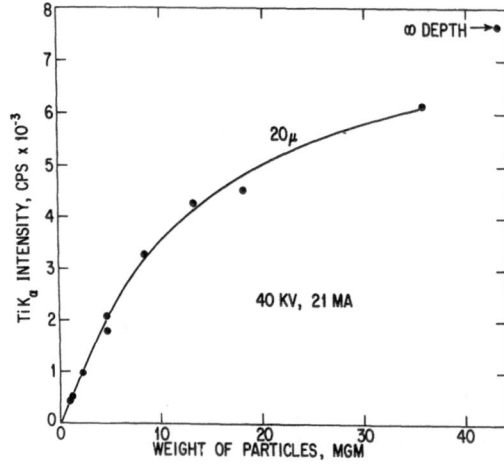

Fig. 11. Fluorescent intensity of titanium metal particles on film.

TABLE III
Precision of Results*

Sample, %	No. of reading pairs	Std. dev., %	Std. dev. expressed as % UO_2	UO_2 in solution, mg/ml
15.0	9	0.426	0.064	1.5 ± 0.006
	9	0.298	0.045	1.5 ± 0.004
17.0	9	0.495	0.084	1.7 ± 0.008
20.0	9	0.458	0.091	2.0 ± 0.009
	9	0.446	0.089	2.0 ± 0.009
22.0	9	0.436	0.096	2.2 ± 0.009
	9	0.295	0.065	2.2 ± 0.006
24.0	9	0.461	0.110	2.4 ± 0.011
24.0	9	0.473	0.113	2.4 ± 0.011
25.0	9	0.474	0.118	2.5 ± 0.012

*From Silverman et al., Anal. Chem. 29:1762 (1957).

additional advantage is achieved, namely, the state of inhomogeneity or particle division which influences the effective absorption of solid samples, i.e., the "particle effect," is eliminated. Consideration will now be given to methods of reducing absorption effects through sample dilution.

Chemical Solution of Sample

Chemical solution affords a very effective means of minimizing matrix interaction effects and eliminating small-order inhomogeneities and surface effects in metallic samples. A method for determining uranium in fuel elements composed of stainless steel and uranium dioxide which involves a chemical solution technique has been described by Silverman et al. [20]. It was found that large amounts of iron, chromium, or nickel in the solution do not affect the determination of uranium.

A standard deviation corresponding to 0.5 to 1% of the uranium measured was observed. Examples of precision are reflected by the data of Table III. A comparison of X-ray with chemical results on the same samples is given in Table IV.

Jones and Ashley [12] determined nickel, chromium, molybdenum, and niobium (columbium) in stainless steels by solution X-ray spectrography. The first three were determined in solution, but since niobium is difficult to maintain in complete solution it was separated and determined independently. Sulfuric acid in fixed quantity was specified for the solution matrix and no internal standard was necessary. Scintillation counter detection with pulse-height discrimination was employed in the measurement of fluorescent intensity. Correction for background was necessary for attaining the required precision for molybdenum, but not for chromium or nickel. The standard deviation of X-ray values from accepted values of NBS standard steel samples shown in Table V was less than 1% of the amount of each element, and the precision was within 1%.

TABLE IV
Comparison of X-ray and Chemical Methods*

| Lab. No. | UO₂, % | |
	X-ray	Wet chemical
1	17.6	17.3
2	15.8	15.9
3	15.4	15.5
4	16.2	16.3
5	16.5	16.2
6	15.2	15.2
7	15.0	15.1
8	15.8	16.0
9	16.0	15.8

*From Silverman et al., Anal. Chem. 29: 1762 (1957).

Fusion Methods

Fusion of a solid sample prior to X-ray analysis has several advantages. One important requirement is that flux material absorption of the characteristic X-rays from the sample be low so as to avoid drastic fluorescent intensity loss, and this usually requires the use of a low-atomic flux. Even then, intensity loss of low-atomic-number elements may present a problem.

A method of sample preparation through 100-fold dilution of the sample in borax, followed by fusion of the mixture, has been applied in the analysis of minerals and ores by Claisse [4]. The dilution reduces matrix effects to a negligible level and fusion eliminates the particle-size effects which are

TABLE V
Chromium and Nickel in Standard Stainless Steel
Samples*

| Sample | % Chromium | | % Nickel | |
	Nominal	X-ray	Nominal	X-ray
NBS 121b	17.68	17.59	11.14	11.12
		17.54		11.10
		17.58		11.12
		Av. 17.57		Av. 11.11
NBS 846	18.37	17.94	9.10	9.14
		17.73		8.96
		17.71		8.98
		Av. 17.79		Av. 9.03
BCS 246	18.8	18.52	12.1	12.36
		18.68		12.30
		18.73		12.29
		18.72		12.27
		Av. 18.66		Av. 12.30

*From Jones and Ashley, Anal. Chem. 31:1629 (1959).

so prominent in analyses of ores. Claisse concludes that because of the particle effect, accurate analyses cannot be made on any substance in the powder form. Other workers have found that the magnitude of the particle-size effect depends considerably upon the average particle size distribution and the specific identity of the components in the system under study [9]; i.e., some systems in powder form can be dependably and accurately analyzed in that state [6]. Moreover, Heidel and Fassel [10] found that in the analysis of scandium ores, dry grinding of the sample with a matrix containing silicon carbide and vanadium pentoxide in the ratio 5:3:3 provided better precision than borax fusion. In the fused preparations having a dilution ratio of 5:3:92 (borax) the net intensity of vanadium was reduced threefold and of scandium ninefold in comparison with dry grinding.

The accuracy of the X-ray borax fusion technique on sulfide ores of known composition is indicated by a mean deviation of 0.1 to 0.2% from the known value. Comparisons are shown in Table VI.

The influence of matrix was dramatically shown by Claisse through diluting a fixed amount of iron in powder and in fused matrices. In low-atomic-number matrices, e.g., alumina, silica, or borax, the measurements agreed closely and even exactly with theory; in the oxides of the higher-atomic-number elements, tantalum, chromium, and tin, the measured values were about one-third the theoretical values, as shown in Table VII. It was observed that the addition of a small amount of potassium pyrosulfate in fusion further reduced the matrix effect.

Powder Dilution

The powder dilution technique requires the selection of materials which have low-absorption properties for the X-rays to be measured. Gunn [7] applied a matrix dilution technique using lithium carbonate—starch as a carrier for elements covering a broad range in atomic number from 20 to 42. A 1:1 weight ratio of lithium carbonate and starch provides a diluent carrier which has high purity, low absorption, good mixing properties, and

TABLE VI
Comparison of Assay Results for Sulfide Ores*

% Iron			% Zinc			% Lead			% Copper		
Given	Found	Dev.	Given	Found	Dev.	Given	Found	Dev.	Given	Found	Dev.
11.8	11.6	0.2	(x)	(x)		0.89	0.80	0.1	0.26	0.22	0.04
11.8	12.0	0.2	11.7	11.7	0.0	49.1	49.0	0.1	0.07	0.27	0.20
17.6	17.6	0.0	13.0	13.3	0.3	35.6	35.8	0.2	0.07	0.08	0.01
24.8	24.8	0.0	14.0	13.9	0.1	17.4	18.2	0.8	0.08	0.24	0.16
27.7	27.7	0.0	5.5	5.7	0.2	26.1	26.0	0.1	0.12	0.24	0.12
29.8	29.8	0.0	8.8	8.7	0.1	5.74	5.74	0.0	0.45	0.45	0.00
34.2	34.7	0.5	2.6	2.6	0.0	1.05	1.05	0.0	0.16	0.00	0.16
36.5	36.5	0.0	3.3	3.9	0.6	1.30	1.60	0.3	(x)	(x)	
36.9	36.8	0.1	6.9	6.9	0.0	3.44	3.55	0.1	(x)	(x)	
40.2	40.2	0.0	4.7	4.7	0.1	1.40	1.35	0.05	1.40	1.40	0.00
38.6	38.9	0.3	4.5	4.3	0.1	0.78	0.90	0.1	0.91	0.95	0.04
Mean deviation		0.1	Mean deviation		0.15	Mean deviation		0.2	Mean deviation		0.1

*From Claisse, Norelco Reporter 4:3 (1957).

TABLE VII
Matrix Effect on the Analysis of a Sample Containing 20% Fe_2O_3*

| | | % Fe_2O_3 found | |
20% in Fe_2O_3	Nondiluted powders	Sample fused in borax	Sample fused in borax and $K_2S_2O_7$
Al_2O_3	20.0	20.0	20.0
SiO_2	19.0	19.7	20.1
Ta_2O_5	7.4	19.2	19.8
Cr_2O_3	5.1	19.4	19.7
SnO_2	5.6	18.3	19.2

*From Claisse, Norelco Reporter 4:4 (1957).

is suitable for preparing briquetted specimens for X-ray inspection. The finely powdered sample is diluted 20-fold in this carrier. In applying a dilution technique of this type to elements differing considerably in atomic number, some compromise between minimizing matrix effects and reducing the intensity of low-atomic-number elements through dilution has to be made. A universal or ideal dilution method completely applicable to all composition types is not known. Obviously, the technique becomes more precise and the degree of compromise less significant for samples in which the number of components is few or the atomic-number range of the elements is narrow. In the analysis of alumina-supported molybdenum—cobalt catalysts the powder dilution technique reduces absorption—enhancement effects to negligible levels and the X-ray analyses are in excellent agreement with chemical results.

The effectiveness of powder dilution also was tested by analyzing blends containing a number of elements differing somewhat in atomic number. The results for Blend A shown in Table VIII are for metals ranging from calcium (A. N. 20) to molybdenum (A. N. 42). Preliminary considerations indicated that chromium would be affected by the proximity of other interfering lines; the high value obtained for this element appeared to bear this out. An explanation for the low value for molybdenum may be that the briquetted wafer was not of infinite depth to the MoK_a X-rays. Without chromium, the average deviation of analysis from synthesis is 0.6% in this sample of broad range in elements. Variations from synthesis for Blend B range from 0.2 to 2.2 for components in the sample at about the 15% level.

The influence of elements in the sample other than those used to establish the calibrations is of interest. Interaction effects of elements of lower or of somewhat higher atomic number than the group of elements selected for study should be more detectable than of those measured in the intermediate range. The results of measuring blends of this type are shown in Table IX.

The deviations as a whole are somewhat greater than observed for calibration blends. The presence of a large amount (46%) of highly absorptive lead in the same composition produces a negative relative error ranging from as much as 20% for iron to 61% for molybdenum.

The presence of sodium chloride produces a relative positive error of 12 and 16% for iron and molybdenum, respectively. The absorptivity of

TABLE VIII
Analysis of Synthetic Blends*

Element	% added	% found	Difference
Blend A: 10% of each metal oxide			
Ca	4.0	4.6	+0.6
Fe	7.0	7.0	0.0
Mo	6.7	4.5	−2.2
Cu	8.0	9.0	+1.0
Zn	8.0	7.7	−0.3
V	5.6	5.7	+0.1
Mn	6.3	7.0	+0.7
Cr	6.8	10.5	+3.7
Co	7.1	6.9	−0.2
Ni	7.9	8.0	+0.1
Blend B: 20% ZnO, As_2O_3, Cr_2O_3; 15% NaCl, $CaSO_4$; 10% Fe_2O_3			
Zn	16.1	14.3	−1.8
As	15.1	17.3	+2.2
Cr	13.7	13.5	−0.2

*From Gunn, Anal. Chem. 29:184 (1957).

sodium chloride for their radiations is low, which produces high results in their measurements. A negative error of 20% is observed for calcium, however. The very high interpolated absorption coefficient of chlorine for calcium radiation (K_α 3.35 A), offers an explanation for the low calcium value.

For the blend containing components with several atomic numbers (sodium chloride, silica, alumina), the absolute errors for iron and molybdenum are of the order of those observed in calibration, i.e., the composition effect is minor. Again the value for calcium is low, probably because of chlorine absorption.

These measurements clearly indicate that the unknown should contain elements that are relatively close to the calibration standards in atomic number. For samples containing extremes in atomic number, i.e., where elements of low and high absorptivity coexist, reference standards should be

TABLE IX
Influence of Elemental Composition*

Composition of sample	Added	Found	Difference	% Rel. error
50% PbO, 50% Fe_2O_3	Fe 35.0	28.0	− 7.0	−20
50% PbO, 50% MoO_3	Mo 33.3	13.0	−20.3	−61
50% PbO, 50% CaO	Ca 35.8	28.0	− 7.8	−22
50% NaCl, 50% Fe_2O_3	Fe 35.0	39.3	+ 4.3	+12
50% NaCl, 50% MoO_3	Mo 33.3	38.5	+ 5.2	+16
50% NaCl, 50% CaO	Ca 35.8	29.8	− 6.0	−20
30% NaCl, 26% SiO_2, 14% Al_2O_3,	Fe 7.0	8.5	+ 1.5	+18
10% Fe_2O_3, 10% MoO_3, 10% CaO	Mo 6.7	7.5	+ 0.8	+11
	Ca 14.3	9.0	− 5.3	−37

*From Gunn, Anal. Chem. 29:184 (1957).

prepared containing the same elements in approximately the same concentrations as in the unknown to ensure a reasonable degree of accuracy.

COMPENSATION BY COMPARISON OR INTERNAL STANDARDIZATION

The second general method of solving absorption and enhancement problems is to compensate for them by comparison or by internal standardization. This empirical approach usually involves no attempt to minimize absorption—enhancement effects or to correct for them through knowledge of the fundamental characteristic absorption properties of the sample; rather, the approach is one of experimental analogy. This is, of course, not unique to X-ray spectrography since other fields of analytical spectroscopy also employ it. For example, the process of excitation in conventional quantitative optical emission is so complex and difficult to control that comparison and internal standardization are the only practical methods which can be used.

Compensation by Family of Curves

Davis and Van Nordstrand [5] have encountered elemental interaction effects in determining additive elements (barium, calcium, and zinc) in lubricating oils, even at concentrations less than 1% of the blended composition. The effect of 0.6% barium on 0.1% zinc in an oil was to lower the zinc intensity by about 30%, for example. A family of curves was found to be a satisfactory solution for this problem. To correct zinc determinations made in the presence of barium, a family of curves was prepared by increasing the concentration of barium in sets of standards in steps of 0.1%. Interpolation between curves is made to obtain the correct value for zinc on an unknown. Typical curves are shown in Fig. 12.

The X-ray method of additive analysis requires 3 to 12 min per element, depending on the level of precision selected. The accuracy was found to be at least equal to that of chemical methods. Comparisons between X-ray and chemical results are given in Table X.

Compensation by Internal Standard

The use of an internal standard added to the substance to be analyzed is a well-recognized means of compensating for matrix or interaction effects. The element used as an internal standard should be close in atomic number to the element measured and homogeneously dispersed in the sample containing it. In this way the absorption properties and fluorescent yields of the standard and unknown will be similar. The amount of added internal standard also is significant. Ideally, the fractions of unknown and internal standard element should be about the same and the intensity ratio of the lines should be unity near the midconcentration range of the element being measured. Practically, because of a number of considerations, this ideal may be difficult to achieve. Usually, depending on the absorption characteristics involved, the concentration of the added internal standard should not exceed 10% of the total blended composition [3]. More than this fractional amount may modify the matrix to such an extent as to produce errors

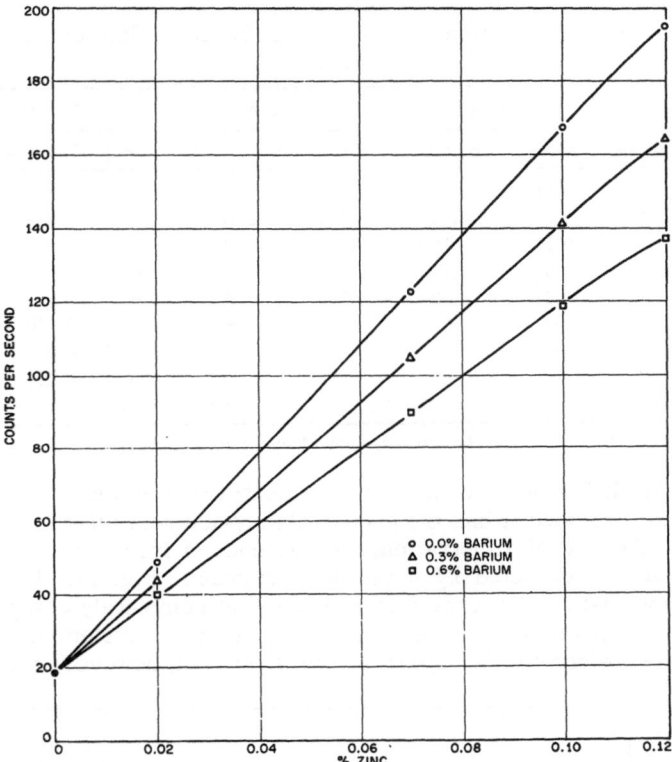

Fig. 12. Calibration curve for zinc using a molybdenum tube at 50 kV and 30 mA with helium. [From Davis and Van Nordstrand, Anal. Chem. 26:973 (1954).]

in the results. The effects of absorption and enhancement on line pairs is of paramount importance in the selection of an internal standard. Adler and Axelrod [1] concluded that neighboring elements in the periodic series make good internal standards for one another by compensating for absorption and enhancement, except when there is a strong line between the edges involved or when there is an edge between the compared lines. In the selection of a substance for use as an internal standard these conclusions should serve as a guiding principle to be applied to any specific problem or system of interest to the analyst.

A method for the determination of columbium (present name—niobium) and tantalum in rare-earth monazite ores in which solution of the sample and reprecipitation is effected has been reported by Mortimore et al. [15]. The purpose of sample dissolution and precipitation is twofold: to minimize particle effects and to incorporate an internal standard intimately with the sample. Zirconium is used as the internal standard for columbium, and hafnium for tantalum.

These workers point out that the influence of other elements in the matrix which absorb selectively may be compensated in two ways: (1) a family of curves for each element may be plotted in which each curve

TABLE X
X-ray and Chemical Analyses for Barium, Calcium, and
Zinc*

	% Barium		% Zinc		% Calcium	
Sample	X-ray	Chemical	X-ray	Chemical	X-ray	Chemical
1	0.51	0.51	0.068	0.068	–	–
2	0.22	0.22	0.101	0.100	–	–
3	0.11	0.12	0.089	0.090	–	–
4	0.08	0.10	0.086	0.086	–	–
5	0.24	0.23	0.103	0.100	–	–
6	–	–	–	–	0.021	0.020
7	–	–	–	–	0.099	0.097
8	–	–	–	–	0.196	0.195
9	–	–	–	–	0.586	0.582

*From Davis and Van Nordstrand, Anal. Chem. 26:973 (1954).

represents a different amount of an element; (2) an internal standard may
be employed, provided it has the proper absorption relationship to the other
elements.   By way of illustration, the influence of iron on the columbium–
tantalum ratio is depicted by a family of curves in Fig. 13.   These curves
clearly show that the internal standard does not completely compensate for
absorption effects in an iron matrix.   The measured wavelengths, 0.746 and
1.522 A, are relatively far apart in the spectrum, however, and their re-

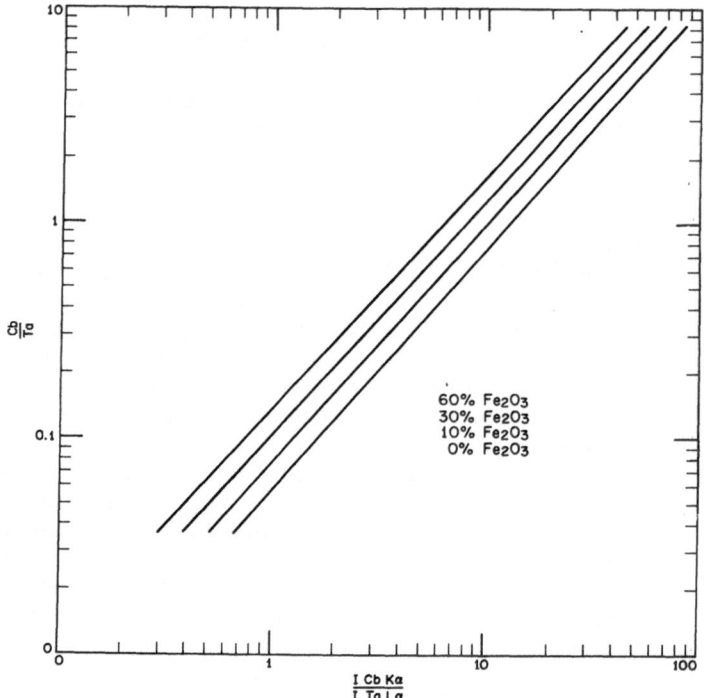

Fig. 13. Effect of iron on columbium–tantalum ratio.

spective mass absorption coefficients for iron are 46 and 285; this probably accounts for the lack of complete compensation.

MATHEMATICAL CORRECTION

The basic premise of quantitative X-ray fluorescence analysis is that the intensity of characteristic fluorescent X-rays emitted by an element under arbitrary conditions of excitation is proportional to the quantity of the element which emits them. The absolute amount of characteristic radiant energy emitted by an element per unit time could be calculated approximately from fundamental physical relations, but at the present state of development this approach is much too involved and tedious for general use. In practice the intensity of radiation emitted by an element in an unknown is not of itself an absolute parameter for measuring concentration; rather, its value in terms of relative internal consistency or in comparison with that of a reference substance—a common practice in several types of instrumental analyses—is used as a measure of concentration. Absorption-enhancement effects have a highly significant bearing on the emitted intensity of the element. Successful efforts have been made to correct for these effects, both from theoretical and empirical standpoints.

Theoretical Calculation of Intensity

Nokes [16] has shown how the relative fluorescent intensity emitted by an element in a binary iron—nickel alloy can be calculated by using fundamental expressions in which the intensity is related to the absorption constants of the elements and the thickness of the sample. The experimental data employed for illustration was that previously reported by Koh and Caugherty [11]; the relative intensity of nickel for selected concentrations is used for comparison. Nokes points out that the fundamental integral expression relating these parameters cannot be evaluated. Hence, simplifying assumptions are made to accomplish the calculations. An imaginary depth below the surface is assumed to be the origin or virtual source of the fluorescence. The absorption coefficient values of nickel for different composition ratios of the alloy are used in an exponential equation to evaluate the depth t of the virtual source. The incident value I_0 is obtained by multiplying the relative intensity of pure nickel by the weight fraction of nickel in the sample.

Example: The alloy considered is 50% nickel. The linear absorption coefficient of Ni K_α in this alloy is 1877 and that of pure nickel 514. The intensity of Ni K_α emitted by the alloy was found experimentally to be 34 relative to 100 for pure nickel.

$$I = I_0 \exp [- \mu t]$$

Then
$$\ln (34/50) = - (1877 - 514)t$$

$$t = 2.83 \times 10^{-4} \text{ cm}$$

By calculations based on nine composition ratios, Nokes found the average value of t to be 2.75×10^{-4} cm. The average value of t can be employed to calculate nickel intensity I for a given composition. For 30% nickel,

$$l = 30 \exp\left[-(2424)(2.75 \times 10^{-4})\right]$$

This value is normalized by multiplying it by (100/86.8) to give 17.8. (The same calculation for 100% nickel gave 86.8.) Nokes' comparison of intensities for different iron-nickel compositions is shown in Table XI.

Thus, reasonable agreement between experimental observation and Nokes' [16] calculation from theory is shown. Nokes also made further applications to ternary alloys in which enhancement ("negative absorption") occurs by applying absorption–enhancement coefficients with promising results.

Experimental Prediction of Intensity

Mitchell [14] has proposed a practical experimental means for predicting the intensity of an element in any matrix. Essentially, the method consists of introducing the element or its compound in known concentration into a matrix composed of another substance and measuring its fluorescent intensity. A few key matrix materials are selected and a plot is made relating the atomic number of the matrix to the intensity emitted by the element under investigation. The observed intensity obviously results from and accounts for matrix interaction, i.e., absorption–enhancement effects. The variation in intensity is, in general, inversely proportional to the mass absorption coefficient. However, exceptions are found near the absorption edge of the measured element. The matrix elements having emission lines just on the short-wavelength side of the absorption edge of the measured element enhance its intensity, thus partially compensating for the matrix absorption loss. This is shown for the element tin in Fig. 14 for various matrices. The highest intensity for Sn K_α in a matrix greater than A.N. 25 is not shown in the matrix of lowest absorption, A.N. 47, silver, but in A.N. 54, xenon, several atomic numbers higher. Extrapolations located the peak at A.N. 54. A reminder that the K-absorption edge of tin is at 0.425 A and the K_α emission line of xenon at 0.416 A explains the enhancement of the characteristic tin emission in theoretical mixture with xenon. Mitchell applied this concept to several other matrix mixtures. The validity of this

TABLE XI
Intensities for Different Iron-Nickel Composition

% Nickel	I_{obs}	I_{calc}	Difference
10	6	5.1	−0.9
20	13	11.0	−2.0
30	20	17.8	−2.2
40	28	25.6	−2.4
50	34	34.4	+0.4
60	43	44.6	+1.6
70	53	55.9	+2.9
80	66	68.8	+2.8
90	82	83.4	+1.4
100	100	100	0

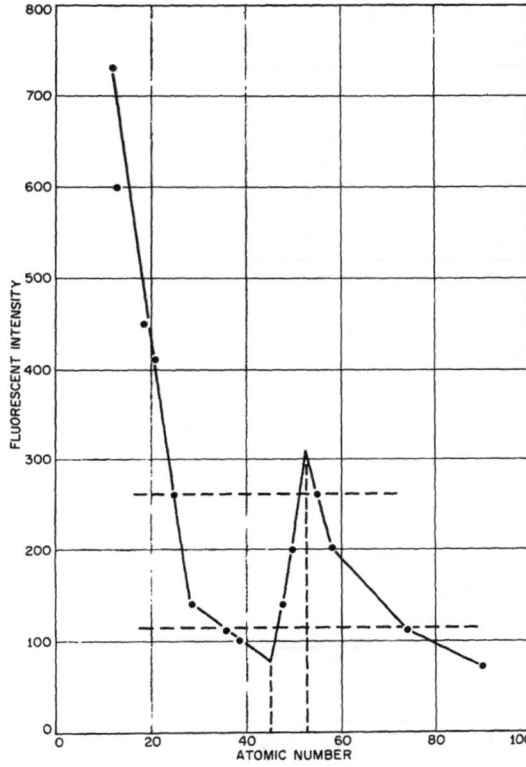

Fig. 14. Fluorescent intensity–atomic number pattern for 10% SnO . [From Mitchell, Anal. Chem. 33:917 (1961).]

approach in predicting the fluorescent intensities of elements in matrices of varying atomic number is shown in Table XII. The agreement between predicted and observed intensities is remarkably good.

Empirical Absorption–Enhancement Indices

A calibration method for determining four components—chromium, iron, nickel, and molybdenum—in a quaternary metal alloy was developed by Beattie and Brissey [2] in terms of theoretical considerations of absorption

TABLE XII
Accuracy of Estimated Fluorescent Intensities*

Element	Matrix		Estimated intensity	Measured intensity
	Substance	A.N. of metal		
10% SnO$_2$	Ni$_2$O$_3$	28	185	191
	V$_2$O$_5$	23	400	407
	Sm$_2$O$_3$	62	170	169
10% MoO$_3$	CeO$_3$	58	95	95
	Sm$_2$O$_3$	50	135	139
10% Fe$_2$O$_3$	V$_2$O$_5$	23	185	183
	CuO	29	608	610

*From Mitchell, Anal. Chem. 33:917 (1961).

TABLE XIII
Chemical Analyses of Binary Alloys*

Analysis for	Binary partner			
	Cr	Fe	Ni	Mo
Cr	—	48.24	48.07	74.28
Fe	50.83	—	51.53	65.33
Ni	48.19	46.65	—	53.70
Mo	23.53	34.44	46.27	—

*From Beattie and Brissey, Anal. Chem. 26:980 (1954).

TABLE XIV
Intensity Ratio for Binary Alloys*

Radiating element	Binary partner			
	Cr	Fe	Ni	Mo
Cr	1.000	1.760	1.815	1.841
Fe	3.360	1.000	1.613	1.835
Ni	2.860	3.670	1.000	1.954.
Mo	3.770	2.787	2.461	1.000

*From Beattie and Brissey, Anal. Chem. 26:980 (1954).

TABLE XV
Absorption Parameters*

Radiating element	Absorbing element			
	Cr	Fe	Ni	Mo
Cr	1.000	0.721	0.813	2.660
Fe	2.482	1.000	0.676	1.582
Ni	1.863	2.420	1.000	1.108
Mo	0.877	0.9435	1.260	1.000

*From Beattie and Brissey, Anal. Chem. 26:980 (1954).

TABLE XVI
Intensity Ratios for Ternary and Quaternary
Alloys*

Alloy No.	$[R_a = I_{aa}/I_{as}]$†			
	Cr	Fe	Ni	Mo
1	3.610	1.918	14.34	—
2	3.535	6.620	2.152	—
3	2.477	3.665	5.373	—
4	5.04	2.346	6.640	14.34
5	4.01	1.942	14.72	41.35
I_{aa}‡	597	2795	3270§	1076

*From Beattie and Brissey, Anal. Chem. 26:980 (1954).
†Intensity in counts per second on pure element corrected
for background.
‡Intensity from a 0.75 × 0.75 in. sample area. Bent mica
crystal was 0.0016 in. thick.
§Nickel intensities were determined with a W-target X-ray
tube operating at 24.1 mA and 48.5 kVp. All other elements
were determined under 49.6 mA and 48.5 kVp.

enhancement effects. A set of simultaneous equations contains empirical corrections for these effects, the unknowns in them representing composition. Coefficients of the equations are based on the internal consistency existing between intensity measurements of each pure metal element and the intensity measurements of the binary alloys of the four metals. For the latter, known composition values must be available.

Equations for the expressions from which the absorption enhancement indices are calculated are

$$A_{ab} = (W_a/W_b)(R_a - 1) \qquad (6)$$

and

$$A_{ab} = (W_b/W_a)(R_b - 1)$$

where A_{ab}, A_{ba} is the absorption parameter for the absorption–enhancement of a to b, b to a, and thus for other pairs.

W_a, W_b, etc. are the fractional weights of elements in an alloy. W_a/W_b, W_b/W_a, W_a/W_c, etc. are the weight ratios of the elements in an alloy and R_a, R_b, etc. are the ratios of the intensity of the pure metal element to that of the element in a compound. Then,

$$- (R_a - 1) W_a + A_{ab} W_b + A_{ac} W_c + \ldots = 0 \qquad (7)$$

$$A_{ba} W_a - (R_b - 1) W_b + A_{bc} W_c \ldots = 0 \qquad (8)$$

$$A_{ca} W_a + A_{cb} W_b - (R_c - 1) W_c + \ldots = 0 \qquad (9)$$

$$W_a + W_b + W_c + \ldots = 1 \qquad (10)$$

and so on for additional components.

(Note that the sum of the weight fractions of the metal elements in the alloy must equal unity.)

After the absorption parameters are evaluated experimentally, they are inserted into equations (7)-(9). For an unknown alloy, the intensity ratio for each element to that of the pure element is found. This is also inserted into equations (7)-(9), and a simultaneous solution is found through the use of algebraic and determinant manipulations to obtain the weight values for the alloy.

By application of the absorption enhancement equations to the analysis of five typical alloys these investigators found that the relative deviation from chemical analysis averaged 3.4% for the four metal elements, the maximum deviation being 8.6%.

ACKNOWLEDGMENT

The use of data reported in the literature by various workers which illustrate different methods of solving absorption and enhancement problems in X-ray emission spectrography is gratefully acknowledged.

REFERENCES

1. I. Adler, and J. M. Axelrod, Spectrochim. Acta 7:91 (1955).
2. H. J. Beattie, and R. M. Brissey, Anal. Chem. 26:980 (1954).
3. L. S. Birks, X-ray Spectrochemical Analysis (Interscience Publishers, Inc. New York, 1959).
4. F. Claisse, Norelco Reporter 4:3 (1957).
5. E. N. Davis, and R. A. Van Nordstrand, Anal. Chem. 26:973 (1954).
6. G. V. Dyroff, and P. Skiba, Anal. Chem. 26:1774 (1954).
7. E. L. Gunn, ibid. 28:1433 (1956).
8. E. L. Gunn, ibid. 33:921 (1961).
9. E.L. Gunn, Advances in X-ray Analysis, Vol. 4, W.M. Mueller, ed. (Plenum Press, Inc., New York, 1958) p. 247.
10. R. H. Heidel, and V. A. Fassel, Anal. Chem. 33:913 (1961).
11. P. K. Koh, and B. Caugherty, J. Appl. Physics 23:427 (1952).
12. R. W. Jones, and R. W. Ashley, Anal. Chem. 31:1629 (1959).
13. H. A. Liebhafsky, H. G. Pfeiffer, E. H. Winslow, and P. D. Zemany, X-ray Absorption and Emission in Analytical Chemistry (John Wiley & Sons, Inc., New York, 1960).
14. B. J. Mitchell, Anal. Chem. 33:917 (1961).
15. D. M. Mortimore, P. A. Romans, and J. L. Tews, Appl. Spectroscopy 8:24 (1954).
16. G. E. Nokes, A.S.T.M. Special Technical Publication No. 157, Philadelphia, Pa., 57, 1953.
17. T. N. Rhodin, Jr., Anal. Chem. 27:1857 (1955).
18. M. L. Salmon, Advances in X-ray Analysis, Vol. 5, W. M. Mueller, ed. (Plenum Press, Inc., New York, 1961) p. 389.
19. J. Sherman, Spectrochim. Acta 7:283 (1955).
20. L. Silverman, W. Houk, and L. Moudy, Anal. Chem. 29:1762 (1957).
21. G. Von Hevesy, Chemical Analysis by X-rays and Its Applications, (McGraw-Hill Book Co., New York, 1932).

X-Ray Spectrometric Analysis of Alloyed Copper

T. J. Cullen

U. S. Metals Refining Company
Carteret, New Jersey

Pure copper has properties which are desirable in only a few applications. Over the years, a series of alloyed coppers has been developed which changes only certain physical characteristics while maintaining the electrical and heat conductivities. These alloys contain only 0.01 to 3.0% of the alloying element(s). Production quality control of these alloyed coppers presents the problems of rapid and accurate analysis. At U.S. Metals Refining Company, these alloys are produced in a continuous casting operation. Master alloys are added, at a predetermined rate, in a hopper between an oxygen-free furnace and a continuous casting apparatus. This situation leads to sampling problems which are circumvented by taking drilled samples. The drillings are briquetted and placed in a rotating sample holder. Surface irregularities are compensated by the use of scattered radiation intensity ratio techniques. Interelement effects of ternary and quaternary alloys are discussed, and sensitivities of various elements in copper are given.

For the purpose of this paper, a differentiation between the terms "alloyed copper" and "copper-based alloys" is made. Alloyed copper is essentially pure copper to which small amounts of elements are added to obtain certain properties, while having a minimum effect on other desirable properties. Copper-based alloys are those alloys, such as brasses, bronzes, and monels, which have properties characteristic of the alloy rather than of the copper.

Pure copper, such as Oxygen-Free High-Conductivity Brand copper (99.99+% copper), produced at the United States Metals Refining Company, exhibits properties different from oxygen-bearing copper of the same degree of purity, excepting the oxygen content. It may be considered that oxygen-bearing copper is an oxygen alloy of copper.

The main difference between OFHC-Brand copper and oxygen-bearing copper is that, in the oxygen-bearing copper, a eutectic of $Cu-Cu_2O$ is formed. The eutectic has a low solubility in copper and is precipitated in the grain boundaries, thus inhibiting grain growth. The effect of smaller grain structure can be detected in the ductility and fatigue endurance tests. The OFHC-Brand copper shows high ductility and fatigue endurance. Also, oxygen-bearing copper ruptures when drawn to a fine wire, while OFHC-Brand copper can be drawn.

Some applications require certain properties of OFHC-Brand copper, while requiring properties of greater tensile strength or lower creeping characteristics at elevated temperatures.

Silver-bearing OFHC-Brand copper which contains 15 o.p.t. silver is used in electrical motors and generators because oxygen-bearing and OFHC-Brand copper will cause failures due to creeping or embrittlement of the copper at the operating temperatures. Tellurium-bearing OFHC-Brand copper, containing 0.5% Te, has the property of being free-machining. Tellurium–copper produces small chips when drilled or machined, rather than long spirals which tend to foul machinery. Chromium–titanium–tin OFHC-Brand copper has the ability to produce a hard copper when heat-treated.

These alloys are produced at United States Metals Refining Company by melting electrolytically pure copper cathodes, adding master alloys, and casting under an inert gas atmosphere. Obtaining a sample from such a process is difficult. One cannot obtain a furnace sample without the danger of furnace contamination by air. Thus, the sample must be obtained from the cast product. In order to ensure rapid quality control, drillings or cut sections must be obtained, drillings being the most rapid method. Melting and recasting the drillings is time-consuming and may alter the composition of the sample.

Since only two- or three-significant figure results are needed for process control, it was decided to determine if briquetted drillings could be used as the sample form for X-ray spectrographic determinations of alloying elements. Since the briquetted drillings have gross surface irregularities, it was thought that rotation of the briquette and/or the use of scattered background [1] or coherently scattered X-ray tube target characteristic radiation internal standardization [2, 3] would give suitable results.

Scattered radiation internal standardization has been shown to compensate for several instrumental and sample effects. If the scattered radiation wavelength is properly chosen, absorption and enhancement effects are corrected. Particle size, sample surface to X-ray tube target distance variations, and power fluctuations can be detected and compensated.

A simple experiment was designed to determine the best technique for analyzing the briquetted drillings. A single sample of tellurium–copper alloy was chosen and analyzed by four techniques—measurement of the Te K_α

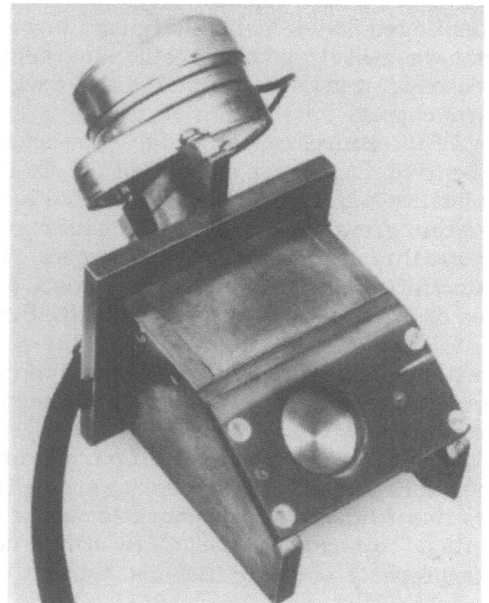

Fig. 1. Rotating sample drawer.
(Courtesy Spex Industries, Inc.)

radiation only, and measurement of the TeK_α and WL_{γ_1} radiation intensity ratios on both static and rotating samples.

Tellurium–copper was specifically chosen because it is a free-machining type of alloy; the tellurium, by metallurgical design is concentrated in the grain boundaries of the alloy, thus a segregation effect is also present.

A General Electric XRD-5 X-ray spectrograph was operated at 50 kV and 50 mA, using a tungsten-target X-ray tube, lithium fluoride analyzing crystal, 0.010 Soller slit, and a scintillation counter. A Spex Industries sample turntable was used in both the rotating and static sample measurements (Fig. 1). The turntable rotated at 25 rpm. Different areas of the sample were measured statically by switching the turntable on and off between measurements, thus selecting areas at random. A minimum of 50,000 counts was taken for each measurement, so that deviations above two significant figures of the final concentration could be attributable to the sample or technique and not the statistical counting variations.

Approximately 10 g of the drillings were briquetted at 15,000 psi pressure in a $1\frac{1}{4}$-in.-diameter die, producing a briquet $\frac{1}{4} \pm \frac{1}{16}$ in. thick. Ten briquets of the same sample, containing 0.60% Te were produced and the intensity measurements made.

Table I shows the results and degrees of deviations encountered from the intensities obtained. The best results were obtained from the measurement of the TeK_α-WL_{γ_1} ratio on a rotating sample.

Measurements of the ratios on samples containing 0 to 1.00% Te gave a straight-line calibration when the ratios were plotted against the tellurium concentration. Variations of ±0.01% Te were obtained, which is the same degree of variation as obtained from routine chemical analysis.

Continuously cast billets of alloyed copper may show variations of concentrations of alloying elements from two sources: first, from the rate of addition of the master alloy, and second, from the natural segregation due to metallurgical design or segregation occurring during solidification while

TABLE I
Deviations
(Briquetted Drillings, 0.60% Tellurium)

	Static		Rotating	
	$TeK_{\alpha},$ counts/sec	$\dfrac{TeK_\alpha}{WL_{\gamma_1}}$	TeK_α counts/sec	$\dfrac{TeK_\alpha}{WL_{\gamma_1}}$
	320	4.36	321	4.36
	312	4.35	318	4.35
	309	4.28	310	4.35
	330	4.39	320	4.37
	322	4.38	332	4.37
	300	4.29	331	4.36
	315	4.30	328	4.36
	332	4.38	335	4.34
	360	4.43	335	4.39
	311	4.35	319	4.37
Avg.	321.1	4.351	324.9	4.362
Std. Dev. %	4.95	1.17	2.74	0.48

casting.   Control of both types of segregation is obtained by sampling the cast alloy at predetermined intervals and obtaining the drillings from select-ed parts of the billet.

In the case of tellurium–copper alloy, natural segregation occurs from metallurgical design and from the rate of solidification. Since the tellurium is concentrated in the grain boundaries of the copper grains, the size of the crystals will determine the locations of the tellurium. When a billet is cast, the outside surface is cooled and solidified first, thus producing small grains.   The production of the small grains will squeeze the tellurium to-ward the center of the billet. The center of the billet will solidify last, and with slow solidification, larger crystals are formed. Thus, the tellurium concentration, through a billet, will show variations. Figure 2 shows this variation on an 8-in.-diameter billet. Tellurium concentration was within the specifications of 0.50 ± 0.10% Te across the diameter, thus the process was in control with regard to addition of the master alloy and casting rate.

As can be seen from Fig. 2, a representative sample can be obtained by taking drillings from a point midway from the outside surface to the center. The average analysis of this midsection is 0.557% Te, while the average of the entire area, by integration of the averages of the sections over the areas, is 0.556% Te.   Reproducibility tests, tellurium determinations on ten drill holes taken in the center of the middle section, show variations of ±0.02% Te from the average.   Repetitive determinations on the same alloy briquet show precision of ±0.01% Te.

Zirconium–copper alloys do not exhibit segregation as do the free-ma-chining tellurium–copper alloys.   Rotation of the briquetted drillings show improvement because the surface irregularities are averaged out. Compar-ison of results with photometric determinations are excellent (Table II). The same technique and instrumental parameters as the tellurium deter-

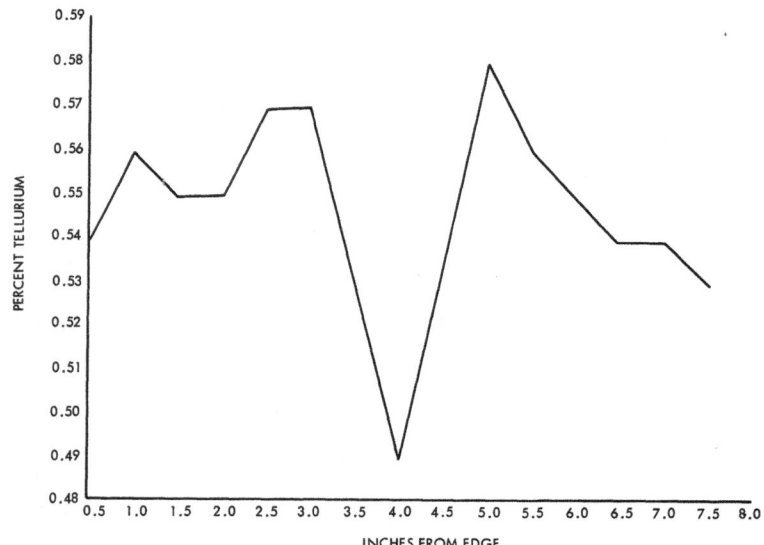

Fig. 2. Variation of tellurium concentration across an 8-in. billet.

TABLE II
X-Ray and Chemical Determinations of Zirconium

% Zirconium			Difference, % Zr	
Static	Rotating	Chemical	Static	Rotating
0.090	0.085	0.085	0.005	0.000
0.085	0.081	0.081	0.004	0.000
0.099	0.095	0.099	0.000	0.004
0.106	0.104	0.103	0.003	0.001
0.131	0.137	0.138	0.007	0.001
0.099	0.103	0.103	0.004	0.000
0.106	0.106	0.107	0.001	0.001
0.082	0.087	0.087	0.005	0.000
0.175	0.170	0.170	0.005	0.000
0.182	0.186	0.134	0.002	0.002
0.106	0.095	0.099	0.007	0.004

*Photometric Alizarin Red S determinations, average of two determinations on two separate portions of samples reported.

mination were used for zirconium. ZrK_α and WL_{γ_1} intensity ratios were found to be adequate.

The use of scattered background intensity at 0.53 A for the ratio measurement in both the tellurium and zirconium determinations gave results of quality equal to that obtained from the use of the coherently scattered X-ray tube characteristic radiation. However, in the case of an addition of a second alloying element to zirconium–copper, for example arsenic, the ratio of ZrK_α/WL_{γ_1} was still valid, while the ratio of $ZrK_\alpha/B.G.$ 0.53 A was decreased.

The effect of interelemental absorption of radiations can best be studied using chromium–copper and chromium–titanium–tin–copper alloys. The effect of titanium and tin on the intensity of chromium radiations is explainable by mass absorption calculations. It is interesting to note that in making mass absorption coefficient calculations on background or coherently scattered radiation term (the denominator) has been shown theoretically and experimentally to be empirically corrected for matrix effects.

Table III shows the results of mass absorption coefficient correction using two alloys. The first is a 0.50% Cr alloy of copper, and the second is a 0.50% Cr, 1.7% Ti, and 2.7% Sn alloy of copper. The sums of the mass absorption coefficients times the concentrations of the elements, including cop-

TABLE III
Mass Absorption Coefficient Corrections
0.50% Cr-Cu Alloys

Alloy	Ratio CrK_α /B.G. 0.53A	$\Sigma(\mu \times c)$	Corrected ratio
Cr–Cu	0.785	158.65	—
Cr–Ti–Sn–Cu	0.688	179.49	0.778

Ratio 0.778 = 0.49% Cr

μ—mass absorption coefficient.
c—concentration.

TABLE IV
Determination of Silver in Various Types of Copper
Briquetted Drilling Technique

Type of copper	Ag, o.p.t.	Ratio AgK_α/B.G. 0.53A	Ag found
OFHC-Brand	0.2	1.201	0.0
OFHC-Brand	15.0	1.450	15.2
OFHC-Brand	60.0	2.179	60.0
Oxygen-bearing	12.5	1.400	12.3
Oxygen-bearing	27.5	1.648	27.7
Anode	45.0	1.942	45.5
Anode	78.2	2.460	77.8
Anode	90.2	2.602	90.5
Blister	100.5	2.822	100.0
Blister	110.3	3.002	111.0

per and chromium, are 158.65 and 179.49, respectively. To correct the ratio, CrK_α/B.G. 0.53 A, of the Cr–Ti–Sn–Cu alloy to the ratio calibration of the Cr–Cu alloy, one multiplies the ratio obtained from the Cr–Ti–Sn–Cu alloy by the correction factor of 1.1314, which is obtained from the ratio of the mass absorption corrections of the two alloys.

Silver-bearing copper is one of the most important alloyed coppers produced. Both oxygen-bearing and OFHC-Brand silver coppers are produced. Silver is also found in anode, blister, and tough-pitch copper. It was desired to determine silver in all types of copper by one method and one calibration. Briquets of millings, drillings, or filings were found to be adequate samples to obtain this calibration. Table IV shows results obtained from the various samples. A background ratio of $Ag K_\alpha$/B.G. 0.53 A was used. An accuracy of ±0.35 o.p.t. was obtained; this is ±12 ppm.

Many different alloys have been studied using this technique. Both production and experimental alloys have been studied and analyzed successfully. Of the light elements determined, of special note are sulfur and phosphorus. With a quartz crystal, helium atmosphere, a flow-proportional counter, and a thin-window chrome X-ray tube run at 50 kV and 35 mA, sulfur–copper alloys can be analyzed when the sulfur content is about 0.2%. Phosphorus at the 0.005% level can be detected, but the low intensities obtained eliminate the possibility of routine phosphorus determinations. Improvements in instrument design may soon make the determination of phosphorus in phosphorus–copper alloys (0.005–0.3% P) possible. Titanium with the TiK_α at 2.75 A can be measured with a scintillation counter tube, and in the case of Cr–Ti–Sn–Cu alloys, the titanium is determined successfully in this manner.

In the course of studying the various alloys, a limit of detection of the elements in copper has been determined. Table V gives the limits of detection obtained using the scintillation counter, lithium fluoride crystal, tungsten-target X-ray tube and power parameters of 50 kV and 50 mA. Rotated briquets of the alloys and the ratio technique were used.

The low limit of detection of silver is explainable from advantageous excitation by the tungsten-target X-ray tube and the efficient counting by the scintillation counter. Limits of detection from solid polished samples are

TABLE V
Limits of Detection of Elements in Copper Briquetted Copper Alloys

Element	Ratio	Limit of detection, %
Ti	TiK_α/B.G. 0.53A	0.05
Cr	CrK_α/B.G. 0.53A	0.01
Mn	MnK_α/B.G. 0.53A	0.01
Fe	FeK_α/B.G. 0.53A	0.007
Zn	ZnK_α/WL_{γ_1}	0.005
As	AsK_α/WL_{γ_1}	0.005
Sb	SbK_α/B.G. 0.53A	0.005
Sn	SnK_α/B.G. 0.53A	0.005
Ag	AgK_α/B.G. 0.53A	0.00035
Cd	CdK_α/B.G. 0.53A	0.001
Pb	PbK_α/WL_{γ_1}	0.01
Ni	NiK_α/B.G. 0.53A	0.01

lower than those limits of detection from briquetted drillings. Solid pieces have a greater effective surface area for excitation X-rays to strike and for the fluorescent radiations to leave. Also, statistical counting variations arise from only one measurement rather than two measurements. Limits of detection of various elements in copper obtained from solid samples have been reported by many authors. This limit of detection varies from instrument to instrument, even of the same manufacture. Careful alignment of Soller slits and adjustment of crystals and operating voltages of counter tubes are absolutely necessary for comparison of data. Because of these factors, the data presented in Table V should not be considered the optimum results obtained, but merely a guide to what can be expected from similar conditions.

X-ray spectrographic results have been found to be accurate and rapidly obtained in the quality control of continuously cast alloyed coppers. Single determinations take less than 5 min from the time the sample is received until the result is reported. The use of scattered radiation ratio methods reduce the necessity to run standards. Only one or two check standards are used to ensure that the system is under control. It has been our experience that the calibrations have been extremely stable, and show only slight, if any, change even when X-ray tubes are changed.

The use of briquetted drillings, filings, or millings has another advantage in that the sample can be submitted in almost any form and still be used with a minimum of effort.

REFERENCES

1. G. Andermann and J. W. Kemp, Anal. Chem. 30: 1306 (1958).
2. T. J. Cullen, Anal. Chem. 33: 1342 (1960).
3. T. J. Cullen, Anal. Chem. 34: 812 (1962).

Additional Papers on X-Ray Spectroscopy

- "Microprobe Analysis of Minerals and Segregates," I. Adler, U.S. Geological Survey, Washington, D.C.

- "Application of the Electron Microprobe to the Operation of an Industrial Plant," W. W. Welbon and R. J. Debeer, General Electric Company, St. Petersburg, Florida.

- "Auxiliary Equipment for X-Ray Process Control Installations," J. W. Kemp, Applied Research Laboratories, Inc., Glendale, California.

- "X-Ray Fluorescence Applied to On-Stream Analysis," W. R. Kiley, Philips Electronic Instruments, Mount Vernon, New York.

- "A Digital Computer for Optical and X-ray Quantometers," W. E. Fowler and A. J. Hartwick, Applied Research Laboratories, Inc., Glendale, California.

- "X-Ray Fluorescence Control Analysis in the Manufacture of Magnetic Alloys," J. P. Kapetan, Western Electric Company, Chicago, Illinois.

- "X-Ray Spectroscopy in Biology and Medicine IX. Microtechniques and Instrumental Sensitivity," J. C. Mathies and P. K. Lund, The Swedish Hospital, Seattle, Washington.

- "Assay of Iodine Content of Thyroid Extract by X-ray Absorption Edge Analysis," D. E. Flick, Eli Lilly and Company, Indianapolis, Indiana.

- "Recent Progress in Soft X-Ray Spectroscopy in the Wavelength Region of 13-150 Angstroms," J. E. Holliday, United States Steel Corporation, Monroeville, Pennsylvania.

Infrared–Raman Spectroscopy

Radio-Frequency Powered Raman Sources

E. R. Lippincott, F. X. Powell, J. A. Creighton,
and D. G. Jones

Department of Chemistry
University of Maryland
College Park, Maryland

The construction and performance of high-intensity electrodeless lamps, suitable as light sources for the excitation of Raman spectra, are described. The lamps are powered by a 3-kW RF oscillator operating at about 5 Mc, and intense light emissions have been obtained from mercury, helium, rubidium, cesium, potassium, cadmium, and thallium. As Raman sources, these lamps offer advantages over conventional discharge lamps because of their ease of construction and their interchangeability within their RF power source.

INTRODUCTION

Several types of light sources have been developed for the excitation of Raman spectra. These sources, or lamps, fall into two categories: those which utilize electrodes to convert electrical energy into radiant energy, and those which do not. The first type is exemplified on the one hand by the Toronto arc, developed by Walsh and co-workers [1], and on the other hand by the high-voltage helium discharge lamps developed by Stammreich [2]. The second type includes the toroidal lamps powered by microwave energy introduced by Ham and Walsh [3] and the larger lamps recently described by Lippincott et al. which obtain their power from a radio-frequency oscillator.

Because of the relative simplicity in design and construction of a supply capable of delivering a large amount of power, lamps with electrodes can usually be built with high luminous output. This type of lamp, however, is restricted in the types of material which can serve as plasma media within the lamp. Furthermore, for each plasma medium a special design of lamp envelope and power supply is usually necessary. For example, the Toronto arc technique is confined almost exclusively to low-voltage DC discharges through mercury vapor, while the high-voltage technique employs an AC discharge through low-pressure helium. On the other hand, the electrodeless lamps have been shown to be capable of maintaining a discharge through a wide variety of substances using the same power supply. Toroidal lamps are simple to construct and outgas, and with the elimination of electrode sputtering and failure of metal-to-glass seals the lamps have a long life. When powered by a microwave or RF source they are extremely bright, comparable in intensity to lamps with electrodes. Unfortunately, because of the high cost of large microwave sources, the microwave-powered lamps that have been described have been small and their total light output relatively low. This restriction is not applicable to lamps powered by RF energy, and the lamps described here and elsewhere [4], which derive their power from a 3-kW oscillator of inexpensive design, have a light output comparable to that of a large Toronto arc source.

THREE-KILOWATT OSCILLATOR

The schematic diagram for a 3-kW oscillator is shown in Fig. 1. The DC power supply for this unit is not shown; however, the one presently used by the authors will supply zero to 3000 V, depending on the Variac setting of the supply, and has a capability of 1 A. The design of the DC supply is not critical: it should be controlled by a Variac and should have a ripple not greater than about 30% of the DC voltage. In operation, the positive side of this supply is connected to the junction point of C_8 and L_6 and the negative side is grounded.

As shown, the schematic is that of a tuned plate—tuned grid oscillator. L_1C_1 comprises the tuned grid circuit, S_1 is the coarse grid current control, and C_1 is the fine control. Grid current should not exceed 150 mA; for operation at 1 to 1.5 kW, the grid current should be adjusted to 100–110 mA.

The inductor—resistor networks L_2R_2 through L_5R_5 are parasitic suppressors. The 3-kW oscillator will not deliver power to the toroidal lamp unless these networks are present. Particularly robust components are required for C_9 and L_6. C_9 is a Sangamo type G13 mica capacitor, 0.015 μF at 4000 V DC. The Sangamo G series or Sprague CC series are sufficiently rugged for this application. For L_6 the authors use two air inductors of 100 turns, each 10 in. long and 2.5 in. in diameter, in series with two inductors of 230 turns of cotton-covered wire each wound on a 1-in.-diameter Pyrex tube 10 in. long. The entire inductor is contained in a Lucite box well separated from grounded objects to which arcing may occur, and is cooled by a fan.

The inductor L_7 is the oscillator load coil and is made of 18 turns of $1/4$-in.-diameter copper tubing. The coil is 6 in. long and has an internal

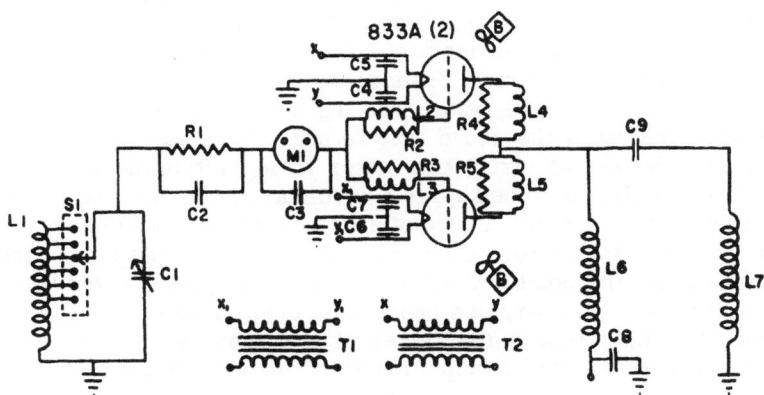

Fig. 1. Schematic diagram of 3-kW oscillator. (B) Rotron muffin fans, (C_1) 10—100 pF, (C_2) 0.005 μF, (C_3) 0.002 μF, (C_4,C_5,C_6,C_7) 500 pF, (C_8) 0.02 μF, 6kV, (C_9) see text, (L_1) 15 turns No. 14 wire 2 in. length and diameter, tapped 1 $1/2$, 3, 6, 9, 12, and 15 turns, (L_2,L_3,L_4,L_5) 4 turns No. 14 wire $3/4$ in. length and diameter, (L_6) see text, (L_7) 19 turns $1/4$-in. copper tubing 6 in. long 4 $3/4$ in. diameter, (M_1) 250 mA DC meter, (R_1) 1 K 50 W, (R_2,R_3,R_4,R_5) 47 Ω 2 W carbon, (S_1) Switch Centralab No. 2500, (T_1,T_2) Transformer Stancor P-6461.

diameter of $4\frac{3}{4}$ in. With this coil the resonant frequency of the oscillator is 4 to 5 Mc.

TOROIDAL LAMPS

The excitation of an electrical discharge and the withdrawing of power from the oscillator may be achieved by inserting a low-pressure gas, contained within a toroidal lamp envelope, inside the oscillator load coil.* At low RF power the gas in the toroid glows weakly as a result of dielectric loss. As the RF power is tuned up, the conductance of the gas increases through the generation of a greater number of ions and electrons, and at sufficiently high RF power the gas behaves as a shorted low-resistance single turn to the inducing load coil. When this occurs, large amounts of power can be fed into the gaseous discharge via transformer action, with collisions between electrons and the gas serving to dissipate the energy. Calculations indicate that the circulating current induced by this mechanism in lamps such as are described here is of the order of 60 A, and the brightness of the discharge is consequently very high.

High-intensity discharges excited by RF energy have so far been obtained by the authors in the following gases or vapors: mercury, cesium, helium, argon, potassium, cadmium, and thallium. The design of the toroidal envelope used to contain the discharges has been dictated largely by the required degree of cooling of the plasma. It is generally desirable, in order to obtain sharp spectral lines and low background continuum and to minimize line self-reversal, to maintain the walls of the lamp at as low a temperature as possible. In the case of mercury, helium, and argon it has been found possible to run lamps completely submerged in water. With the less volatile metals, such efficient cooling results in too low a vapor pressure to maintain a discharge, and in the case of potassium, rubidium, and cesium, forced air cooling has been used. For cadmium and thallium all cooling has been kept to a minimum.

The most useful toroid design, suitable for mercury, rubidium, cesium, and potassium, and employing forced air cooling, is shown in Fig. 2. It consists of an inner Pyrex lamp chamber surrounded by an outer jacket through which air is circulated. This double-toroid design is unsuitable for cooling by water. Where water cooling is desirable, a single toroid of dimensions similar to the above may be used, the entire toroid being submerged in a beaker of circulating water fitting within the oscillator load coil. Where cooling is to be kept to a minimum, as in the case of cadmium and thallium lamps, a single toroid of silica or Vycor is suitable.

The toroidal lamps (or in the case of the submerged lamps, the surrounding beaker) are usually wrapped with asbestos paper coated with moist titanium dioxide. When dry this produces a highly reflecting surface which increases the light flux at the center of the toroid, and in the case of the cadmium and thallium lamps helps maintain a high lamp temperature.

---

*It is important that the lamp be inserted directly into the load coil (L_6 in Fig. 1) rather than into some other parallel LC circuit remote from the oscillator and drawing power from L_6 along a transmission line. In such an arrangement, for efficient power transfer it is necessary to tune the remote LC circuit to the resonant frequency of the oscillator. Since the condition of the lamp plasma markedly affects the mutual inductance of this LC circuit, a slight change in the plasma may detune the LC circuit, causing the plasma to change further and the lamp regeneratively to extinguish.

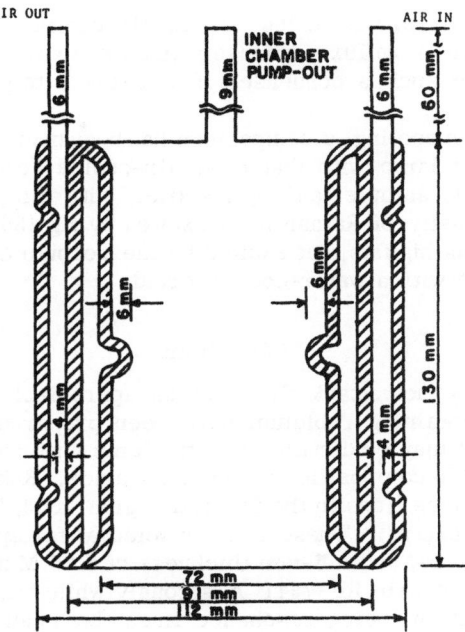

Fig. 2. Toroidal lamp, longitudinal cutaway.

The initiation of a plasma discharge requires that the substance which is to serve as the discharge medium be in a gaseous state. Toroids containing permanent gases require no treatment prior to ignition; ignition can be accomplished by inserting the toroid into the oscillator load coil and advancing the DC supply voltage until ignition occurs. Most metals, however, require preheating prior to ignition. Preheating can be realized by two methods: (a) insertion of the toroid into a furnace and (b) injection of an inert gas which, when converted into a plasma, will serve to vaporize the metal. Argon is suitable as a starter gas, since it readily produces a plasma, but once the metal is vaporized it is usually necessary to pump the argon away in order to obtain a lamp spectrum free from argon lines. In practice the authors have found the use of sealed-off lamps and the furnace preheating technique the more convenient.

Pyrex is slowly attacked and darkened by metal vapors, particularly by potassium and to a lesser extent by rubidium. This attack can be hindered by coating the glass, for example, with boric acid and potassium fluoride (Corning K-coat), but the coating is difficult to apply effectively, and except with potassium it is often more convenient to clean and re-evacuate the lamps when necessary.

TYPES OF LAMPS

Mercury

In order to obtain a low spectral background and high intensity of Hg 4358 A, it is necessary to keep the vapor pressure of mercury in the toroid

low.   With double toroids using forced air cooling it has been the usual
practice to attach a reflux condenser directly to the lamp, while without
forced air cooling such a condenser is necessary to prevent quenching of
the discharge.

With forced air cooling, a comparison has been made of the performance
of an RF-powered lamp with that of an air-cooled Toronto arc. Using the
same power input, sample, and optics (i.e., with simply the lamps interchanged) the intensity of Raman lines excited by Hg 4358 A from the toroid
was very nearly as high as that excited by the Toronto arc. No comparison
has yet been made with a water-cooled toroid.

Rubidium

Rubidium does not attack Pyrex at an appreciable rate, and uncoated
double toroids containing rubidium have been prepared for routine Raman
work.   The metal may be distilled into the lamp under vacuum or prepared
in situ by thermal decomposition of rubidium azide. Rubidium gives rise to
two intense resonance lines in the far-red region (7800, 7947 A) suitable for
exciting Raman spectra.   These are sufficiently far apart to be separable
with chemical filters, and a 5-mm thickness of 0.25 M neodymium chloride
can be used to absorb Rb 7947 A strongly while transmitting 90% of Rb
7800 A.   In addition, being resonance lines, both can be very selectively
absorbed by a rubidium vapor filter. Thus, in addition to a neodymium
chloride prefilter, the authors have used a rubidium vapor postfilter cell,
8 cm long and heated to about 250 C, to absorb selectively the primary line
(Rb 7800 A) from the Raman spectrum.

In addition to the red lines, rubidium has two weaker resonance lines in
the blue region (4201, 4215 A).   While not as intense as lines of similar
wavelength from other sources, they have been found useful, for example,
in investigations of the low-lying lattice vibrations of solids, where the
strong primary line scattered by the solid can be very selectively absorbed
by a rubidium vapor postfilter.

Because of resonance line self-reversal, forced-air-cooled rubidium
lamps are advantageously run at low power input. With the lamps described here, increase of input power above 400 W results in a decrease in
the intensity of resonance emission.   More efficient cooling to keep the
rubidium vapor pressure low and minimize self-reversal is desirable.

Helium

Because of the high ionization potential of helium, the full 3-kW capacity
of the oscillator described here is necessary to maintain a plasma. However, the possibility exists of maintaining a plasma at a lower input by decreasing the number of turns in the oscillator load coil, thereby increasing
the transformer ratio for power transfer into the lamp. With a load coil
of 7 turns, $3\frac{1}{2}$ in. long and $4\frac{1}{4}$ in. in diameter, the resonant frequency of the
oscillator is about 10 Mc, and in a toroid 8 cm long of inner and outer di-
ameter 6.0 and 8.7 cm, and completely submerged in water, it is possible
to maintain a helium plasma with a power input of 1500 W.

Helium has intense emission lines at 5875 and 6678 A which are suitable
for the excitation of Raman spectra. However, in the region of He 5875 A,

weak band systems due to the presence of excited diatomic species He_2^* in the discharge often interfere with the Raman spectrum. The intensity of these bands decreases with fall of pressure, and in the high-voltage helium lamps of Stammreich a helium pressure of 3 mm Hg is normally used. Excited by an RF field, a plasma can be maintained at helium pressures as low as 0.1 mm Hg, with consequent very low intensity of He_2^* bands. Below 0.5 mm pressure, however, there is a marked falling off in the intensities of He 5875 and 6678 A, and the authors have adopted 0.5 mm Hg as an optimum helium lamp pressure.

Cesium and Potassium

Cesium attacks glass more slowly than rubidium, and the general construction and behavior of cesium lamps is very similar to those of rubidium. With potassium the attack is more rapid and necessitates a protective coating on the glass. Cesium has intense resonance lines at 9521, 8943 A and at 4555, 4593 A, and potassium at 7664, 7699 A.

Cadmium

Cadmium has an intense emission at 6438 A, and in a silica toroid with no provision for cooling, a plasma may be maintained with a power input of 1500 W. The temperature of the lamp is such that the level of spectral background is high; however, cadmium appears to have no advantage over helium as a discharge medium for Raman excitation in the red region.

Thallium

Thallium is the most refractory metal in which a plasma discharge has been achieved by use of the 3-kW oscillator. A power input of 2.5-kW was necessary to maintain the discharge, and the silica toroid reached a dull red heat. Thallium is of interest because it emits an intense resonance line (5350 A) in the green region of the spectrum.

Because of the ease with which lamps can be interchanged within the oscillator load coil, the RF-powered light sources described above constitute an extremely versatile Raman source. The authors have found that lamps containing respectively mercury, helium, and rubidium provide a satisfactory coverage of the spectrum for routine Raman work. These lamps are relatively easy to prepare, have a long operational life, and give rise to the following lines suitable for Raman excitation: Hg 4358 A; Hg 5461 A; He 5875 A; He 6678 A; Rb 7800 A. With improvements in near-infrared detectors, the cesium lamp may also become a valuable source for routine Raman work.

ACKNOWLEDGMENT

This work was supported in part by grants from the Atomic Energy Commission and Advanced Research Projects Agency, United States Department of Defense.

REFERENCES

1. H. L. Walsh, M. F. Crawford, and W. J. Staple, Nature 164:737 (1949).
2. H. Stammreich, Spectrochim. Acta 8:41 (1956).
3. N. S. Ham and A. Walsh, Spectrochim. Acta 12:88 (1958).
4. F. X. Powell, O. Fletcher, and E. R. Lippincott, Rev. Scientific Instruments 34:36 (1963); F. X. Powell, E. R. Lippincott, and D. Steele, Spectrochim. Acta 17:880 (1961).

The Infrared Determination of Oil in Phosphorus

A. Davis, H. M. Joseph, and L. E. Tufts

Hooker Chemical Corporation
Niagara Falls, N. Y.

The chemical methods for determining oil in the parts per million range in yellow phosphorus are, at best, time-consuming and often very difficult to reproduce. A simple rapid infrared method dependent upon the measurement of the C—H bands of the oil dissolved in a molten sample of yellow phosphorus is routinely used for this determination in our laboratory. This method requires approximately 15 to 30 min per analysis in contrast to 2 to 4 hr for other methods.

INTRODUCTION

The oil content in yellow phosphorus is important to the phosphorus-producing industry. The color of many products produced directly from phosphorus correlates with the oil content of the phosphorus used in its manufacture. The oil content is normally low, ranging from 5 to 200 ppm.

Several chemical methods are now used for this determination. In one, the phosphorus is chlorinated under a blanket of phosphorus trichloride, then hydrolyzed, and the organic material is extracted with carbon tetrachloride. The carbon tetrachloride solution is evaporated and the nonvolatile residue determined by weighing. The resulting residue must be corrected for any chlorination of the oil that took place. This method is long, requiring 3 to 4 hr per sample, and the reproducibility is poor.

A second chemical method utilizes the color resulting from the oxidation of yellow phosphorus with nitric acid. This method, while faster (2 hr) than the chlorination method, is very dependent upon accurately reproducing the oxidation conditions.

A fast direct method of analyzing the oil in yellow phosphorus was needed. Infrared spectroscopy was thought to offer the possibility of a simple solution to this analytical problem. The method described in this paper resulted.

PROCEDURE

A sample of yellow phosphorus is melted under hot (about 60 C) water and transferred to a separatory funnel containing a layer of water. A Beckman 1-cm quartz cell (No. 46006) is suspended in a large test tube. This test tube is continuously purged with CO_2. (Figure 1 is a diagram of the apparatus used to transfer the liquid phosphorus to the cell.) The cell is filled to within approximately 1 cm of the top with liquid phosphorus, and a layer of water is added to protect the phosphorus from air.

A heated cell holder (Fig. 2) is maintained in the spectrophotometer at 70 C. The cell is placed in the cell holder, which in turn is in the sample

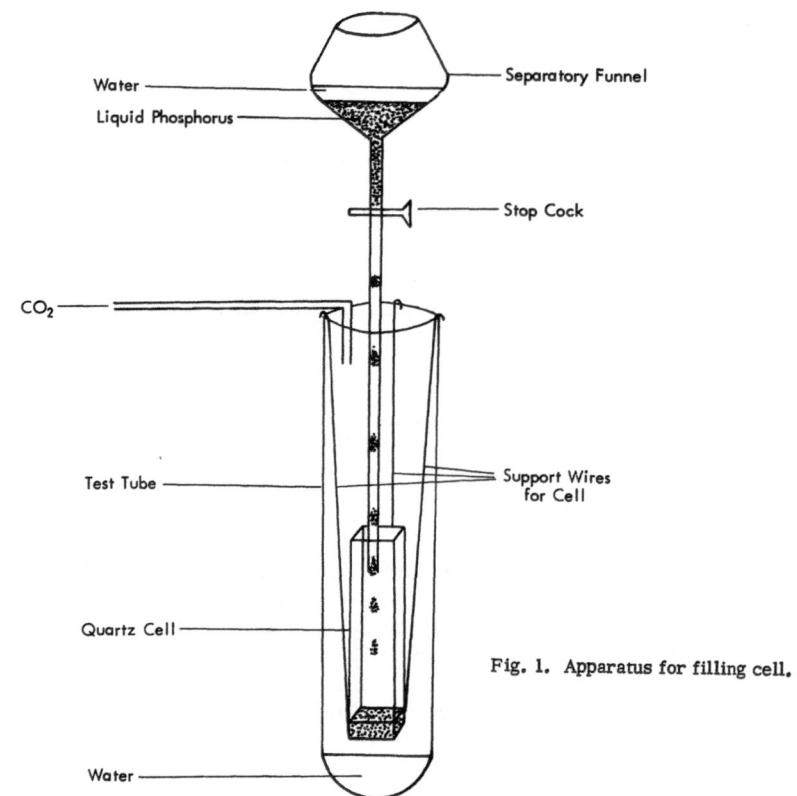

Water

Liquid Phosphorus

Separatory Funnel

Stop Cock

CO_2

Test Tube

Support Wires
for Cell

Quartz Cell

Fig. 1. Apparatus for filling cell.

Water

beam of the infrared spectrophotometer. A second quartz 1-cm cell filled with carbon tetrachloride is mounted in the reference beam. The spectrum is recorded from 2.7 to 3.8 μ. Both a Beckman IR-4 and a Perkin-Elmer 137 have been used for this analysis. The operating conditions for the Beckman IR-4 are listed in Table I, while standard slow scan conditions are used for the P–E 137.

A typical spectrum of phosphorus containing oil is shown in Fig. 3. The C—H stretching due to unsaturated material is located at 3.31 μ, while the C—H stretching due to saturates is at 3.44 μ.

Calibration and standardization is accomplished by use of a mixture containing 25 mg/liter benzene and 25 mg/liter heptane in spectroscopic-grade carbon tetrachloride. The base-line method is used for determining absorptivities at 3.31 and 3.44 μ. The sum of the concentration of saturated and unsaturated organic materials is calculated using the absorptivities for heptane and benzene. This is reported as oil content of the yellow phosphorus sample.

DISCUSSION

The correlation between nitric acid and infrared methods for oil in phosphorus is satisfactory. The correlation between infrared and the chlorination method is inconsistent.

Top View

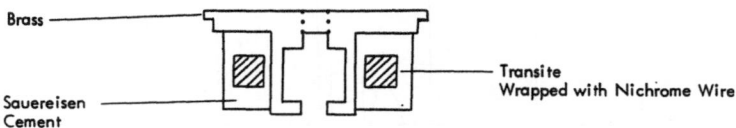

Brass

Sauereisen
Cement

Transite
Wrapped with Nichrome Wire

Front View

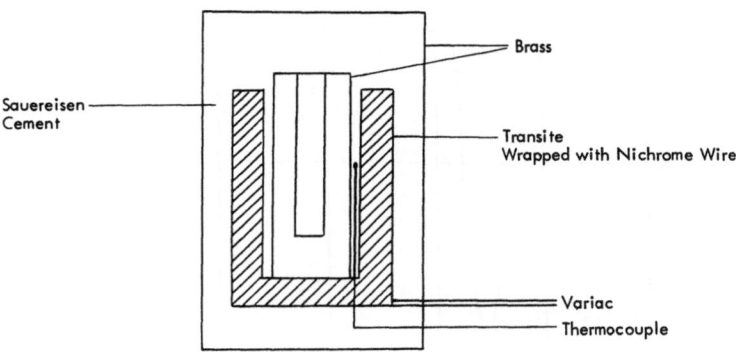

Sauereisen
Cement

Brass

Transite
Wrapped with Nichrome Wire

Variac

Thermocouple

Fig. 2. Heated cell holder.

TABLE I
Condition for Analysis on Beckman IR-4

Speed	0.5 μ/min
Period	8
Slit schedule....................	0.5 mm at 9.3 μ
Gain single-beam—double-beam ratio ..	1 : 3.3
Heated cell holder temperature.......	70°C
Cells........................	Two 1-cm quartz Beckman No. 46006

Reference cell filled with spectrograde carbon tetrachloride

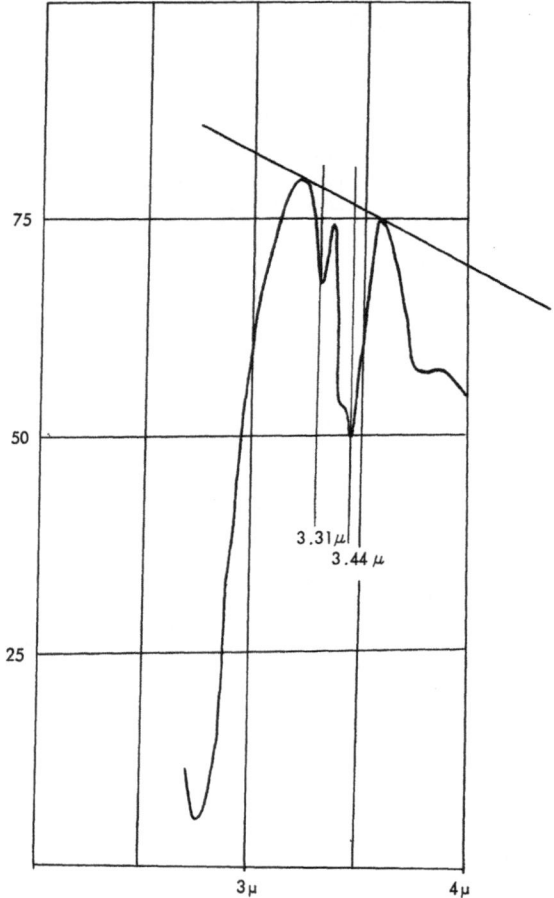

Fig. 3. Spectrum of phosphorus containing 100 ppm oil.

The color of an ammonical solution of phosphorus pentasulfide correlates well with the oil content found by infrared in the phosphorus used as
the raw material. Figure 4 shows infrared analyses for oil in phosphorus
vs. transmittance of an ammonical solution of phosphorus pentasulfide.

Table II shows the reproducibility between laboratories and within a
laboratory for a series of samples of phosphorus. The duplicate analyses
were run from one week to one month after the first analysis was performed.

The coefficient of variation between laboratories is approximately 10%.
While this is high for normal infrared quantitative analysis, it is very satisfactory for this determination.

Several different oils were checked to use as calibration standards. However, the ratio of saturated to unsaturated hydrogen varied from oil to oil.
When the saturated and unsaturated hydrocarbons were determined separately the resulting sum was within experimental error of the actual oil
content.

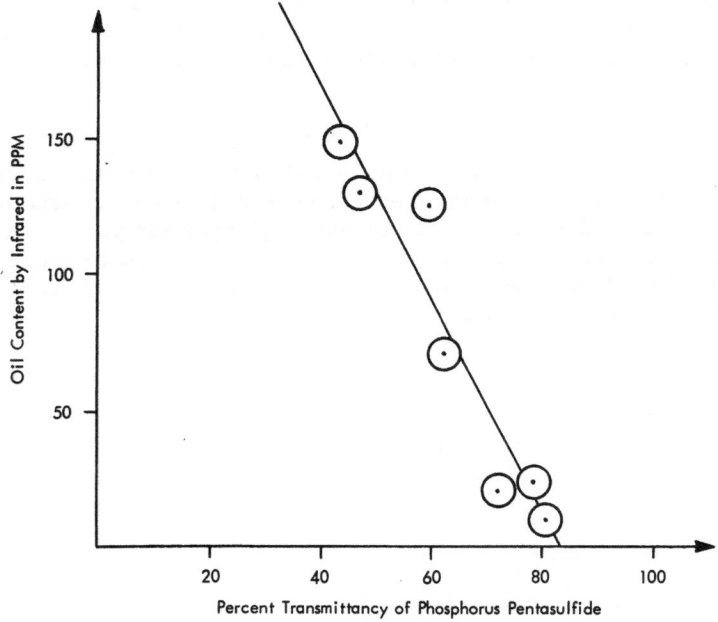

Fig. 4. Infrared analysis for oil in phosphorus vs. transmittance test for phosphorus pentasulfide.

SAFETY

Since an accidental burn involving molten yellow phosphorus can be extremely serious, safety measures should be carefully considered before attempting this analysis.

A safety shower must be available in an easily accessible spot. Several buckets of cold water immediately at hand are very useful to freeze any spilled phosphorus.

TABLE II
Comparison of Phosphorus Results by
Infrared

Oil content, ppm		
Research Department		Columbia, Tenn. Plant
44	26	33
27	22	23
40	34	32
16		16
19		17
20		18
22		27
110		122
140		146
120		128
130		142

An extra person should be available to help in case of an emergency. Both the analyst and the helper should wear protective clothing, which should include a face shield, heavy rubber gloves, and a plastic apron.

CONCLUSION

A simple method for the determination of oil in phosphorus has been developed which depends upon the measurement of the C—H absorption in the infrared. The method is fast, taking about $\frac{1}{2}$ hr per sample, or a group of 4 to 5 samples can be determined within an hour. The resulting oil determination has shown a good correlation with the color of several products produced from the phosphorus.

The Use of Thin-Layer Chromatography
With Infrared Spectroscopy

Marcia K. Snavely and Jeanette G. Grasselli

Chemical and Physical Research Department
The Standard Oil Company (Ohio)
Cleveland, Ohio

Thin-layer chromatography is a fast and economical way of separating mixtures of non-volatile liquids or solids. It requires only a small amount of sample and is applicable to problems in diversified fields. It serves as a useful complement to column and paper chromatography, and, like these methods, it requires some other analytical tool for positive identification of the separated components. In this work infrared spectroscopy has been used with thin-layer chromatography to analyze synthetic mixtures, organic reaction products, and commercial samples. Several techniques for obtaining IR spectra from the thin-layer spots will be described, and special methods of preparing the thin-layer plates will be recommended. The combined use of these two tools holds unique advantages and is certain to be employed even more frequently in the analytical and research laboratories.

INTRODUCTION

One of the limitations of such analytical instruments as infrared, ultraviolet, X-ray, and nuclear magnetic resonance is the difficulty in working with and identifying components of complex mixtures. One of the limitations of such analytical techniques as distillation, fractionation, and chromatography is that fractions separated by these techniques are not specifically identified. Therefore, the combination of analytical tools—the one used for separation or purification of components, the other used for specific identification—has recently gained greater recognition and popularity. The growing number of scientific articles describing the combined use of these techniques and instruments attest to this fact. These articles have discussed the combination of gas chromatography with mass spectroscopy [1-4], ultraviolet [5], and infrared [6-9]; column chromatography with infrared [10]; and thin-layer chromatography with ultraviolet [11-13] and mass spectroscopy [14]. It is the purpose of this paper to describe an additional combination of analytical tools—infrared and thin-layer chromatography—which has been found extremely simple to execute and can be applied to a variety of problems.

Although the beginnings of thin-layer chromatography occurred more than 30 years ago when Ismailov and Schraiber [15] examined the rings produced on layers of adsorbent powder by organic mixtures, it was not until 1956-1960 when Professor E. Stahl from the University of Saarlandes in Germany standardized the method and equipment [16-21] that it became popular, especially in Europe. The method has recently gained favor in the

United States, where equipment is now available from a number of manufacturers.*

The thin-layer method is simple in both design and concept. Flat glass plates are coated with an adsorbent layer of material. A sample is placed on one end of the plate and solvent is allowed to traverse the plate, chromatographically separating the components. As a general rule, any of the adsorbents appropriate for column chromatography can be used provided they are available as powders having a uniform, fine particle size and that this powder can be made to adhere to a glass plate. Silica gel is by far the most popular and is used for the rapid separation of neutral, acidic, and basic substances. It is available with or without a calcium sulfate binder, the latter especially suitable for the thin-layer chromatography of inorganic or hydrophilic compounds. It is also available with a fluorescent indicator, which allows the detection of separated components without recourse to dyes.

Another popular adsorbent is aluminum oxide, which is used for the separation of less polar compounds such as polyene or alkaloids. Diatomaceous earth is also available and is especially suited for partition separations of substances which are strongly hydrophilic or amphoteric. Polyamide powder can be used for separating phenols and phenolic substances, and cellulose powders are available for nucleic acid derivatives and nucleotides.

The glass plates used to hold the adsorbents are generally one of two sizes, 5 by 20 cm or 20 by 20 cm, but can vary from tiny microscope slides [22] to large plates measuring 38 by 48 cm [23].

The adsorbing material is applied to the glass plate as a slurry and allowed to dry into an adherent layer. The thickness of the layer is usually 250 μ but can be varied. Commercial applicators are available which regulate the thickness and provide a uniform layer of adsorbent. The latter is especially important because the uniformity of the layer will affect the distance and rate a sample will travel. Recent articles refer to the preparation of plates without the use of these commercial spreaders. T. M. Lees and P. J. DeMuria [24] describe a method of preparing plates using only adhesive tape and a uniform glass rod. Bravo and Hernandez [11] make their plates merely by placing the suspension on the center of the plate and moving the plate to and fro horizontally to cover the entire surface. There have also been references to a number of homemade spreaders [23, 25, 26], none of which would be too difficult to duplicate.

The coated plates must be activated by drying the plates in an oven above 100 C. The temperature and duration of the drying should be determined by the degree of activity desired. Less hydrophobic materials may require less activated layers, in which case air-dried plates may be used.

After the plates have been prepared and activated, the sample can be applied. Most of the literature references advise using a micropipet to apply the sample solution in a line $3/4$ in. from the end of the plate [27]. A plastic template is available commercially or can easily be constructed to give reproducible placement of samples. If a micropipet proves unsatisfactory, at least two more sophisticated sample applicators [23, 28] have been con-

*Brinkmann Instruments, 115 Cutter Mill Road, Great Neck, N. Y.; Research Specialties Company, 200 South Garrard Blvd., Richmond, California; Arthur H. Thomas Company, Vine Street at 3rd, Philadelphia 5, Pennsylvania; Bio-Rad Laboratories, 32nd and Griffin Avenue, Richmond, California.

structed or a micro syringe can be used. $1\,\mu l$ to 1 ml of sample solution can be applied, containing from 0.01 μg to 50 mg of sample.

After the solvent used to dissolve the sample has evaporated from the starting line, the thin-layer plate is placed in a container which has been filled with sufficient developing agent to immerse the lower edge of the plate. The containers are usually standard rectangular or cylindrical jars of borosilicate glass with ground edges. More reproducible results are obtained if one side of the jar is lined with filter paper wetted with solvent; this ensures a saturated atmosphere. The jar is covered and the solvent allowed to rise until it has reached a height of at least 10 cm, a process which rarely takes more than 30 min. The plate is then removed, air-dried to evaporate the solvent, and, if the sample is colorless, examined under UV light or sprayed with an appropriate reagent to indicate its position on the plate. A list of these indicators and their uses can be found in an excellent review article on thin-layer chromatography by E. G. Wollish, M. Schmall, and M. Hawrylshyn [27].

For the permanent recording of thin-layer chromatograms, there have been a number of methods now proposed, including the use of Polaroid film [30], common letter-copying machines [31], or the Xerox 914 office copier [32]. E. Merck (Germany) has also developed a plastic coating spray called Neaton, which hardens as a clear coating, binding the adsorbent layer and enabling removal of the layer from the glass plate. This is available in the United States from the Brinkmann Instrument Company.

One of the most difficult aspects of the thin-layer technique is the selection of a suitable solvent as the developing agent. Sometimes narrow plates (1.5 by 25 cm) or even microscope slides are used for exploratory work. In essence it is a trial-and-error determination, although the selection follows the principles of adsorption chromatography, the eluting power of solvents increasing with increasing polarity. (As a general rule, the polarity of the solvent should correspond to the polarity of the substance being developed.) Mixtures of solvents are frequently used, the ratio of the components adjusted to give the proper polarity.

Thin-layer chromatography has several unique advantages over other comparable techniques: it is much faster and more sensitive than column or paper chromatography; it is extremely economical; and it is simple to operate. Also, harsh chemicals which might attack paper can be used as indicators. And, when there is no particular method of detection known for an organic substance, the chromatoplate may be heated to 500 C, producing dark spots due to charring.

The versatility of the thin-layer method is demonstrated by the variety of problems to which it is applicable. It has been used to separate hydrocarbons, alcohols, carbohydrates, nitrogen compounds, inorganic substances, plastics, organometallics, proteins, lipids and essential oils, radioactive compounds, and many others. Extensive bibliographies, covering literature up to August, 1962, have been published by the Research Specialties Company, Brinkmann Instruments, and the Arthur H. Thomas Company. The only drawback to these bibliographies is that, since the technique has been used so much more extensively in Europe, many of the articles are in foreign-language journals. These are sometimes difficult to obtain, and they require translation.

There are two major limitations to the applicability of thin-layer. One is that only high-boiling liquids and soluble solids can be separated; the other is that it is not specific—for example, standards have to be run along-side the sample to determine if R_f values are the same, in much the same way as peak retention times are compared in gas chromatography. (R_f value is the same measurement that is used in paper chromatography: the ratio of the distance traveled by the sample spot to the distance traveled by the sol-vent front.)   In those cases where the sample gives an entirely unknown band or spot or when verification of identity is needed, preparative thin-layer chromatography can be used. Preparative thin-layer chromatography merely involves using large enough plates and large enough samples to recover and analyze the separated fractions by another means. The word "preparative" is somewhat misleading, however, since it generally refers in other techniques to amounts larger than those utilized for thin-layer chromatography.

Preparative thin-layer chromatography is not new; there have been a few references to it in the literature. J. H. Dhont and C. DeRooy [13] col-lected 2,4-dinitrophenylhydrazones and subjected them to UV spectropho-tometry, melting point determination, and flash-exchange chromatography. H. L. Bird, et al. [12] used UV to identify and quantitatively analyze steroid samples.   Bravo and Hernandez [11] separated dichlorophene and hexa-chlorophene, collected them quantitatively, and subjected them to UV exam-ination.   Recently, R. D. Spencer and B. H. Beggs [29] described the sep-aration by TLC of four triaryl-s-triazines and the subsequent examination by UV and IR. In all of these references the same technique is used: the band containing the adsorbent material and separated fraction is removed from the plate, extracted with a solvent, and filtered to eliminate interfer-ence from the adsorbent. For UV work the resulting solution can be run directly; for IR work the solution is generally stripped. There is also the possibility of running the solution directly in the infrared, but the interfering bands of the solvent would require differential techniques and possibly or-dinate scale expansion.   Most of the solvents used for extraction are polar and give strong absorptions in many areas of the spectrum; thus, there would be relatively large regions of the spectrum with little or no energy to bring out the bands of the unknown.

In our laboratory several different procedures were attempted in order to find the best possible way of obtaining an IR spectrum from a thin-layer spot.

EXPERIMENTAL

The Perkin–Elmer models 21 and 221-G infrared spectrophotometers were used for this study. The Desaga/Brinkmann apparatus for thin-layer chromatography was used, including adjustable applicator, glass plates, and developing jars. The initial plates were coated with silica gel G, a standard-ized mixture of silicic acid and gypsum made by E. Merck (Germany) and distributed by Brinkmann.   More recently silica gel HF_{254} has been used. This contains an indicator which exhibits strong fluorescence at 2540 A and eliminates the necessity of spray reagents.

Both silica gel preparations, however, contain an impurity. Soxhlet extracts, using acetone as a solvent, give infrared spectra showing a 3.4 μ absorption and a strong, sharp band at 7.2 μ (see Fig. 1). These bands appear in many of the sample spectra obtained from thin-layer spots extracted with polar solvents. Caution must be exercised in interpreting spectra with bands in these regions. Other silica gel is available* which is reported to have a higher degree of purity, but further testing will have to confirm this.

The thin-layer plates were cleaned carefully in a bath of approximately 400:1 parts water to concentrated hydrofluoric acid. The hydrofluoric acid wash was necessary because it was impossible to get the plates clean enough to ensure good adhesion using common detergents or even chromic acid cleaning solution.

After the plates were thoroughly cleaned and dried, they were coated with a 250-μ layer of silica gel and distilled water (ratio of slurry, 1 to 2). The plates were then activated by putting them in a 110 C oven to dry for at least an hour.

TEST MIXTURES

Triphenylphosphine–Tetracyclone Mixture

In order to determine the best method of running the thin-layer spots on IR after removal from the plates, a standard mixture of approximately 50% tetracyclone and 50% triphenylphosphine was prepared. This mixture was dissolved in acetone (a 25% solution), and 70 μl were placed in a line $^3/_4$ in. from the bottom of a 200 by 200 mm plate with a microsyringe. The plate was then put into a developing jar which contained $^1/_2$ in. of a 1:1 mixture of carbon tetrachloride and chloroform. The lid was placed on the jar, and the plate allowed to develop until the solvent front had reached 10 cm. Since the plate coating was a nonfluorescent silica gel, the band positions were indicated by spraying the edges and a center strip of the plate with $KMnO_4$. A homemade cardboard guide which covered the entire plate, except for $^1/_2$ in. strips at either side and in the middle, was used for the spraying so that the remainder of the plate was not damaged. Two nicely separated bands were obtained with R_f values of 0.89 and 0.71, respectively. The lines between the sprayed portions were scraped off the plates with a spatula, and 5 ml of acetone were added. (The choice of a solvent here is no problem since the unknown materials will always be soluble in the original solvent used to apply the sample to the thin-layer plates.) The samples were then filtered with ultrafine Büchner funnels (pore size: 0.9-1.4 μ) under vacuum. Neither centrifuging nor using a fine Büchner funnel (pore size: 4-5.5 μ) removed all traces of the silica gel. The ultrafine Büchner funnel, however, was entirely satisfactory when used with a strong vacuum so that the filtering process was not too time-consuming.

Three different plates of the tetracyclone and triphenylphosphine mixture were run. The three solutions obtained after filtration were run on IR in different manners to determine the best method. One of the solutions was evaporated to dryness and the remaining solid run as a mineral oil mull. Another was evaporated to dryness on sodium chloride crystals, producing a thin film which could be run directly on infrared. The third was placed directly on top of 300 mg KBr and allowed to evaporate. When the KBr dried

*Adsorbosil—distributed by Applied Science Laboratories, Inc., P. O. Box 140, State College, Pa.

(these were allowed to stand overnight), pellets were made from the powders. The resulting spectra obtained on the PE-221G for the three methods
are applied to Band I are shown in Fig. 2. The mull spectrum showed
mainly the strong bands of mineral oil. Only two other extremely weak
bands were detected, at 690 and 740 cm^{-1}. These bands can be attributed
to the triphenylphosphine by comparison with references of the phosphine
and tetracyclone, but if the spectrum had been that of an unknown it would
have been almost impossible to identify. Likewise, the spectrum of the film
obtained from benzene was extremely poor, giving no more information
than the mull spectrum. However, the KBr spectrum was extremely clear
and well-defined, giving bands up to 70% absorption without ordinate scale
expansion and indicating that the recovered triphenylphosphine was purer
than the original.

The mull, film, and pellet spectra obtained from Band II of the sample
are shown in Fig. 3. The mull and film spectra, as in the triphenylphosphine
spectra, showed only a series of weak bands in the 650 to 800 cm^{-1} region.
The pellet spectrum was quite a bit stronger but still not the quality of the
triphenylphosphine spectrum. However, by using ordinate scale expansion
of 5×, the bands became strong enough to compare and match with a KBr
reference of tetracyclone.

This experiment showed, then, that if a completely unknown band from
thin-layer is to be run on infrared, the best method is to extract with a
suitable solvent, filter through an ultrafine fritted glass Büchner funnel,
and deposit the solution on a layer of KBr in a glass vial. The solvent is
then stripped from the KBr, and a pellet made of the remaining dry powder.

Hydrazone Mixture

In order to further test the procedure, a 1:1 standard mixture of the
hydrazones of crotonaldehyde and 2-hydroxyadipaldehyde was dissolved in
10 ml acetone (25% solution) and placed on either side of a 200 by 200 mm
thin-layer plate. It is possible to use both the upper and lower halves of
thin-layer plates for separations if two interrupting lines are scraped with
a spatula near the center of the plate to prevent overlapping of the solvent
fronts or sample bands of the two samples. The hydrazone mixture was
placed in a developing chamber containing chloroform. Separation was extremely good; the two bands were scraped from the plates, dissolved in
5 ml acetone, filtered, and placed upon 300 mg KBr. The resulting spectra
are shown in Fig. 4. The spectrum of Band I (Fig. 4a) is shown both with
the vertical scale normal and expanded five times. This band matched
exactly a literature reference of crotonaldehyde hydrazone. Band II gave a
sharp and well-resolved spectrum which matched exactly a reference of
2-hydroxyadipaldehyde hydrazone.

APPLICATIONS

Commercial Motor Oil Additive

To test the thin-layer method on a different type of sample, a commercial motor oil additive was separated. This additive was sold as a blending
agent for compounding heavy-duty automotive lubes. It was a thick, black,

viscous material, almost gummy in nature, which lent itself perfectly to the thin-layer, infrared technique. It was therefore dissolved in n-hexane (20% solution), placed on the thin-layer plate, and developed with methanol. Two bands with R_f values of 0.8 and 0.05 were detected and scraped from the plate. Methanol was used to extract the first band, hexane the second. Figure 5 shows the original spectrum of the additive obtained as a smear between sodium chloride plates and the spectra of the two thin-layer bands after extraction and running as KBr pellets. The strong absorptions of Band II at 8.5, 9.6, and 9.9 μ indicated the presence of a sulfonate. An exact match was found with barium sulfonate, commonly used as a high-temperature detergent in heavy-duty motor oil additives. A weak band at 7.7 μ in the spectrum of Band II indicated the presence of another component, barium phenate. The spectrum of Band I was very similar to a reference of zinc dithiophosphate, an inhibitor commonly put into motor oil additives to reduce bearing corrosion and oxidation. The only differences were small shifts of the bands in the 10- and 14-μ regions. To determine the origin of these shifts, a reference sample of zinc dithiophosphate was dissolved in methanol and the solvent allowed to evaporate on 300 mg KBr; this was then made into a pellet. No shifts were detected. However, when the zinc dithiophosphate was run on a thin-layer plate, using methanol as both the developing agent and extracting solvent, the shifts did occur, and a spectrum exactly like that of Band I of the sample was obtained. Therefore, the material must change somewhat on the silica-gel column, although the nature of these changes is not understood. The band shifts in this case were not severe enough to cause difficulties with the identification, but spectroscopists should be aware that changes on the silica-gel column can and do occur with certain materials.

Commercial Plasticizer

Figure 6 shows the spectrum of a commercial plasticizer run as a smear between salt crystals. When run on thin-layer plates with chloroform as the developing agent, two major spots were obtained. The KBr remaining after evaporation of the chloroform used for extraction was too wet to make into pellets. Therefore, smears were made between salt plates of the wet KBr. The resulting spectra are shown in Figs. 6b and 6c. Band I was a perfect match with dibutyl fumarate. Band II was identified as diallyl phthalate. A standard mixture of these two materials was then prepared, and from the infrared spectrum the concentrations in the original plasticizers were calculated as 75% dibutyl fumarate and 25% diallyl phthalate. Without the separation by thin-layer, the infrared analysis would have been much more difficult, and the major component of the sample might not have been specifically identified.

Organometallics

One of the fields of research in our laboratory to which the combination of infrared and thin-layer has been most usefully applied is that of organometallics, where the synthesis of new and different compounds which require purification and separation goes on continuously. In this area all the separation and extraction of the thin-layer bands was performed by the organometallics project, which then submitted only the extracted sample to infrared for analysis. Once the identity of the thin-layer bands had been established, organic reactions could be followed and monitored using the thin-layer plates [30]. This, we feel, is the most fruitful combination of these techniques, similar to the manner in which gas chromatography and IR are used in our laboratory [9]. The chemist, knowing his systems, is best equipped to do the separating, and once positive identifications are established by a corollary tool such as IR; he can continue to pursue the analysis in routine fashion.

The use of thin-layer chromatography with infrared in the field of organometallics requires the introduction of an additional feature. Many of the synthesized compounds involving carbon to metal ligands are unstable in air, either spontaneously oxidizing or rearranging to form different compounds. When this is the case and an infrared spectrum is desired, most of the sample preparation and separation must be done in a dry box. The IR spectra are obtained from mulls or solutions because it is almost impossible to prepare KBr pellets while working within the confining limits of the dry box. Also, pellets are undesirable because of the inaccuracies introduced in the extremely important and sensitive band positions.

The solutions from the thin-layer spots are usually centrifuged instead of filtered, since filtering would require additional bulky equipment in the dry box. After the bands are scraped from the plates, a solvent added, and the samples centrifuged, the supernatant liquid is removed with a medicine dropper. The IR spectrum is then obtained directly, or the solvent removed and the sample carefully mulled in the dry box. The silica gel is not completely removed by centrifuging, but it does not interfere in the nitrile and metal carbonyl stretching regions from 4 to 5.5 μ, where most of the significant differences in these spectra occur.

A typical example was a derivative prepared from monoacrylonitrile tungsten pentacarbonyl and cyclopentadiene [30, 31]. Two stereo-isomers were possible, the exo- and endocyclic forms. These isomers were separated by thin-layer chromatography in the dry box using a mixture of hexane and benzene as the developing agent. Acetone was the extracting solvent, and the IR spectra of the two bands in the nitrile region resulted in identification of the isomers.

The reaction product of a metal hexacarbonyl with ligands such as triphenylphosphine (TPP) may also form geometric cis and trans isomers [30, 32, 33]. A sample of $(TPP)_2M(CO)_4$ was applied to the thin-layer plates, and two bands were obtained. They were removed and extracted with CCl_4. Band I was identified as the trans and Band II as the cis form by absorption in the metal carbonyl region [32, 33].

FUTURE

The combination of thin-layer chromatography and infrared spectroscopy holds great promise for the future. One modification of the method, which could be utilized immediately, is the use of foamed plastic to hold KBr pellets. The macrotechnique was described by W. H. Shumaker in 1961 [34] but it has recently been extended to include micro KBr pellets containing only 50 μg of sample in 5-10 mg KBr [35]. Good spectra are obtained without the use of beam condensers or scale expansion. This technique will probably be used for most future thin-layer bands extracted for IR in our laboratory because it holds several distinct advantages over the macro-technique: less sample can be used on the thin-layer plates eliminating the danger of overloading the plates; duplicate plates will not have to be run in order to get enough sample to deposit on 300 mg KBr; and the minor constituents of a mixture can be identified more easily. The biomedical field will especially benefit from this technique since micromethods are generally a necessity. Extra precaution must be taken to ensure purity of the sample, however, because the contamination of 50 μg of sample can occur rather easily.

Other innovations will be introduced into the thin-layer, infrared technique in the near future. More sophisticated apparatus will undoubtedly be developed and produced commercially. There is already evidence for this with controlled sample applicators described by Ritter and Meyer [23] and Morgan [28]. Ritter and Meyer also illustrate a small "vacuum cleaner" made of glass and connected to a vacuum line for removing the bands from the plates. This would eliminate the dust produced from scraping, and the sample could then be collected quantitatively. No doubt there will be other improvements in the design and manufacture of the equipment. What is really needed in the field is the development or discovery of a material which has the adsorption qualities of silica gel but which is as transparent as KBr in the infrared. Then samples could be run directly from the thin-layer plates without the extraction procedure. With this aim in mind, our laboratory attempted separations on a 25 μ thickness of Millipore filter paper. This filter paper, as shown by H. L. Sloane [36] in a paper presented at the 1962 Pittsburgh conference, can be used differentially in the infrared to give a remarkably flat baseline with high energy throughout the spectral range. However, the adsorbing properties of Millipore are not nearly as effective as its infrared transmitting properties. The DeSaga test mixture made up of 0.01% 4-dimethylaminoazobenzene, Indophenol, and Sudan Red G was placed on a strip of Millipore. This test mixture, which separates beautifully on any silica-gel layer, did not separate at all on the Millipore. Thus, its use would be limited to the simplest of separations, probably ones which could be done more easily in another manner.

Not only will the apparatus for thin-layer chromatography become more sophisticated in the future, but there will be new innovations in the method and theory as well. S. M. Rybicka [37] has described a concentration gradient development which he used to separate mono-, di-, and triglycerides. He found that no single solvent mixture gave complete separation of these species without some components remaining on the baseline and others reaching the solvent front. Therefore, he developed a technique in which the concentration of the developing solution was gradually increased as the thin-layer plate was being developed. This technique permits a great number of separations which were previously impossible.

Another new development in theory involves the use of unidimensional multiple chromatography, the repeated irrigation with solvents in the same direction. John A. Thoma [38] has described how one can predict, with a simple equation, the number of irrigations producing a maximum separation between two solutes. This ability to predict should help considerably in difficult separations, and the repeated irrigations should in no way affect the suitability of the thin-layer bands for infrared analysis.

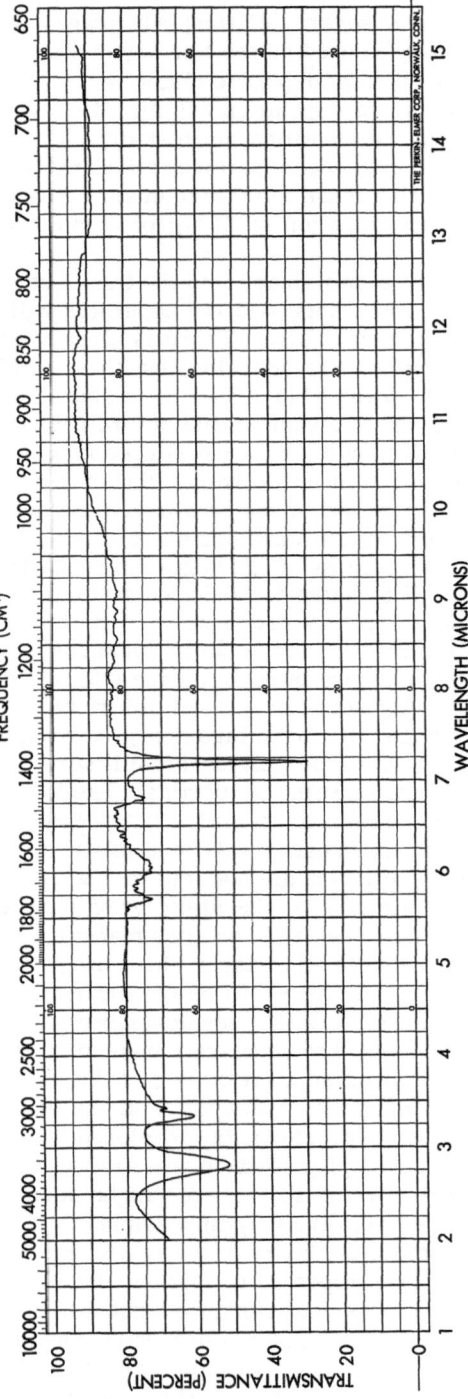

Fig. 1. Silica–gel contaminant—KBr pellet.

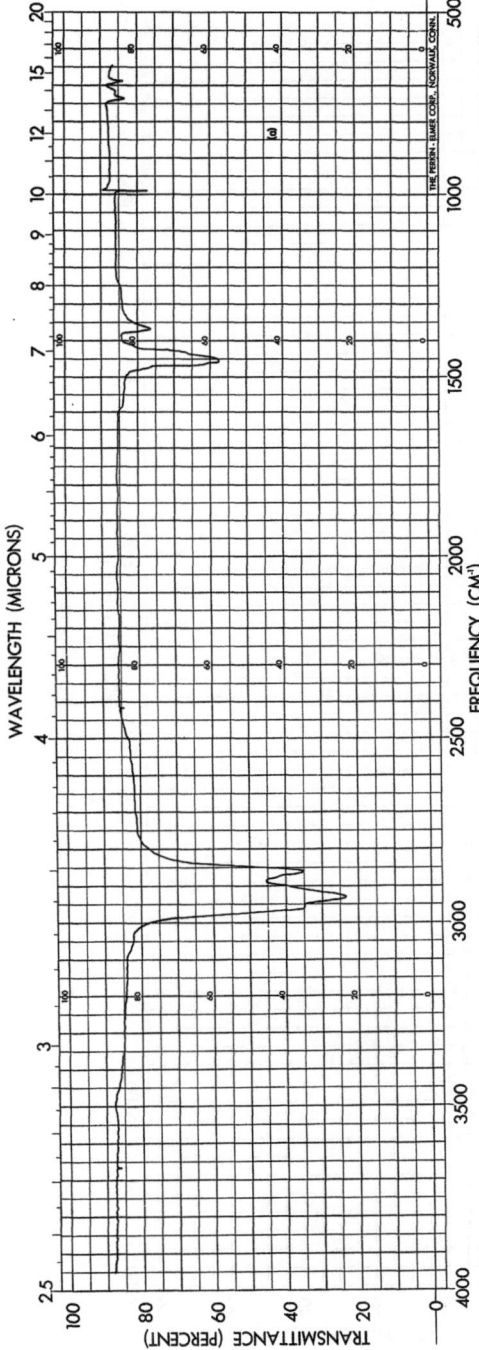

Fig. 2a. Band I of triphenylphosphine—tetracyclone test mixture—mull.

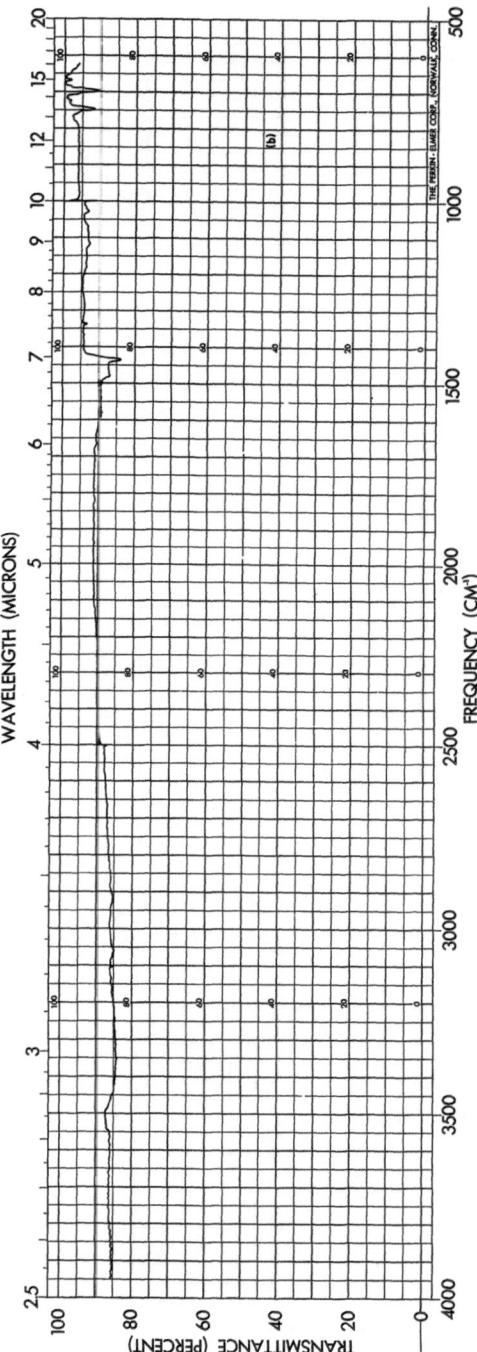

Fig. 2b. Band I of triphenylphosphine—tetracyclone test mixture—film.

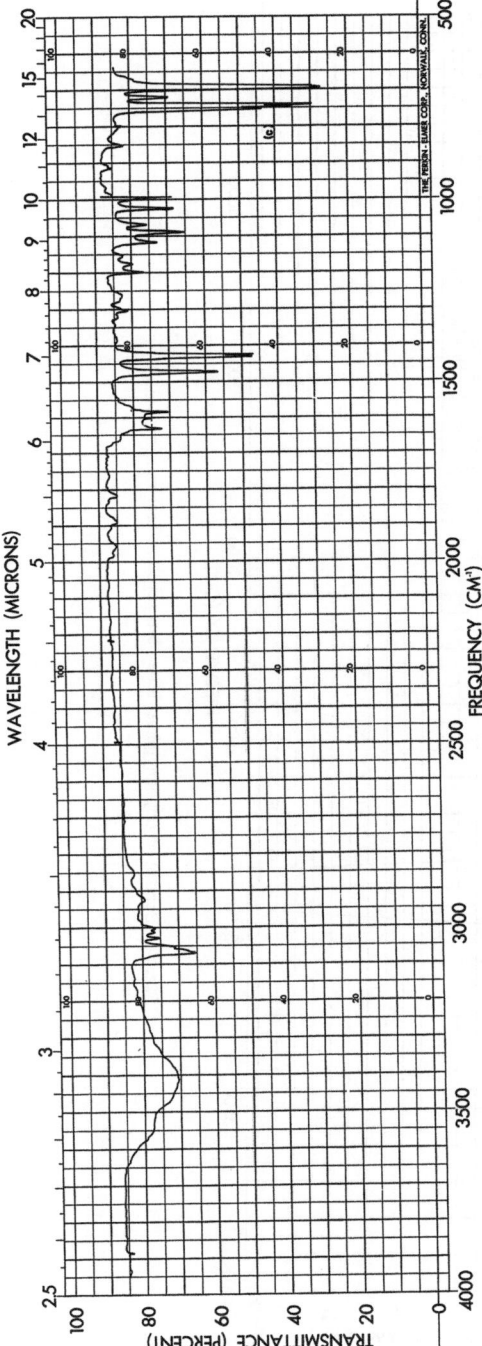

Fig. 2c. Band I of triphenylphosphine—tetracyclone test mixture—KBr pellet.

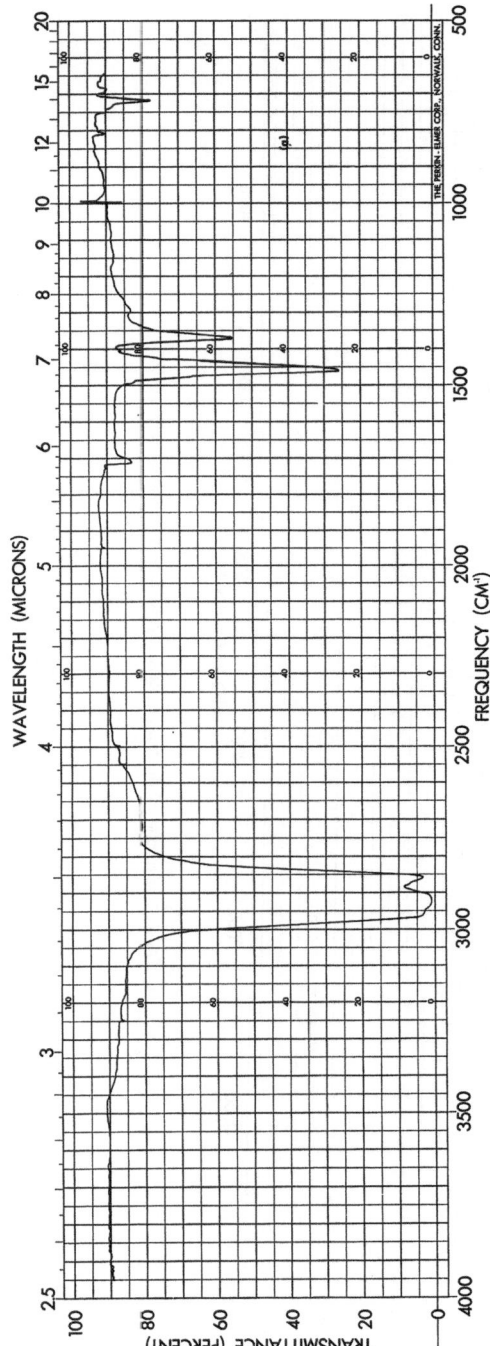

Fig. 3a. Band II of triphenylphosphine–tetracyclone test mixture—mull.

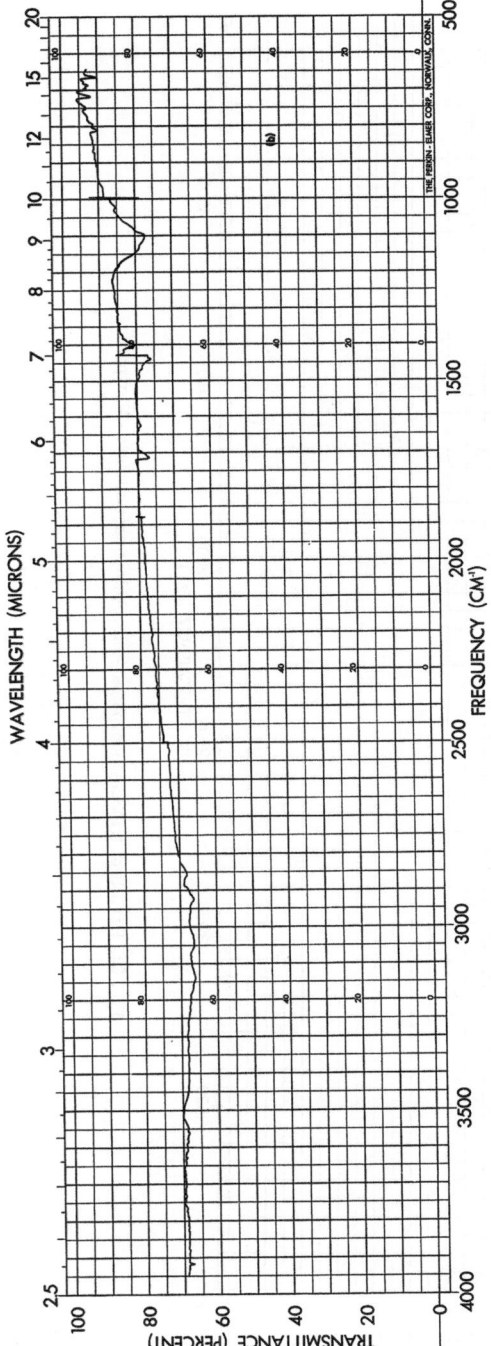

Fig. 3b. Band II of triphenylphosphine—terracyclone test mixture—film.

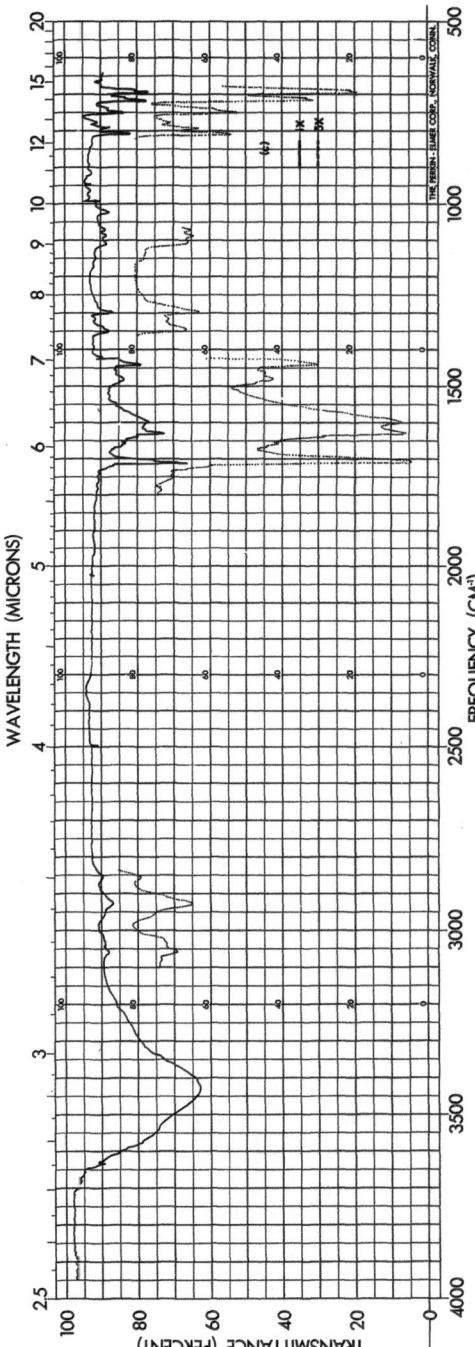

Fig. 3c. Band II of triphenylphosphine—tetracyclone test mixture—KBr pellet.

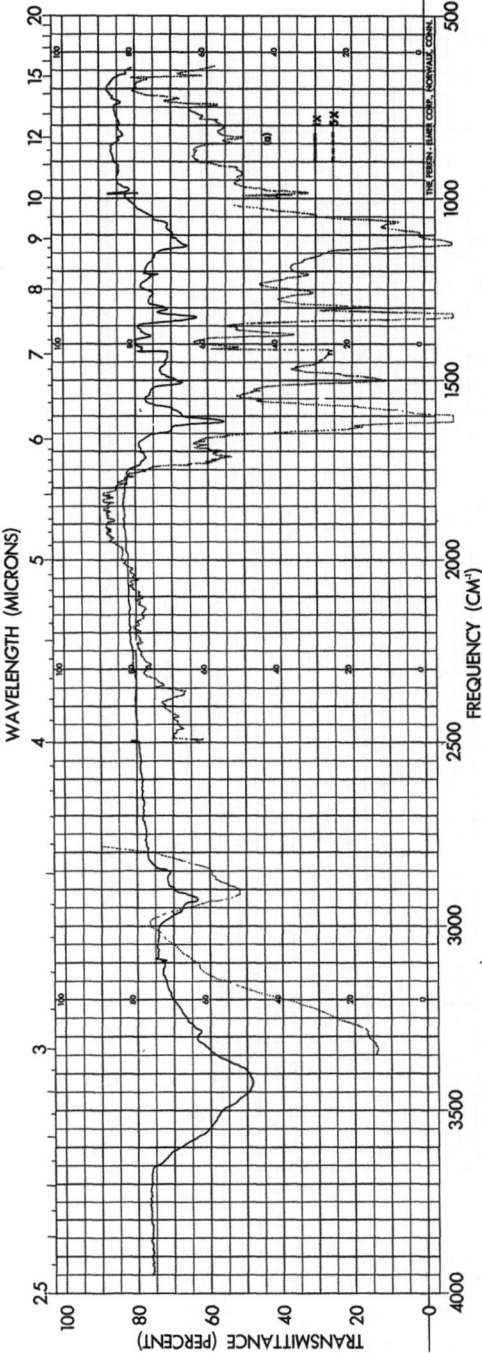

Fig. 4a. Thin–layer spots of hydrazone test mixture—Band I: crotonaldehyde 2,4-dinitrophenyl hydrazone.

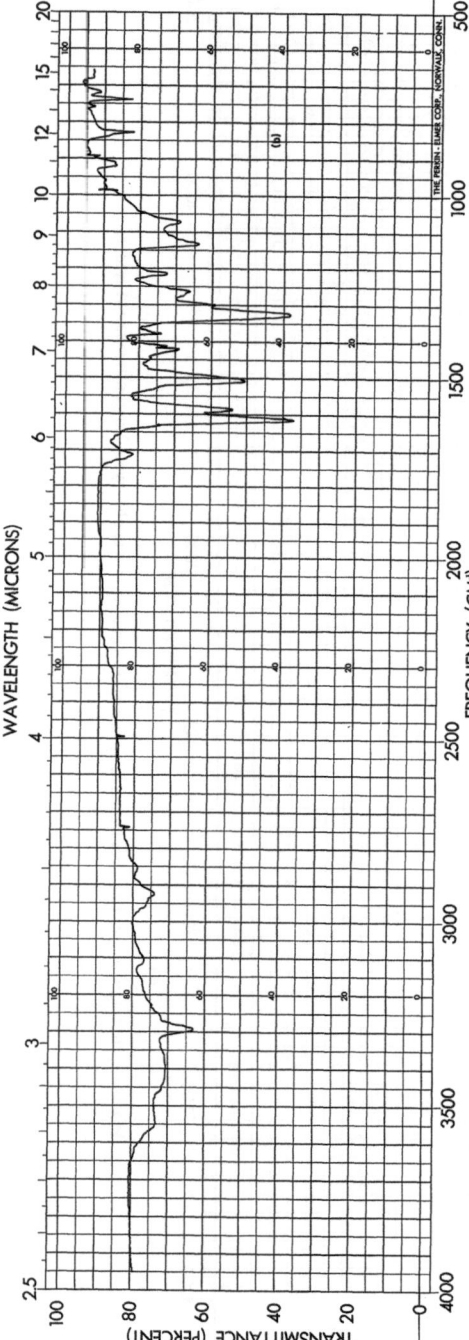

Fig. 4b. Thin-layer spots of hydrazone test mixture.—Band II: 2-hydroxadipaldehyde 2,4-dinitrophenyl hydrazone.

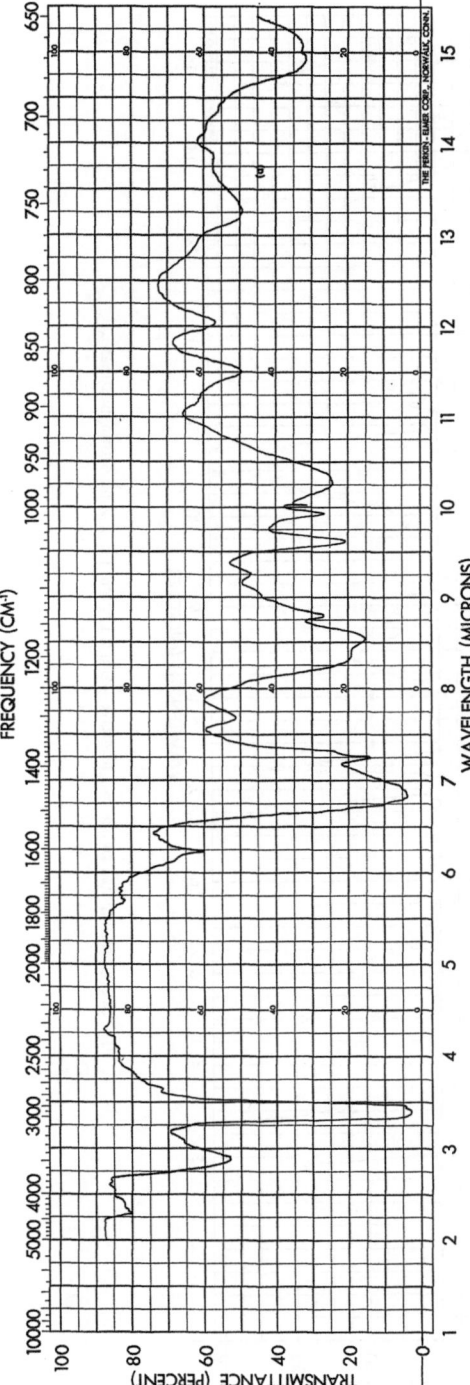

Fig. 5a. Commercial motor oil additive—smear between NaCl plates.

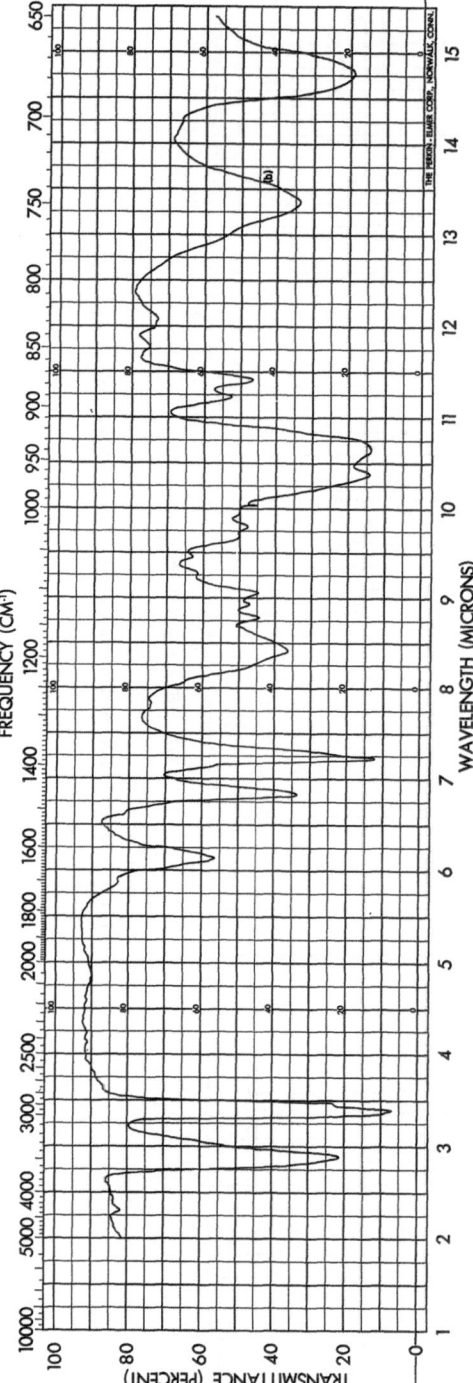

Fig. 5b. Commercial motor oil additive—Band I from TLC, zinc dithiophosphate.

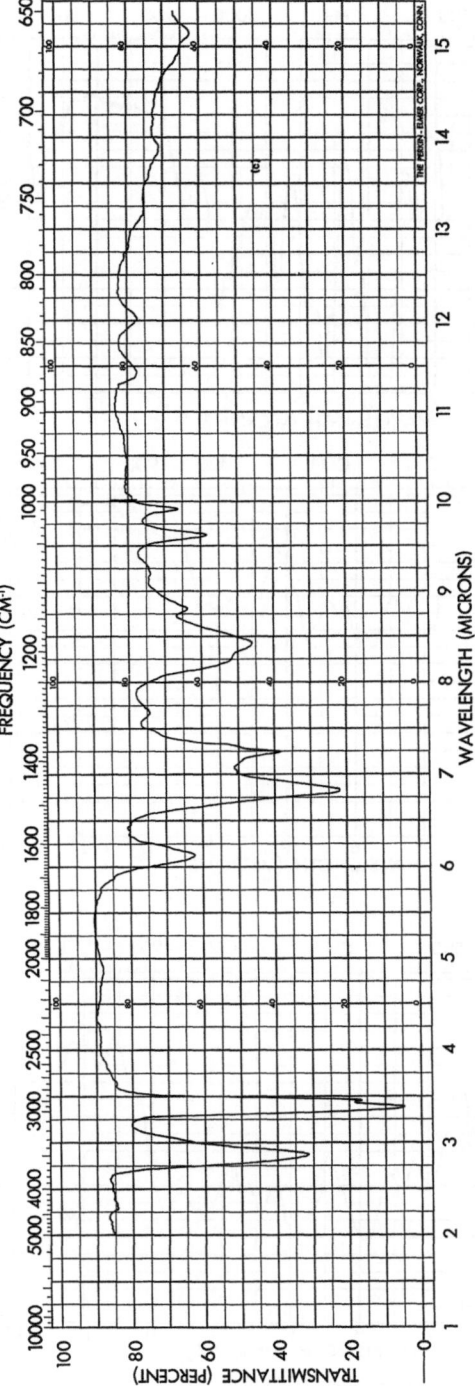

Fig. 5c. Commercial motor oil additive—Band II from TLC, barium sulfonate and barium phenate.

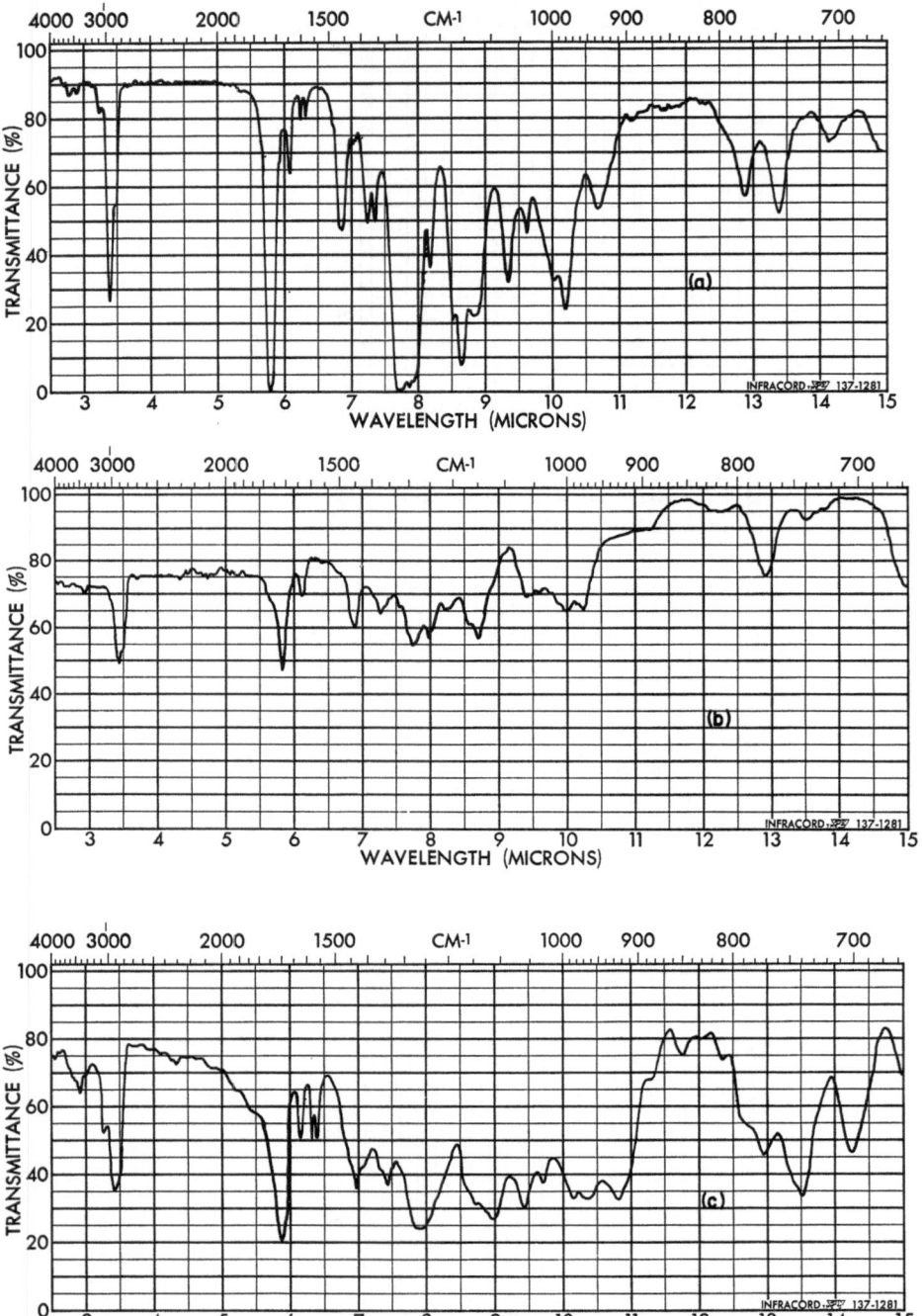

Fig. 6. Commercial plasticizer: (a) Smear between NaCl plates: (b) Band I from TLC, dibutyl fumarate: (c) Band II from TLC, diallyl phthalate.

SUMMARY

A few of the applications illustrating the versatility of combining infrared and thin-layer chromatography have been given. Thin-layer can be used to separate complex mixtures of high-boiling liquids or solids and simplify infrared identifications. Conversely, infrared can qualitatively or quantitatively identify unknown bands from thin-layer chromatography. The combination of these two techniques complements the combination of gas chromatography and infrared, which analyzes volatile liquids or gases.

The technique combining these two methods of analysis is quite simple, involving only extraction, filtration, and stripping. The thin-layer development requires about thirty minutes, the infrared spectrum twenty minutes. Stripping of the solvent is the most time-consuming element of the combination, and if greater speed is desired, this can be done under nitrogen or vacuum.

The large number of analytical problems to which the combination of thin-layer and infrared can be applied has been described, but the number is certain to increase with improvements in thin-layer apparatus and refinements in the technique combining the two analytical tools.

REFERENCES

1. R. S. Gohlke, Anal. Chem. 31:535 (1959).
2. L. P. Lindeman and J. L. Annes, ibid. 32:1742 (1960).
3. A. A. Ebert, Jr., ibid. 33:1865 (1961).
4. R. A. Brown and E. R. Quiram, Appl. Spectroscopy 17:33 (1963).
5. W. I. Kaye, Fifth International Instruments and Measurements Conference, Stockholm, September, 1960.
6. J. Haslam, A. R. Jeffs, and H. A. Willis, Analyst 86:44 (1961).
7. W. S. Gallaway, T. Johns, D. G. Tipotsch, and W. F. Ulrich, Pittsburgh Conference on Analytical Chemistry and Applied Spectroscopy, March, 1958.
8. R. V. Helm, D. R. Latham, C. R. Ferrin, and J. S. Ball, Anal. Chem. 32:1765 (1960).
9. J. G. Grasselli and M. K. Snavely, Appl. Spectroscopy 16:190 (1962).
10. M. Cachia, J. Appl. Chem. 8:291 (1958).
11. R. Bravo and F. Hernandez, J. Chromatog. 7:60 (1962).
12. H. L. Bird, Jr., H. F. Brickley, J. P. Comer, P. E. Hartsaw, and M. L. Johnson, Anal. Chem. 35:346 (1963).
13. J. H. Dhont and C. DeRooy, Analyst 86:74 (1961).
14. K. Heyns and H. F. Grützmacher, Angew. Chem. Intern. Ed. Engl. 1:400 (1962).
15. N. A. Ismailov and M. S. Schraiber, Farmatsiya 3:1 (1938).
16. E. Stahl, Arch. Pharm. 292:411 (1959).
17. E. Stahl, Chemiker Ztg. 82:323 (1958).
18. E. Stahl, Fette, Seifen, Anstrichmittel 60:1027 (1958).
19. E. Stahl, Parfue. Kosmetik. 39:564 (1958).
20. E. Stahl, Pharm. Rundschau. 1 No. 2 (1959).
21. E. Stahl, Pharm. Weekblad. 92:829 (1957).
22. R. Wasicky, Anal. Chem. 34:1346 (1962).
23. F. J. Ritter and G. M. Meyer, Nature 193:941 (1962).
24. T. M. Lees and P. J. DeMuria, J. Chromatog. 8:108 (1962).
25. G. R. Duncan, ibid. 8:37 (1962).
26. E. A. Mistryukov, ibid. 9:311 (1962).
27. E. G. Wollish, M. Schmall, and M. Hawrylyshyn, Anal. Chem. 33:1138 (1961).
28. M. E. Morgan, J. Chromatog. 9:379 (1962).
29. R. D. Spencer and B. H. Beggs, Pittsburgh Conference on Analytical Chemistry and Applied Spectroscopy, March, 1963.
30. J. M. Augl and A. A. Buss, Cleveland Section A. C. S., Meeting-In-Miniature, December 19, 1962.
31. D. P. Tate, J. M. Augl, A. A. Buss, Inorg. Chem. 2:427 (1963).
32. R. Poilblanc and M. Bigorgne, Bull. Soc. Chim. France, 1301 (1962).
33. D. P. Tate, J. M. Augl, A. A. Buss, B. L. Ross, and J. G. Grasselli (in manuscript).
34. W. H. Shumaker, Chemist-Analyst 50:22 (1961).
35. W. H. Shumaker (private communication).
36. H. L. Sloane, Pittsburgh Conference on Analytical Chemistry and Applied Spectroscopy, March, 1962.
37. S. M. Rybicka, Chem. Ind. 7:308 (1962).
38. J. A. Thoma, Anal. Chem. 35:215 (1963).

The Characterization of Saturated Aliphatic Esters in the 15–40 Micron Region*

J. J. Lucier

Department of Chemistry
University of Dayton
Dayton, Ohio

and

F. F. Bentley

Aeronautical Systems Division, United States Air Force
Wright-Patterson Air Force Base
Dayton, Ohio

Saturated aliphatic esters have very characteristic spectra in the 15–40 μ region. Acyl and alkyl components can be identified separately. A correlation chart has been prepared showing characteristic bands for formates, acetates, propionates, butyrates, iso-butyrates, valerates, isovalerates, hexanoates, heptanoates, methyl esters, ethyl esters, propyl esters, isopropyl esters, n-butyl, isobutyl, secondary butyl and tertiary butyl esters, and neopentyl esters. Band assignments in this region are discussed for some simple esters.

INTRODUCTION

The chemical literature contains a considerable amount of information on the infrared spectra of aliphatic esters. Most of this information is in the 4000-650 cm^{-1} region (rock-salt region). The information which has been most useful in the salt region involves the carbonyl stretching frequency (1735-1750 cm^{-1}), the alpha methylene C—H scissor mode (1400-1430 cm^{-1}), and the C—O stretching frequency (1150-1270 cm^{-1}). Information obtained from these bands along with information obtained from related molecular vibrations sometimes enable the analyst to deduce the gross structure of a given molecule. However, it is not always possible to distinguish between closely related structures, and specific characterization with respect to acyl and alkyl parts of the molecule cannot always be made. The additional information obtainable from the 650-250 cm^{-1} (CsBr or 15-40 μ) region often gives a more specific "fingerprint" of the molecule since it is concerned with bending or deformation modes of the entire carbon and oxygen skeleton. In many cases there are skeletal deformation frequencies which enable the chemist to identify specific alkyl or acyl groups, and if desired, to use these bands for quantitative analysis. This work attempts to show how such spectra may be used to characterize saturated aliphatic esters.

*Published in Spectrochimica Acta in abbreviated form. This work was supported by the Aeronautical Systems Division, United States Air Force.

The spectra—structure correlations given in this study as based on the infrared spectra from 15 to 40 μ of over 100 aliphatic esters (formates, acetates, propionates, butyrates, valerates, hexanoates, heptanoates, mixed esters, etc.). Although the band assignments for the formates, acetates, and other classes of esters are reviewed and discussed in regard to new data obtained in this study, primary emphasis has been given to developing empirical information of value to the analyst in characterizing molecules which do not vary widely in structure. The data presented are intended to supplement the information available in the rock-salt region. Characteristic absorption [3] in the latter region indicates the presence of an ester, but they are relatively insensitive to the nature of the acyl or alkyl substituents. Spectra—structure correlations in the long-wavelength region not only confirm the presence of esters, but also enable the chemist to characterize the acyl and alkyl substituent of a given ester.

A detailed discussion of the long-wavelength spectra of the various classes of esters studied together with spectra—structure correlation charts and examples of the spectra used in the study are given in the following pages of this article.

APPARATUS AND PROCEDURE

The instrument used in these studies was a Perkin—Elmer Model 221 spectrophotometer equipped with a CsBr interchange. The instrument calibration was checked at frequent intervals using the 667.3 cm^{-1} CO$_2$ (14.986 μ) band. The instrument was also calibrated using a mixture of indene, camphor, and cyclohexanone [1,13]. The accuracy of the recorded spectra is believed to be better than ±0.05 μ, which is within 1.5 cm^{-1} in this region of the spectrum. The spectra were scanned at a rate of 1 μ/min from 665 to 400 cm^{-1} (15-25 μ) and 0.5 μ/min from 400 to 250 cm^{-1} (25-40 μ). The instrument was purged of water vapor and carbon dioxide by means of high-purity nitrogen. Generally, the instrument conditions were those previously described by Bentley [2]. All samples were examined in the liquid state or as nujol mulls. Cesium bromide cells with path lengths of 0.1 to 0.2 mm were used to record the spectra of liquid compounds.

RESULTS AND DISCUSSION

Formates

In the infrared spectra of a series of n-alkyl formates from ethyl to n-amyl formate (Fig. 1) two absorption bands were observed, one at 476-459 and one at 340-333 cm^{-1}. There was also a weak but sharp band at 621 cm^{-1}. The band at 476-459 cm^{-1} decreases in intensity and moves to higher frequencies as the chain length increases. Beyond amyl formate this band is too weak to be observed. Absorption in the 340-333 cm^{-1} region is very strong and is confined to the 340-338 cm^{-1} region for n-alkyl radicals larger than methyl. Methyl formate has a very strong band at 333 cm^{-1} but does not have a band in the 476-459 cm^{-1} region.

Many branched alkyl formates (Fig. 2) have absorption bands in the 521-485 cm^{-1} region and the 339-287 cm^{-1} region. Progressive substitution starting on the beta-carbon atom and then on the alpha-carbon atom appa-

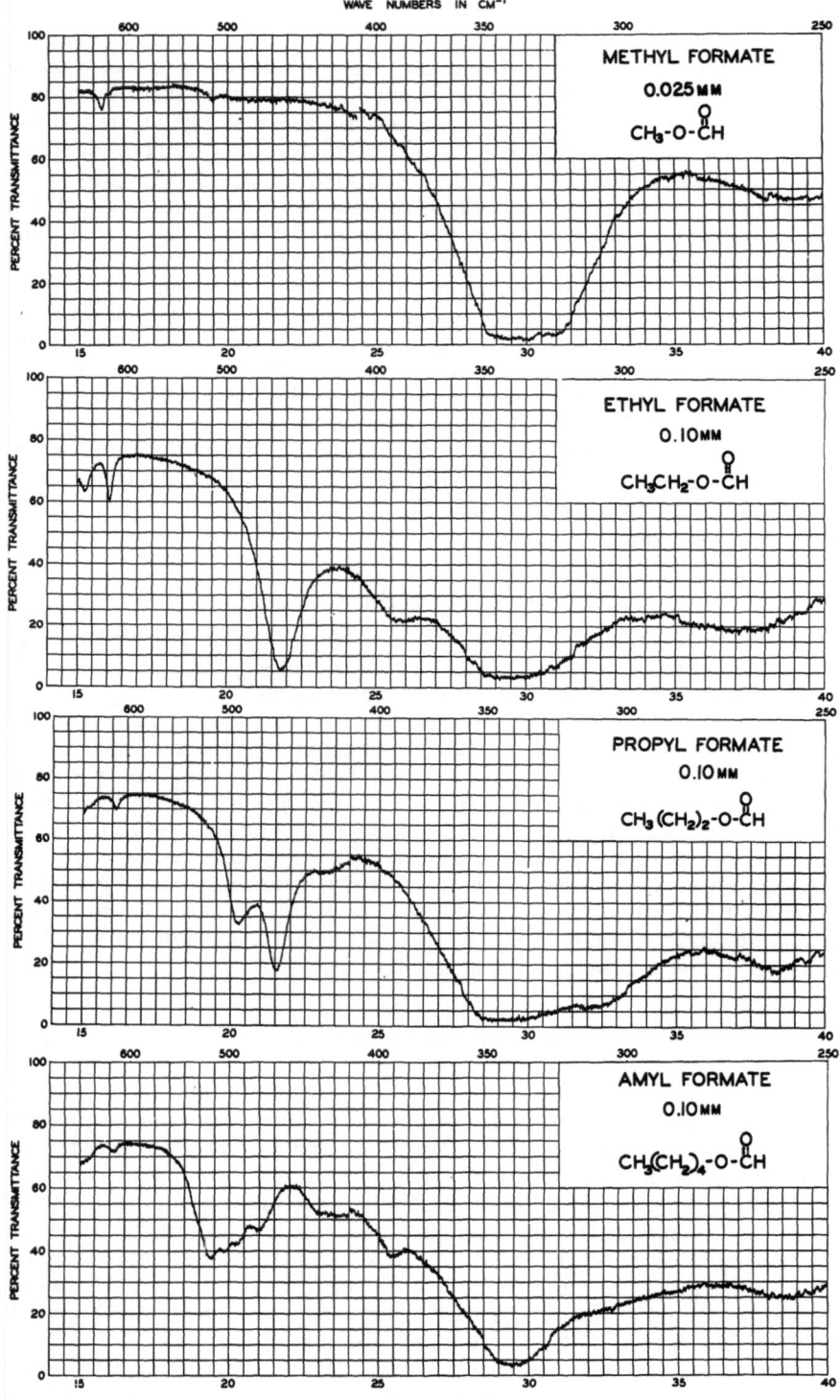

Fig. 1. Infrared spectra from 15–40 μ of methyl, ethyl, n-propyl, and n-amyl formates.

TABLE I

Band Shifts Caused by Progressive Substitution on
Beta- and Alpha-Carbon Atoms of Formate Esters

Ester	Wavelength, μ	Frequency, cm^{-1}
Ethyl formate	29.5	339
Isobutyl formate	30.4	329
Neopentyl formate	32.0	313
sec-Butyl formate	34.0	294

rently causes a frequency shift to lower wavelengths (Table I). This can be seen by starting with the 339 cm^{-1} band of ethyl formate and then noting the same band in isobutyl (328 cm^{-1}), neopentyl (312 cm^{-1}), secondary butyl (294 cm^{-1}), and isopropyl (287 cm^{-1}) formate.

Formate esters of normal alcohols have a weak but sharp band at 621 cm^{-1} except for methyl formate which has a band at 635 cm^{-1}. This band decreases in intensity with molecular weight increase and it is difficult to observe beyond the C_5 formate. The formates of secondary alcohols do not have a band at this position.

Acetates

Two strong absorption bands, the first between 640-623 cm^{-1}, and the second between 610-602 cm^{-1}, are characteristic of all aliphatic acetates (Figs. 3 and 4). A variable band in the 323-306 cm^{-1} region is also characteristic of many acetate esters.

Careful analysis reveals the presence of sharp bands lying within the above-mentioned boundaries. These sharp bands can be used to distinguish the acetates of primary alcohols from acetates of secondary and tertiary alcohols. The first of these distinguishing bands is at 639-633 cm^{-1} for acetates of primary alcohols and at 633-623 cm^{-1} for acetates of secondary and tertiary alcohols. The second distinguishing band is at 608-600 cm^{-1} for acetates of primary alcohols and at 610-608 cm^{-1} for acetates of secondary and tertiary alcohols. It should be noted that this second distinguishing band is somewhat less variable in frequency than the first distinguishing band but is 4 to 10 cm^{-1} higher in frequency for acetates of secondary and tertiary alcohols than for the corresponding primary acetate. For example, n-propyl acetate has a band at 604 cm^{-1} while isopropyl acetate has a band at 610 cm^{-1}. The displacements cited result from substitution of carbon atoms in the alpha position of the alkyl component of the ester. Even highly branched esters such as neopentyl and isobutyl acetates have these bands in the same positions as do methyl and ethyl acetates, since these esters do not have carbon atoms in the alpha position.

Tertiary alkyl groups not only cause the 640-623 cm^{-1} band to shift to lower frequency, but they also cause the band to become much weaker. This can be observed in tertiary butyl acetate and in tertiary amyl acetate. This same weakening of the 640-623 cm^{-1} band can be observed in some secondary alcohol acetates, notably in 1,3-dimethyl butyl acetate and in 5-nonyl acetate.

In the 323-306 cm^{-1} region there is a variable band which is characteristic of many acetate esters. The very strong band at 308 cm^{-1} in methyl

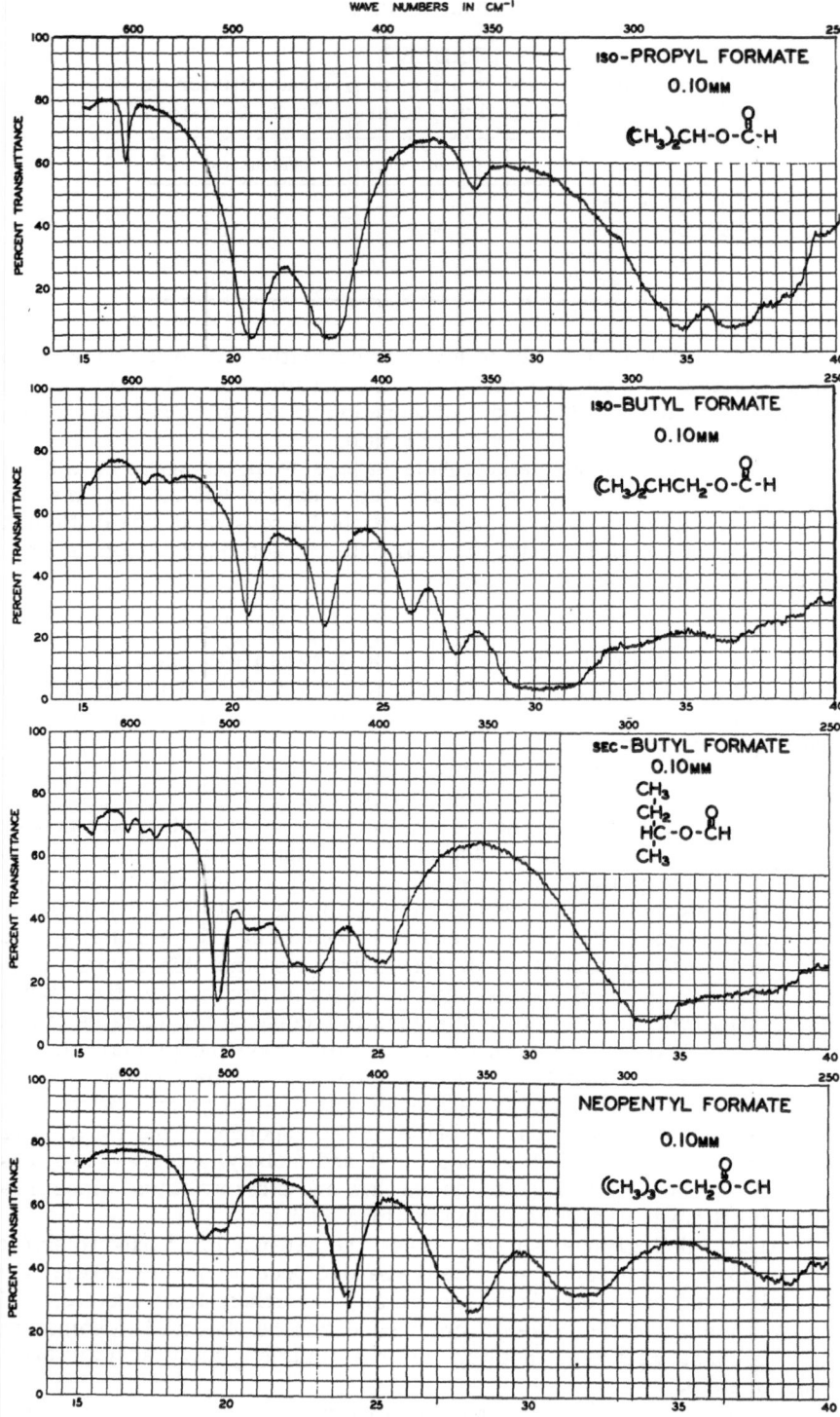

Fig. 2. Infrared spectra from 15–40 μ of isopropyl, isobutyl, secondary butyl, and neopentyl formates.

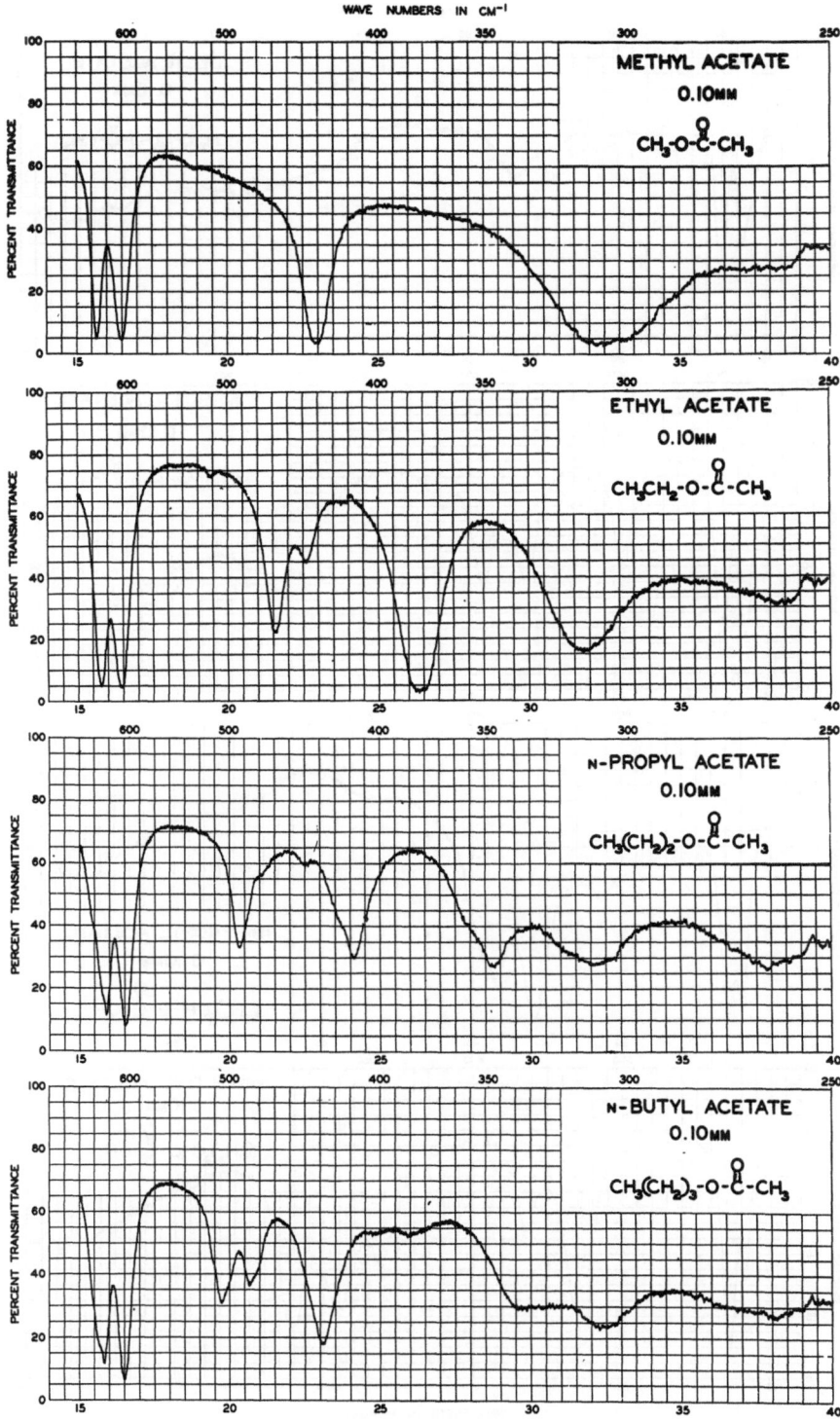

Fig. 3. Infrared spectra from 15-40 μ of methyl, ethyl, n-propyl, and n-butyl acetates.

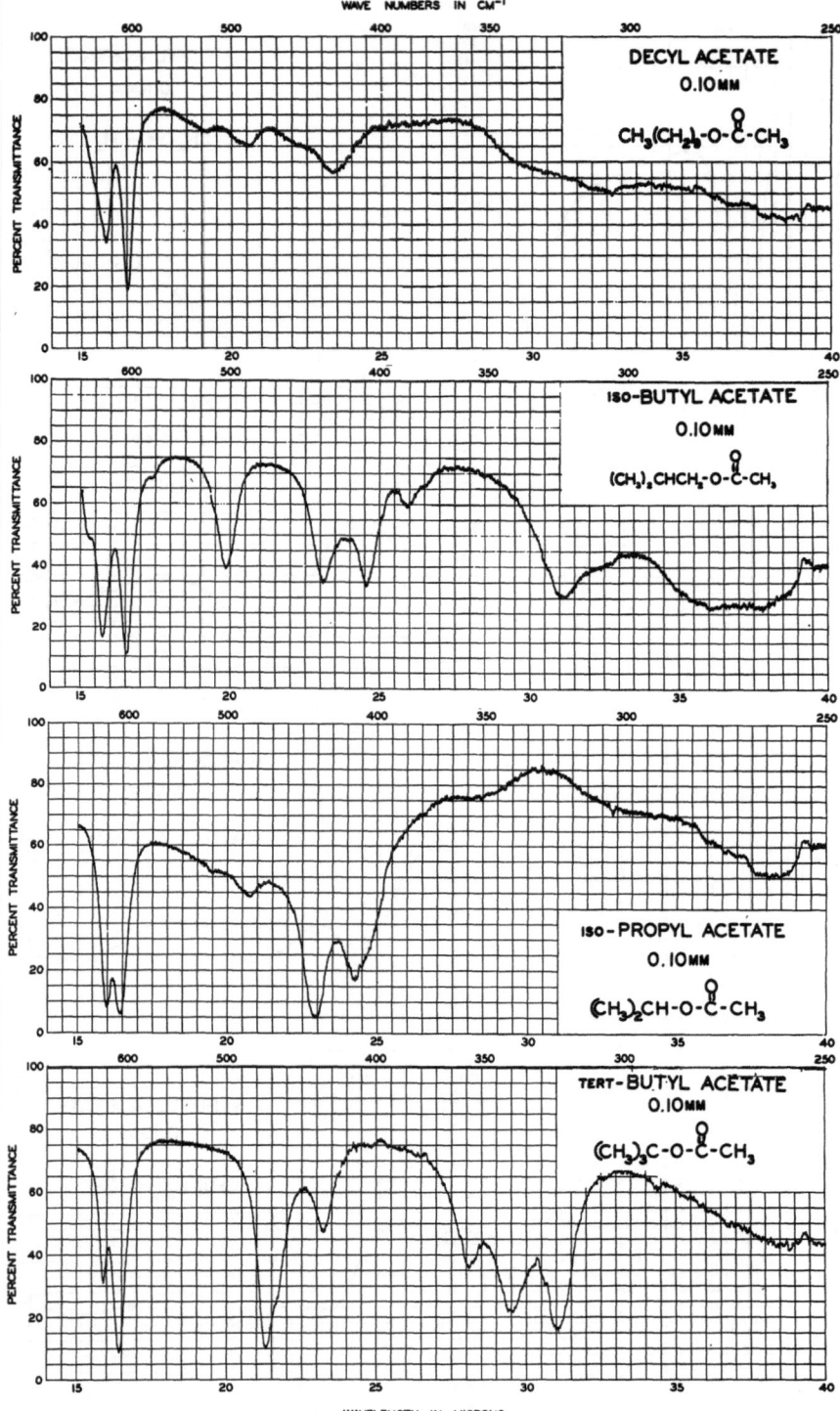

Fig. 4. Infrared spectra from 15–40 μ of n-decyl, isobutyl, isopropyl, and tertiary butyl acetates.

acetate is a good example. This band decreases in intensity as the molecular weight increases until in decyl acetate it is no longer observed. Isopropyl and secondary butyl acetates do not have bands in this region.

Methyl acetate has a strong band at 435 cm^{-1}. This band appears to shift to higher frequencies in an orderly fashion from methyl, through ethyl and propyl, and then splits into two medium to strong bands in n-butyl acetate, and then into four bands in n-amyl acetate.

Propionates

Alkyl propionates have two bands (Fig. 5) which are medium to weak from 617-600 cm^{-1} and 585-575 cm^{-1} with the exception of isoamyl propionate. In all cases there is an apparent interaction between the bands so that they are not cleanly resolved. Most propionates also have a strong band at 351-336 cm^{-1}. Exceptions are isopropyl and secondary butyl propionates, and both of these esters have a strong band at 298 cm^{-1}.

Butyrates

Alkyl butyrates have a weak band at 635-625 cm^{-1} and a medium to strong band at 602-585 cm^{-1} (Fig. 6). All of the propyl and butyl butyrates have a medium to strong band at 505 cm^{-1} to 493 cm^{-1} with the exception of tertiary butyl butyrate. All of the butyrates have a medium to strong band from 467 to 433 cm^{-1} with the exception of propyl butyrate.

Isobutyrates

Isobutyrates have a medium to weak band at 544-521 cm^{-1} and at 515-481 cm^{-1}. Many of the isobutyrates have a weak to medium band at 439-420 cm^{-1}, at 402-377 cm^{-1}, and at 327-317 cm^{-1}.

Valerates

Valerates have a medium to weak band at 641-633 cm^{-1}, and a medium band at 592-585 cm^{-1}. This second band is weak in the case of the isobutyl, tertiary amyl, and neopentyl esters. There is also a medium to weak band at 442-425 cm^{-1} except for the isopropyl and secondary butyl esters. Isopropyl and secondary butyl have a band at 415 cm^{-1}.

Isovalerates

Isovalerates have medium bands at 633-625 cm^{-1} and at 595-588 cm^{-1}. The first of these bands enables the analyst to distinguish this class from normal valerate esters, which have a corresponding weak band at 641-633 cm^{-1}.

Hexanoates

All hexanoates have weak bands at 637 cm^{-1}, at 606-599 cm^{-1}, and at 585 cm^{-1}. However, it must be noted that many esters of higher acids have similar bands. These esters also have a weak band at 346-339 cm^{-1}.

Heptanoates

All heptanoates have a weak band at 641-637 cm^{-1}, at 604-599 cm^{-1} and a medium to weak band at 521-495 cm^{-1}. All of these bands are usually found in hexanoates and in many esters of high-molecular-weight fatty acids.

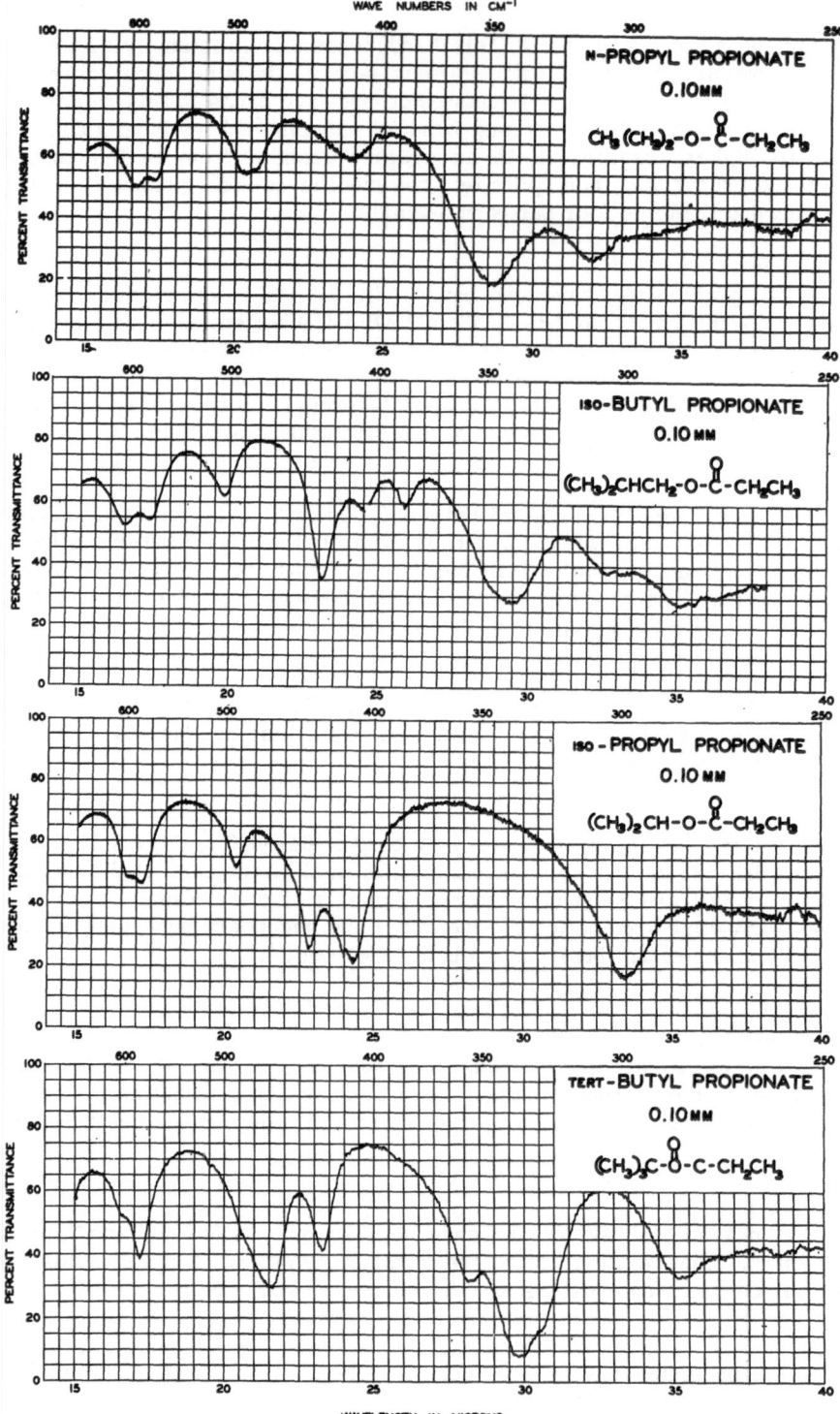

Fig. 5. Infrared spectra from 15–40 μ of n-propyl, isobutyl, isopropyl, and tertiary butyl propionates.

Methyl Esters

Methyl esters have a medium to weak band at 641-633 cm^{-1} except for the propionate.   Except for the formate and the isovalerate, all methyl esters have a weak band at 614-603 cm^{-1}.   For methyl esters of straight-chain acids higher than propionate there is a medium to weak band at 585 cm^{-1}.   There is a strong to medium band at 451-433 cm^{-1} except for the formate and isobutyrate.   The first three of these bands are often found in high-molecular-weight esters such as hexanoates, and consequently must not be relied upon in making characterizations.

Ethyl Esters

Specific ethyl esters may readily be distinguished in this region by "fingerprint" techniques, but they are difficult to characterize as a class. All have variable bands at 389-369 cm^{-1}.

Propyl Esters

Propyl esters have bands of variable intensity at 498-493 cm^{-1} and at 351-339 cm^{-1}.   There is also a band of variable intensity at 606-585 cm^{-1}, with the sole exception of the formate.

Isopropyl Esters

Except for the formate, isopropyl esters have a strong band at 424-412 cm^{-1}.   The formate has a strong band at 431 cm^{-1}.   Straight-chain esters have a band near 436 cm^{-1} except for the valerate.   The isopropyl esters also have a variable band at 606-585 cm^{-1} and at 505-483 cm^{-1}.

n-Butyl Esters

n-Butyl esters have a medium to strong band at 505 ± 2 cm^{-1} and a strong to medium band near 435 cm^{-1}.   They also have a weak band at 348-336 cm^{-1}.

Isobutyl Esters

Isobutyl esters have a medium band at 510-503 cm^{-1} except the formate and isobutyrate.   The formate has a strong band at 486 cm^{-1}.   There is a strong band at 433 cm^{-1} and a variable band at 388-379 cm^{-1}.

Secondary Butyl Esters

Secondary butyl esters of straight-chain acids have a strong to medium band near 590 cm^{-1} except the formate and the acetate.   The isobutyrate and the isovalerate have a weak band at 593 cm^{-1}.   There is a variable band at 500 cm^{-1} except for the formate.   The formate has a strong band at 508 cm^{-1}.   There is also a strong to medium band at 426-415 cm^{-1} except for the formate.

Tertiary Butyl Esters

Tertiary butyl esters have a strong band at 472-463 cm^{-1}, a medium band at 435-424 cm^{-1}, a band of variable intensity near 357 cm^{-1}, and a strong band at 339-327 cm^{-1}.   The formate was not investigated (Figs. 4, 5, and 6).

Neopentyl Esters

Neopentyl esters have a medium to weak band from 544-521 cm^{-1}.   They also have a strong band at 455-452 cm^{-1} except for the formate and the iso-

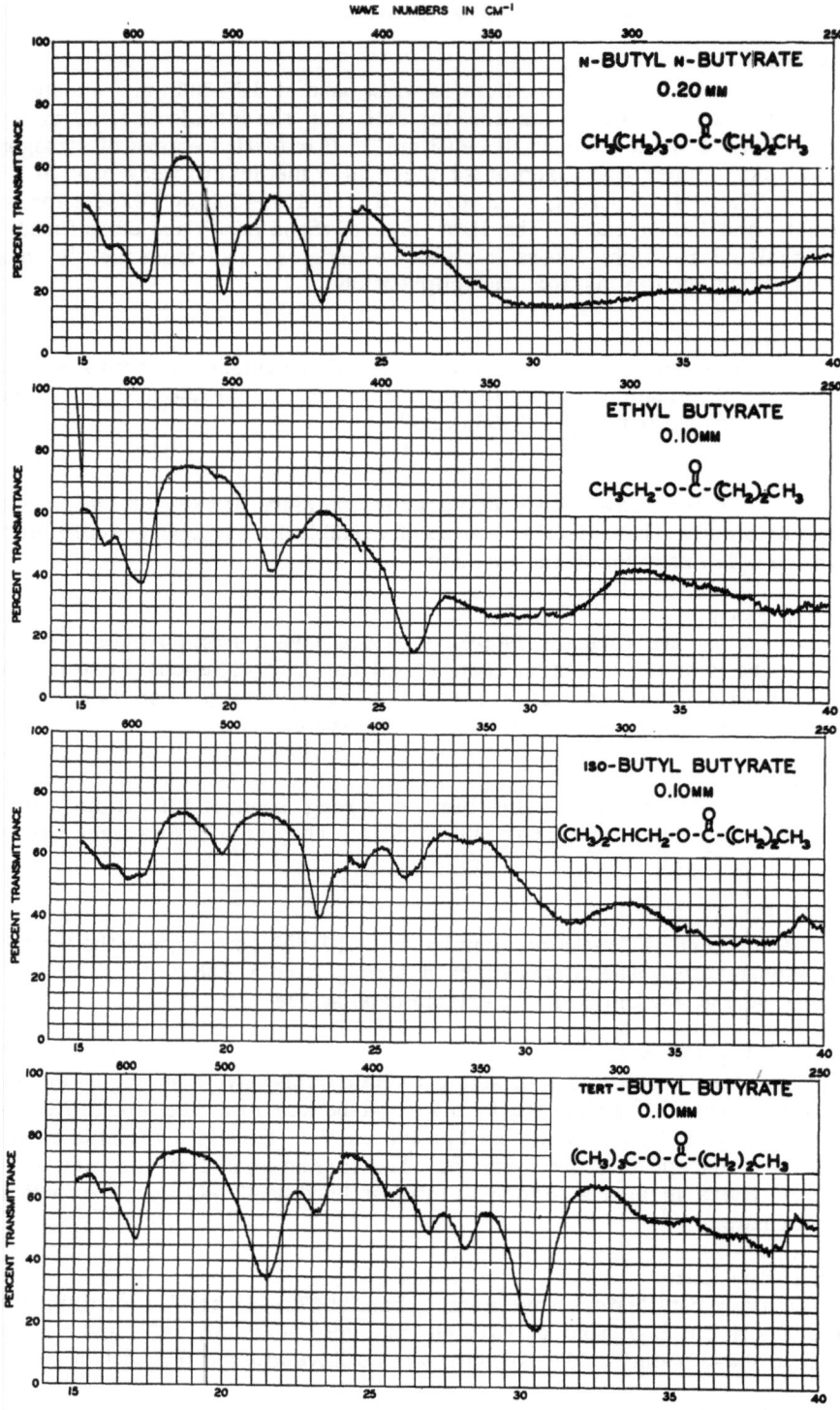

Fig. 6. Infrared spectra from 15–40 μ of n-butyl, ethyl, isobutyl and tertiary butyl butyrates.

butyrate. The isobutyrate has strong bands at 424 cm^{-1} and at 411 cm^{-1}. Homologs of the straight-chain acids up to the valerate have a medium to weak band at 417 cm^{-1}.

CORRELATION SUMMARY

The correlation chart (Fig. 7) is a display of those bands which can best serve to identify the alkyl and acyl components of the esters covered in this study. The chart is a summary of the foregoing discussion on alkyl and acyl correlations in the 15-40 μ region, and reference to this chart will enable the analyst to identify many esters. The chart should be used in conjunction with the text, particularly where bands are listed as variable.

Careful examination of the chart will reveal that each alkyl and each acyl group has a unique set of bands.

Acyl groups larger than heptanoate and alkyl groups larger than neopentyl are not shown since groups of this size do not have spectra which are sufficiently unique to distinguish them from their neighbors. The numbers behind the names of each group indicate the number of samples examined.

BAND ASSIGNMENTS

Formates

Band assignments for methyl formate have been studied by several authors [3-5], notably by Wilmshurst [4]. Earlier authors (Lecomte [8] in 1942 and Thompson [3] in 1945) observed the 635 cm^{-1} band for methyl formate both by infrared and Raman techniques. Lecomte [8] makes a tentative assignment but says it is a doubtful one. Later, Lorenzelli and Möller [11] assigned this band as O—C—O deformation but their assignment is based largely on reference to Lecomte's work. On the other hand, Wilmshurst [4] does not report this band for the liquid or the vapor spectrum and he assigns the O—C—O bending mode to another frequency, viz., 767 cm^{-1}. Consequently, a frequency assignment for the 635 cm^{-1} band in methyl formate is not clear from any published information.

In this work the 635 cm^{-1} band for methyl formate was found to be sharp but weak. Ethyl, n-propyl, n-butyl, and n-amyl formates have a similar band at 621 cm^{-1}. This band becomes weaker as the homologous series is ascended and is very weak in n-amyl formate (path length 0.1 mm). Corresponding bands were not observed in formates of secondary alcohols. Isopropyl formate has a band at 610 cm^{-1}, but it does not correspond in frequency or intensity with the bands just mentioned. Furthermore, secondary butyl and isobutyl formates do not have bands at 635, 621, or 610 cm^{-1}. Thus the band at 635 cm^{-1} for methyl formate and at 621 cm^{-1} for the C_2–C_5 normal formates are characteristic of the straight-chain alkyl formates. The 635 cm^{-1} band was also found in the vapor spectrum of methyl formate. Solutions of methyl formate in diethyl ether gave widely different values for the molecular extinction coefficient at different concentrations, a fact which indicates that the band is not caused by a fundamental vibration of the free molecule.

A strong band was observed for methyl formate at 345 cm^{-1}. This value agrees with the solution spectrum reported by Wilmshurst [4]. In the vapor state this band is at 325 cm^{-1} and Wilmshurst designates it as the C—O—C

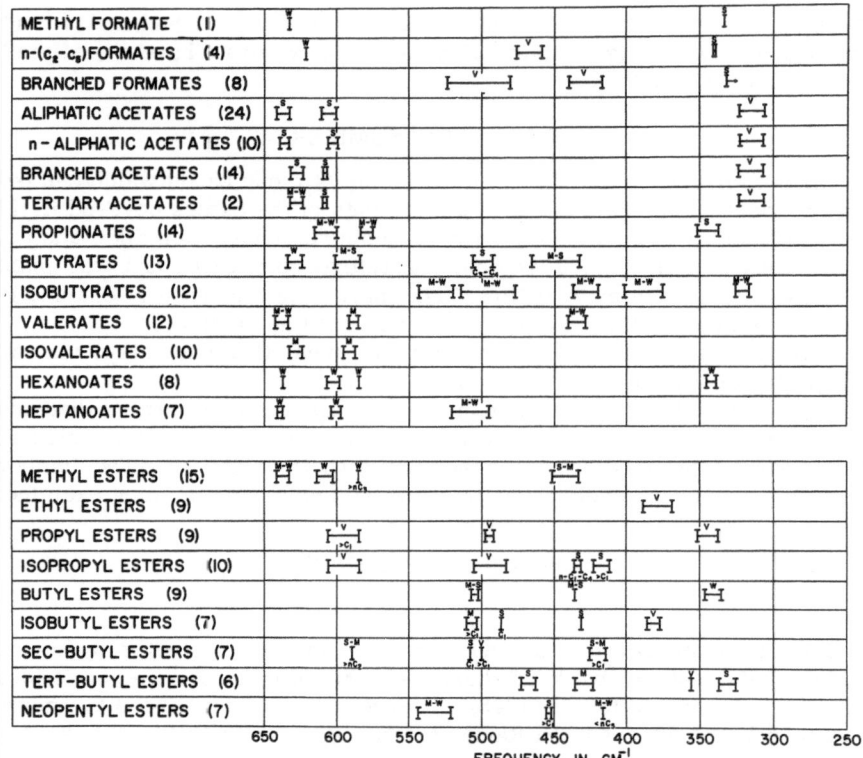

Fig. 7. Correlation summary of saturated aliphatic esters.

bending mode. This last assignment is undoubtedly valid since it has been reported by several workers and has been confirmed by Raman spectra [4, 10].

Acetates

Several workers [2-7] have reported work on methyl acetate and have made vibrational assignments. There is some variation in the frequencies reported for methyl acetate (Table II) but there is no doubt that the authors are describing the same bands. The assignments shown in the table are those of Wilmshurst [4] who found all four bands by Raman as well as by infrared methods. The Raman lines 640, 433, and 303 cm^{-1} are skeletal in-plane bending modes and the 610 cm^{-1} line is a skeletal out-of-plane bending mode. The 640 cm^{-1} band results from O—C—O bending, the 433 cm^{-1} band from C—C—O bending, the 303 cm^{-1} band from C—O—C bending and the 610 cm^{-1} band from CH_3CO_2 out-of-plane bending.

Wilmshurst [4] points out that methyl acetate (as well as methyl formate) has two possible conformations, the cis (I) and the trans (II) (Fig. 8). Lecomte made the same observation in an earlier paper [8]. Recent work on the analysis of the band contours indicates that the cis form is the only one present. This is in agreement with a conclusion reached on the basis of the dipole-moment measurements and electron–diffraction studies [7, 9, 12].

Thompson and Torkington [3] earlier ascribed the 637 cm^{-1} band to deformation of the entire framework while the 606 cm^{-1} band was ascribed to

TABLE II
Infrared Frequencies for the Bending Modes of Methyl Acetate

	This work	Thompson and Torkington [3]	Lorenzelli and Möller [5]	Corval and Lecomte [6]	Wilmshurst [4]	Miyazawa [7]
$-O-C\lesssim^O$ bending	637	640	646	641	639	640
$CH_3C\lesssim^O_{}-O-$ out-of-plane bending	606	612	613	609	615	610
$-C-C\lesssim^O$ bending	434	–	433	436	429	433
C—O—C bending	304	–	305	–	303	303

$CH_3C\lesssim^O$ deformation. These authors point out that methyl ketones have a band at 595 cm^{-1} but that neither diethyl nor dibutyl ketone has a band in this region. Lorenzelli and Möller [5] have made assignments based largely on past and current work of J. Lecomte [8]. The 637 cm^{-1} band is attributed to $-O-C\lesssim^O$ deformation, since essentially the same band appears in acetic acid. The band is slightly weaker in methyl acetate than in acetic acid. The weakening of intensity of this band may be caused by the substitution of the methyl group (methyl acetate) for the hydrogen, (acetic acid) a 15-fold increase in mass.

In this work the spectra of a series of acetate esters were examined. Normal alkyl acetates were found to have two fairly constant bands, one at 633-639 cm^{-1} and one at 600-608 cm^{-1}, as well as two bands (435-505 cm^{-1} and 303-431 cm^{-1}) which vary in a regular manner through the first four members (Table III). These bands are the $C-C\lesssim^O$ (435-505 cm^{-1}) and the C—O—C (303-431 cm^{-1}) bands, and in both cases there is an increase in frequency as the alkyl group gets larger.

In butyl acetate there is an additional strong band at 481 cm^{-1}. Higher homologs of straight-chain acetate esters do not show notable regularity in the bands except for the first two bands, which remain strong and sharp through the C_{18} acetates.

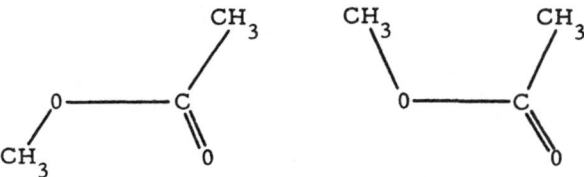

I   II

Fig. 8. Cis and trans conformations of methyl acetate.

TABLE III
Fundamental Acetate Bending Frequencies

Acetate	Wave number, cm^{-1}			
	$-O-C-$ (with O) bending	CH_3C-O- (with O) out-of-plane bending	$-C-C-$ (with O) bending	C$-$O$-$C bending
Methyl	637	605	435	303
Ethyl	633	607	461	378
Propyl	631	607	492	412
Butyl	632	607	505	431

If the methyl group of the acetate radical is replaced by a heavier alkyl radical, both the O—C—O bending mode (631-637 cm^{-1}) and the CH$_3$CO$_2$ out-of-plane bending mode (605-607 cm^{-1}) become broader. This is apparent when these bands for methyl acetate are compared with the same bands for methyl propionate (Fig. 9). If these bands in methyl propionate are then compared with the corresponding bands in ethyl propionate, the bands are further broadened and are barely resolved.

Acetates of Secondary and Tertiary Alcohols

It was observed in this work that the CH$_3$CO$_2$ out-of-plane bending frequency is at a higher frequency in acetates of secondary and tertiary alcohols than in acetates of primary alcohols (Table IV).

At the same time the O—C—O bending mode (640-623 cm^{-1}) shifts to a lower frequency if the acetates of primary alcohols are compared with acetates of secondary and tertiary alcohols (Table V). Isobutyl and neopentyl acetates do not show similar shifts, apparently because they lack carbon atoms in the alpha position. The reason for this apparent interaction is not clear. In this case both bands have different symmetry properties but are close enough in frequency to couple their energies.

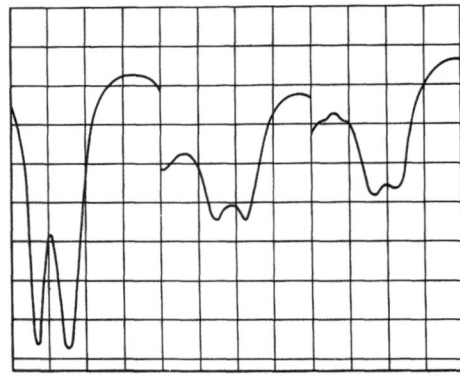

Fig. 9. The in-plane O—C—O and out-of-plane CH$_3$CO$_2$ bending modes for methyl acetate, methyl propionate, and ethyl propionate in 640-600 cm^{-1} region.

TABLE IV
CH_3CO_2 Out-of-Plane Bending in Acetates

n-Propyl	606 cm^{-1}
n-Butyl	601 cm^{-1}
Isobutyl	604 cm^{-1}
Neopentyl	605 cm^{-1}
Isopropyl	610 cm^{-1}
sec-Butyl	611 cm^{-1}
tert-Butyl	608 cm^{-1}

TABLE V
O—C—O Bending in Acetate Esters

n-Propyl	623 cm^{-1}
n-Butyl	629 cm^{-1}
Isobutyl	630 cm^{-1}
Neopentyl	627 cm^{-1}
Isopropyl	620 cm^{-1}
sec-Butyl	620 cm^{-1}
tert-Butyl	623 cm^{-1}

REFERENCES

1. N. Alpert, F. Behnke, and P. A. Strauss, Appl. Spectroscopy 13(5):130 (1959).
2. F. F. Bentley, E. F. Wolfarth, Nora E. Srp, and W. R. Powell, Spectrochim. Acta 13:1 (1958).
3. H. W. Thompson and P. Torkington, J. Chem. Soc. 640 (1945).
4. J. K. Wilmshurst, J. Mol. Spectroscopy 1:201 (1957).
5. V. Lorenzelli and K. D. Möller, Compt. rend. 249:669 (1959).
6. M. Corval and J. Lecomte, Mikrochimica Acta 25 (1955).
7. T. Miyazawa, Bull. Chem. Soc. Japan 34:691-6 (1961).
8. J. Lecomte, J. phys. radium 3:193 (1942).
9. J. M. O'Gorman, W. Shand, and V. Schomaker, J. Am. Chem. Soc. 73:4222 (1950).
10. R. F. Curl, J. Chem. Phys. 30:1529 (1959).
11. V. Lorenzelli and K. D. Möller, Compt. rend. 249:520 (1959).
12. R. J. B. Marsden and L. E. Sutton, J. Chem. Soc 1383 (1936).
13. Jones, Faure, and Zaharias, Revue Universelle des Mines XV 5:417 (1959).

The Urey - Bradley Force Field:

Its Significance and Application

Kazuo Nakamoto

Department of Chemistry
Illinois Institute of Technology
Chicago, Illinois

One of the most fundamental problems in normal coordinate analysis is the selection of a suitable potential field to express the interatomic forces in a molecule. Although the generalized valence force (GVF) field is widely used by many investigators, it encounters serious difficulties in complex molecules since the number of force constants exceeds the number of observed normal vibrations. In order to circumvent this difficulty, Shimanouchi introduced the Urey–Bradley force (UBF) field which consists of stretching and bending force constants, as well as repulsive force constants between nonbonded atoms. The number of force constants in the UBF field is much smaller than that in the GVF field. In addition, the UBF field has the advantages that the force constants have a clearer physical meaning than those of the GVF field, and they are transferable from molecule to molecule. The UBF field, however, does not include any interaction terms between non-neighboring stretching vibrations and between bending vibrations. In some molecules, ignorance of these terms causes difficulties in adjusting force constants to fit the observed frequencies. Recently, a number of attempts have been made to improve the UBF field, and these will be reviewed briefly.

One of the most fundamental problems in normal coordinate analysis is the selection of a suitable potential field which represents the interatomic forces in a molecule. Among several types of potential fields, the general valence force (GVF) field and the Urey–Bradley force (UBF) field are most frequently used. This paper is intended to describe the significance of the UBF field and its application to polyatomic molecules.

In order to see the difference between the GVF and UBF fields, it is useful to compare the potential energy expressions for the H_2O molecule in both fields. With the internal coordinates shown in Fig. 1, the potential energy in terms of the GVF field is written as

$$2V = f_{11}(\Delta r_1)^2 + f_{11}(\Delta r_2)^2 + f_{33}r_0^2(\Delta a)^2 + 2f_{12}(\Delta r_1)(\Delta r_2) + 2f_{13}r_0(\Delta r_1)(\Delta a)$$
$$+ 2f_{13}r_0(\Delta r_2)(\Delta a) \quad (1)$$

Here f_{11} and f_{33} are the stretching force constants of the O—H bond and the bending force constant for the H—O—H angle, respectively; f_{12} and f_{13} are the interaction force constants between stretching and stretching, and between stretching and bending vibrations; f_{33} and f_{13} are multiplied by the equilibrium distance, r_0 to make the dimensions of all the force constants the same.

With matrix notation, equation (1) can be written more conveniently as

$$2V = \tilde{\mathbf{R}}\mathbf{F}\mathbf{R} \quad (2)$$

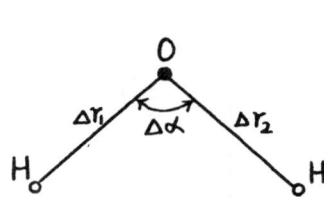

Fig. 1. Internal coordinates of the H_2O
molecule.

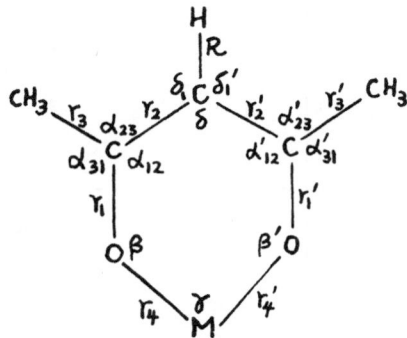

Fig. 2. Internal coordinates of a metal acetylacetonate.

where R is a column matrix of the form

$$R = \begin{bmatrix} \Delta r_1 \\ \Delta r_2 \\ \Delta \alpha \end{bmatrix}$$

$\tilde{R}$ is its transpose

$$\tilde{R} = [\Delta r_1 \; \Delta r_2 \; \Delta \alpha]$$

and F is a matrix whose components are the force constants:

$$F_{GVF} = \begin{bmatrix} f_{11} & f_{12} & r_0 f_{13} \\ & f_{11} & r_0 f_{13} \\ & & r_0^2 f_{33} \end{bmatrix} \tag{3}$$

As is seen above, four GVF force constants are needed to describe the potential energy of the H_2O molecule. Since only three fundamentals are observed, it is impossible to determine all four constants simultaneously. One method to circumvent this difficulty is to calculate the vibrational frequencies of isotopic molecules such as HDO and D_2O assuming the same set of force constants.

This method is satisfactory, however, only for simple molecules. As molecules become more complex, the number of force constants in the GVF field becomes too large to allow any reliable evaluation. For example, consider a metal acetylacetonate [1, 2] whose internal coordinates for the in-plane vibrations are selected as shown in Fig. 2. With these internal coordinates at least thirty GVF constants are needed to describe the potential energy of this molecule (see Table I). Experimentally, however, only fifteen in-plane fundamentals are observed. Thus it is almost impossible to determine these force constants simultaneously.

In order to circumvent this difficulty, Shimanouchi [3] introduced the UBF field [4] which consists of stretching (K) and bending (H) force constants as well as repulsive (F) force constants between nonbonded atoms. It is

TABLE I
The **F** Matrix of a Metal Acetylacetonate

F	r_1	r_1'	r_2	r_2'	r_3	r_3'	r_4	r_4'	R	α_{12}	α_{23}	α_{31}	α_{12}'	α_{23}'	α_{31}'	δ	δ_1	δ_1'	β	β'	γ
r_1	f_1		f_1^2		f_1^3		f_1^4			f_1^{12}		f_1^{31}							f_1^β		
r_1'		f_1		f_1^2		f_1^3		f_1^4					f_1^{12}		f_1^{31}					f_1^β	
r_2'			f_2		f_2^3				f_2^R	f_2^{12}	f_2^{23}					f_2^δ	$f_2^{\delta_1}$				
r_2'				f_2		f_2^3			f_2^R				f_2^{12}	f_2^{23}		f_2^δ		$f_2^{\delta_1}$			
r_3'					f_3						f_3^{23}	f_3^{31}									
r_3'						f_3								f_3^{23}	f_3^{31}						
r_4'							f_4	f_4^4											f_4^β		f_4^γ
r_4'								f_4												f_4^β	f_4^γ
R									f_5								$f_R^{\delta_1}$	$f_R^{\delta_1}$			
α_{12}										f_{12}											
α_{23}											f_{23}										
α_{31}												f_{31}									
α_{12}'													f_{12}								
α_{23}'														f_{23}							
α_{31}'															f_{31}						
δ																f_δ					
δ_1																	f_{δ_1}				
δ_1'																		f_{δ_1}			
β																			f_β		
β'																				f_β	
γ																					f_γ

(Symmetrical)

clearly recognized that the repulsive force acting between nonbonded atoms is certainly not negligible and is at times quite appreciable. The fact that these three force constants have clear-cut physical meaning is an advantage of the UBF field relative to the GVF field, where physical meaning of some of the interaction constants is obscure.

The potential energy of the H_2O molecule in the UBF field may be written as

$$2V = K\{(\Delta r_1)^2 + (\Delta r_2)^2\} + H(r_0\Delta a)^2 + F(\Delta q)^2 \tag{4}$$

where q is the distance between two H atoms. This expression would be satisfactory if Δr_1, Δr_2, Δa, and Δq were mutually independent because the first derivatives of the potential energy could then be equated to zero using the equilibrium condition. In the present case, however, the four coordinates are related by the geometrical relation

$$q = (r_1^2 + r_2^2 - 2r_1r_2 \cos a)^{1/2} \tag{5}$$

Accordingly, the potential energy must be expressed including the first derivatives:

$$2V = 2K'r_0(\Delta r_1 + \Delta r_2) + K\{(\Delta r_1)^2 + (\Delta r_2)^2\} + 2H'r_0^2(\Delta a) + H(r_0\Delta a)^2 \\ + 2F'q_0(\Delta q) + F(\Delta q)^2 \tag{6}$$

The differential of equation (5) gives

$$\Delta q = s\Delta r_1 + s\Delta r_2 + t(r_0\Delta a) \\ + \{t^2(\Delta r_1)^2 + t^2(\Delta r_2)^2 - s^2(r_0\Delta a)^2 - 2t^2(\Delta r_1)(\Delta r_2) + 2ts(\Delta r_1)(r_0\Delta a) + 2ts(\Delta r_2)(r_0\Delta a)\}/2q_0 \tag{7}$$

where

$$s = r_0(1 - \cos a)/q_0 \quad \text{and} \quad t = (r_0 \sin a)/q_0 \tag{8}$$

If equation (7) is inserted in equation (6), the potential energy is now expressed in terms of Δr_1, Δr_2, and Δa only

$$2V = 2(K'r_0 + sF'q_0)(\Delta r_1 + \Delta r_2) + (K + t^2F' + s^2F)(\Delta r_1^2 + \Delta r_2^2) + 2(H'r_0 + tF'q_0)(r_0\Delta a) \\ + (H - s^2F' + t^2F)(r_0\Delta a)^2 + 2(-t^2F' + s^2F)(\Delta r_1)(\Delta r_2) + 2ts(F' + F)(\Delta r_1)(r_0\Delta a) \\ + 2ts(F' + F)(\Delta r_2)(r_0\Delta a) \tag{9}$$

The first derivatives in equation (9) can now be set equal to zero using the equilibrium condition ($\partial V/\partial r_i = \partial V/\partial a = 0$). Thus the **F** matrix becomes

$$\mathbf{F}_{UBF} = \begin{bmatrix} K + t^2F' + s^2F & -t^2F' + s^2F & ts(F' + F)r_0 \\ & K + t^2F' + s^2F & ts(F' + F)r_0 \\ & & (H - s^2F' + t^2F)r_0^2 \end{bmatrix} \tag{10}$$

A comparison of the two **F** matrices (3 and 10) gives the relation between the GVF and UBF force constants:

$$f_{11} = K + t^2 F' + s^2 F$$
$$r_0^2 f_{33} = (H - s^2 F' + t^2 F) r_0^2 \tag{11}$$
$$f_{12} = -t^2 F' + s^2 F$$
$$r_0 f_{13} = ts(F' + F) r_0$$

The number of force constants is four in both fields. It is possible, however, to reduce it to three in the UBF field by taking $F' = -(1/10)F$. This is based on the assumption that the repulsive energy between nonbonded atoms is proportional to $1/r^9$. If it is proportional to r^{-n}, then $F' = -[1/(n + 1)]F$. In any event F' is not large, and a variation in the value of n does not cause a significant difference in the final result. Thus the number of UBF force constants for the H_2O molecule is smaller by one compared with that of the GVF field.

The UBF field for general tetrahedral molecules [3] is written as

$$V = \frac{1}{2} \sum_i^* \left[K_i + \sum_{j \neq i} (t_{ij}^2 F_{ij}' + s_{ij}^2 F_{ij}) \right] (\Delta r_i)^2$$

$$+ \frac{1}{2} \sum_{i<j} [H_{ij} - s_{ij} s_{ji} F_{ij}' + t_{ij} t_{ji} F_{ij} + (\{3\kappa/\sqrt{8}\} r_{ij}^2)](r_{ij} \Delta a_{ij})^2$$

$$+ \sum_{i<j} [-t_{ij} t_{ji} F_{ij}' + s_{ij} s_{ji} F_{ij}](\Delta r_i)(\Delta r_j) \tag{12}$$

$$+ \sum_{i \neq j} [t_{ij} s_{ji} F_{ij}' + t_{ji} s_{ij} F_{ij}](r_j/r_i)^{1/2}(\Delta r_i)(r_{ij} \Delta a_{ij})$$

$$+ \sum_{i \neq j \neq k} (\{\kappa/\sqrt{2}\} r_{ij} r_{ik})(r_{ij} \Delta a_{ij})(r_{ik} \Delta a_{ik})$$

where

$$s_{ij} = (r_i - r_j \cos a_{ij})/q_{ij}$$
$$s_{ji} = (r_j - r_i \cos a_{ij})/q_{ij} \tag{13}$$
$$t_{ij} = r_j \sin a_{ij}/q_{ij}$$
$$t_{ji} = r_i \sin a_{ij}/q_{ij}$$

In addition to the K, H, and F force constants, it is necessary to include κ (internal tension), which originates in the redundancy concerning six angles around the central atom. Thus the number of UBF force constants for CH_4 becomes four [K(C—H), H(HCH), F(H$\cdots$H), and κ], and that for CH_3Cl becomes seven [K(C—H), K(C—Cl), H(HCH), H(HCCl), F(H$\cdots$H), F(H$\cdots$Cl), and

TABLE II
UBF Force Constants for Halo-genomethanes (mdyn/A)

$K(C-Cl) = 1.76$	$K(C-Br) = 1.43$
$H(Cl-C-Cl) = 0.08$	$H(Br-C-Br) = 0.045$
$H(Cl-C-Br) = 0.03$	$F(Cl\cdots Cl) = 0.656$
$F(Br\cdots Br) = 0.493$	$F(Cl\cdots Br) = 0.575$
	$\kappa = 0.3$

κ]. The situation is similar in the GVF field where four constants (f_α, f_d, $f_{\alpha\alpha}$ and f_{dd}) and seven constants (f_D, f_d . $f_{\alpha'}$, f_β, $f_{D\alpha}$, $f_{D\beta}$, and $f_{\alpha\beta}$) are normally used for CH_4 and CH_3Cl, respectively. In more complex molecules, how-ever, the number of force constants in the UBF field becomes much smaller than that of the GVF field. For example, the number of the UBF force constants for a metal acetylacetonate is reduced to eighteen, which is about half the number for the GVF field [1].

Another advantage of the UBF field is the transferability of force constants from molecule to molecule. For example, the force constants obtained for CCl_4 and CBr_4 (Table II) give excellent agreement for CCl_3Br, CCl_2Br_2, and $CClBr_3$. As will be shown later, there are several other series of molecules in which the UBF force constants have excellent trans-ferability. This transferability of the UBF force constants is highly im-portant in normal coordinate analyses of complex molecules. With this property, it is possible to estimate a priori the values of force constants for complex molecules from those previously obtained for simpler mole-cules. Again a metal acetylacetonate may be used as an example. Figure 3 shows that bond distances and bond angles in the acetylacetonato- Fe(III)

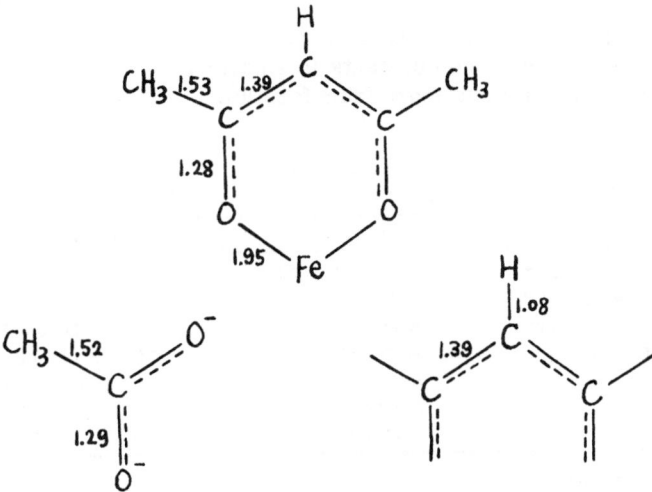

Fig. 3. Bond distances in tris-(acetylacetonato)-Fe(III), the acetate anion and benzene.

TABLE III

Comparison of Calculated
and Observed Frequencies of
bis-(Acetylacetonato)-Cu (II)
(cm^{-1})

		Calculated	Observed
A_1	λ_1	1544	1554 1534
	λ_2	1264	1274
	λ_3	936	937
	λ_4	659	684
	λ_5	457	458
	λ_6	272	—
	λ_7	197	—
B_2	λ_8	1580	1580
	λ_9	1489	1464
	λ_{10}	1226	1190
	λ_{11}	901	937
	λ_{12}	628	654
	λ_{13}	373	—
	λ_{14}	302	—

complex [5] are similar to those in the acetate ion [6] and in benzene [7].
This result suggests that the UBF force constants obtained for the latter
two compounds may be used for the metal complex with slight modification.
In fact, the agreement in frequency such as shown in Table III was obtained
by using the set of force constants shown in Table IV, where most of the
values were taken from those obtained for the acetate ion [8] and for benzene [8].

In conclusion, the advantages of the UBF relative to the GVF field are
summarized as follows:

1. Force constants have clearer physical meaning.
2. The number of force constants is smaller.
3. Force constants are more transferable.

TABLE IV

Force Constants of bis-(Acetylacetonato)-Cu(II) (mdyn/A)

Stretching	Bending	Repulsive
$K(C \cdots O)$, 6.90*	$H(CHCO) = 0.24$	$F(C \ldots O) = 0.50$*
$K(C \cdots C)$, 5.35[†]	$H(RCC) = 0.25$	$F(C \ldots R) = 0.40$
$K(C—R)$, 3.60*	$H(RCO) = 0.31$*	$F(O \ldots R) = 0.50$*
$K(Cu—O)$, 2.20	$H(COCu) = 0.05$	$F(C \ldots Cu) = 0.05$
	$H(OCuO) = 0.05$	$F(O \ldots O) = 0.05$
	$H(HCC) = 0.20^+$	$F(C \ldots C) = 0.37$[†]
	$H(CCC) = 0.33^†$	$F(C \ldots H) = 0.54^+$

*Acetate ion.
[†]Benzene.

It should be noted, however, that the UBF field ignores

 4. Interactions between non-neighboring stretching vibrations.*
 5. Interactions between bending vibrations.

The ignoring of these interaction force constants can cause serious difficulty in fitting calculated frequencies to observed ones. These disadvantages of the UBF field can be partly remedied, however, by introducing new interaction force constants which are not included in the simple UBF field. These new force constants should have clear physical meaning as well as reasonable transferability.

Recently, Shimanouchi and his co-workers [8] conducted an extensive study on the transferability of the UBF force constants. According to their results, molecules can be classified into three classes:

 I. Molecules for which force constants are completely transferable.
 II. Molecules for which force constants are not strictly transferable, but for which the UBF field is applicable in individual cases.
 III. Molecules for which the UBF field is applicable upon introduction of new force constants.

Molecules belonging to Class I are illustrated in Fig. 4. In each series of molecules of these types, force constants are completely transferable, and the calculated repulsive force constants are reasonably close to those expected from the Lennard–Jones type of potential. If molecules of Type I contain an H atom, however, the force constants are no longer transferable within the same series of molecules, but the UBF field is applicable to each molecule (Class II). This is illustrated in Table V where the force constants are slightly different in each compound.

Molecules belonging to Class III can be further classified into three groups:

 1. Molecules having central atoms which utilize d-orbitals for bonding.
 2. Molecules containing nonlocalized electrons.
 3. Ethane and other hydrocarbons.

*The interaction between neighboring stretching vibrations is expressed in terms of the repulsive force constant [see equation (10)].

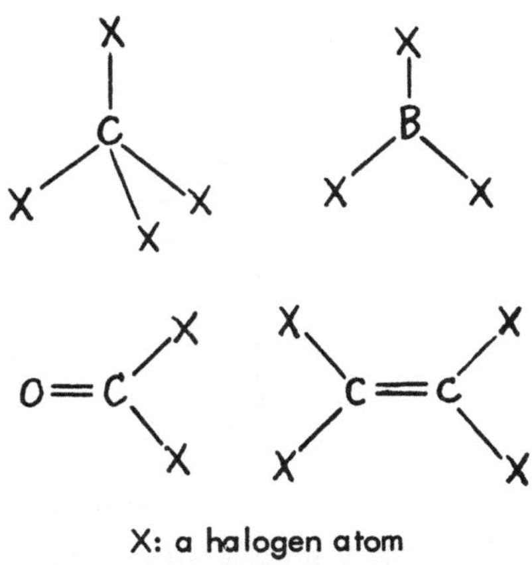

Fig. 4. Molecules belonging to Class I.

X: a halogen atom

TABLE V
Force Constants of Chloromethanes (mdyn/A)

	CH_4	CH_3Cl	CH_2Cl_2	$CHCl_3$	CCl_4
$K(C-H)$	4.76	4.74	4.49	4.24	–
$H(HCH)$	0.42	0.42	0.39	–	–
$F(H...H)$	0.08	0.06	0.04	–	–
$K(C-Cl)$	–	2.06	1.94	1.92	1.81
$H(ClCCl)$	–	–	0.06	0.08	0.08
$F(Cl...Cl)$	–	–	0.70	0.70	0.66
$H'(HCCl)$	–	0.18	0.08	0.02	–
$F(H...Cl)$	–	0.60	0.68	0.73	–
κ	−0.01	−0.06	0.10	0.23	0.26

When the central atom uses d-orbitals for bonding, interactions between two stretching or between two bending vibrations may occur through d-orbitals. Typical examples are given by molecules of the SiX_4 type (X = halogen), for which an interaction force constant P between two stretching vibrations is needed [8]. Trigonal bypyramidal molecules such as PX_5 and SbX_5 (X = F or Cl) also need an interaction force constant (l) between two angles perpendicular to each other [9]. Complex ions such as $[PtX_6]^{2-}$ need both of these force constants [10]. In trans-$[Pt(NH_3)_2X_2]$, long-range interaction force constants between two HNH angles and between two HNPt angles were found to be necessary in addition to the interaction constants between two bonds in the trans position [11]. All these new interaction constants are illustrated in Fig. 5.

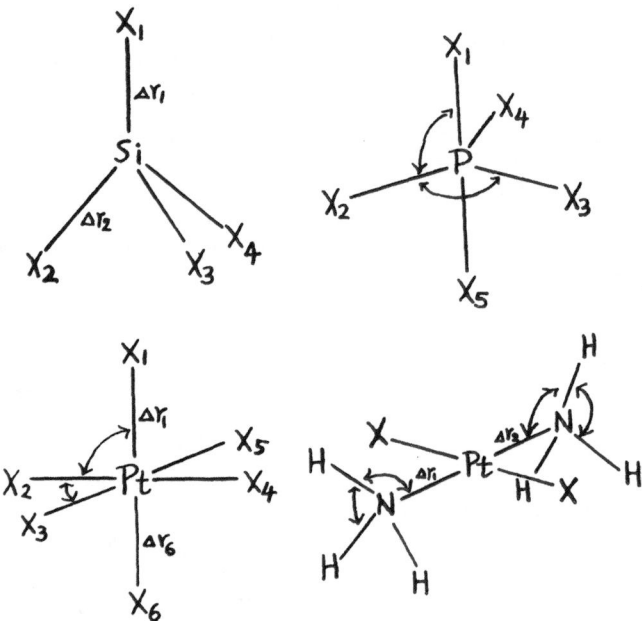

X: a halogen atom

Fig. 5. Stretching–stretching and bending–bending interactions.

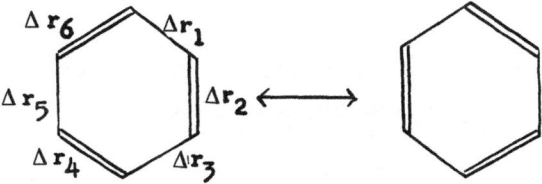

Fig. 6. The Kekulé interaction.

Molecules of group (2) are those for which several resonance formulas can be written. It is known that the simple UBF field does not give satis-factory results for inorganic ions such as NO_3^- and SO_4^{2-}. The same is true for organic molecules like benzene and naphthalene. This difficulty can be avoided if one considers interactions between two bonds which al-ternate between single and double bonds in various resonance formulas. In the case of benzene, the complete potential function may be written as

$$2V = 2V_{\text{UBF}} + 2\rho\left(\sum_{}^{6}\Delta r_i \Delta r_{i+1} + \sum_{}^{3}\Delta r_i \Delta r_{i+3} - \sum_{}^{6}\Delta r_i \Delta r_{i+2}\right) \qquad (14)$$

where ρ is called the Kekulé constant (see Fig. 6.). The physical origin of this interaction has been discussed by Scherer and Overend [12].

It has long been recognized that the simple UBF fails to give good agreement between calculated and observed frequencies for the rocking and wagging vibrations of ethane and other hydrocarbons. In order to improve the agreement, it is necessary to include the bending–bending interactions [13-15] shown in Fig. 7. This modification also gives satisfactory results for polymers such as polyvinylchloride and polypropylene. However, the physical origin of these interactions is still obscure, and must be clarified in the future.

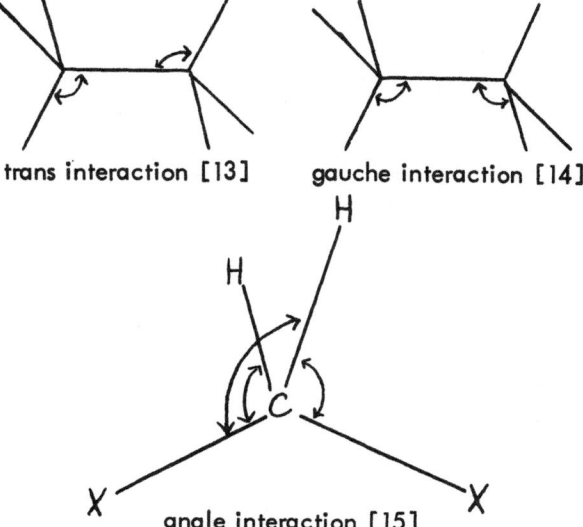

trans interaction [13]   gauche interaction [14]

angle interaction [15]

Fig. 7. Various bending–bending interactions.

REFERENCES

1. K. Nakamoto and A. E. Martell, J. Chem. Phys. 32:588 (1960).
2. K. Nakamoto, Infrared Spectra of Inorganic and Coordination Compounds (John Wiley, New York, 1963).
3. T. Shimanouchi, J. Chem. Phys. 17:245, 734, and 848 (1949).
4. H. C. Urey and C. A. Bradley, Phys. Rev. 38:1969 (1931).
5. R. B. Roof, Acta Cryst. 9:781 (1956).
6. W. H. Zachariasen and H. A. Pettinger, Acta Cryst. 12:526 (1959).
7. I. L. Karle, J. Chem. Phys. 20:65 (1952).
8. T. Shimanouchi, "Force Constants of Small Molecules," Private Publication, Univ. of Tokyo, Sept., 1962; S. Mizushima and T. Shimanouchi, Infrared Absorption and the Raman Effect (Kyoritsu, Tokyo, 1958).
9. R. A. Condrate and K. Nakamoto, presented at Mid-America Spectroscopy Symposium, Chicago, May, 1963.
10. J. Hiraishi, I. Nakagawa, and T. Shimanouchi, International Symposium on Molecular Structure and Spectroscopy, Tokyo, Sept., 1962.
11. K. Nakamoto, P. J. McCarthy, R. A. Condrate, and J. Fujita (to be published).
12. J. R. Scherer and J. Overend, Spectrochim. Acta 17:719 (1961).
13. J. R. Scherer and J. Overend, J. Chem. Phys. 33:1681 (1960).
14. H. Takahashi and T. Shimanouchi, International Symposium on Molecular Structure and Spectroscopy, Tokyo, Sept., 1962.
15. I. Nakagawa and S. Mizushima, J. Chem. Phys. 21:2195 (1953); 22:759 (1953).

The Urey-Bradley Force Constants of Trigonal Bipyramidal XY_5-Type Molecules*

Robert A. Condrate and Kazuo Nakamoto

Department of Chemistry
Illinois Institute of Technology
Chicago, Illinois

Normal coordinate analysis has been carried out for the trigonal bipyramidal XY_5-type molecules, and the results have been applied to $SbCl_5$, SbF_5, PCl_5, and PF_5 using the Urey–Bradley force field. It has been found that the introduction of an angle–angle interaction force constant to the simple Urey–Bradley force field is necessary to obtain good agreement between calculated and observed frequencies. The results provide consistent band assignments and force constants for the four molecules studied.

INTRODUCTION

Although the infrared and Raman spectra of trigonal bypyramidal XY_5-type molecules have already been obtained by several investigators, there appears to be a controversy about the band assignments for several fundamentals of certain molecules. A typical example is PCl_5, for which the σ_6 mode was assigned to 190 cm^{-1} by Siebert [1], to 335 cm^{-1} by Wilmshurst and Bernstein [2], and to 273 cm^{-1} by Miller [3]. The purpose of this paper is to solve such conflicting band assignments through a systematic normal coordinate analysis of $SbCl_5$, PCl_5, SbF_5, and PF_5, for which the monomeric trigonal bipyramidal structure is well established. The present work should be contrasted with the previous normal coordinate analyses [1, 4-8], which were based on limited experimental data.

In carrying out the normal coordinate analysis, it was soon found that the simple Urey–Bradley force field fails to give good agreement between the calculated and observed frequencies of these molecules. The addition of an angle–angle interaction force constant to the simple Urey–Bradley force field was necessary to improve the results. It should be mentioned that this interaction force constant has already been introduced by Hiraishi et al [9] for octahedral XY_6-type molecules which utilize d-orbitals for bonding. Recently, Shimanouchi [10] formulated a general rule that all molecules utilizing d-orbitals for bonding need an angle–angle interaction force constant in the Urey–Bradley field. Therefore, the inclusion of this additional force constant in the present study can be well justified, since the bonding in trigonal bipyramidal molecules presumably involves d-orbitals. The excellent agreement between calculated and observed frequencies provides in turn an indication of validity of Shimanouchi's generalization.

---

*To be submitted by Robert A. Condrate to the faculty of Illinois Institute of Technology in partial fulfillment of the requirements for the degree of Doctor of Philosophy.

Fig. 1. Microdensitometer tracing vs. frequency for PCl₅. Asterisk denotes band is due to POCl₃ impurity.

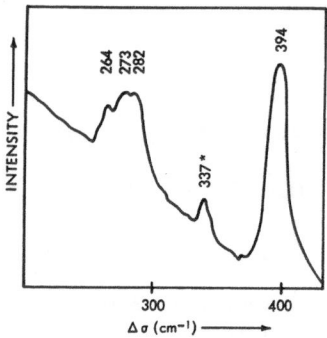

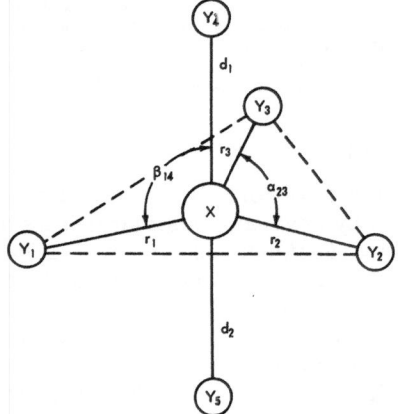

Fig. 2. Structure and internal coordinates of the trigonal bipyramidal XY_5-type molecule.

EXPERIMENTAL

Most of the spectral data used for the normal coordinate analyses were taken from the literature. The Raman spectrum of PCl₅ was obtained, however, to confirm some of the calculated frequencies. A saturated benzene solution of phosphorous pentachloride (Baker Analyzed Reagent) was analyzed in a Hilger E-612 two-prism Raman spectrograph. Mercury light of wavelength 4358 A was used for exciting the Raman transitions. A solution of Dupont rhodamine 5 GDN-Extra dye mixed with an ethanol solution of paranitrotoluene was used to filter out the other mercury lines. Eastman 103a-J plates were employed to record the spectra. In addition to a number of Raman lines due to benzene, the following lines were observed for PCl₅: 585 cm⁻¹ (medium strong), 394 cm⁻¹ (very strong), 282 cm⁻¹ (strong), 273 cm⁻¹ (strong), 264 cm⁻¹ (medium strong). The transitions below 200 cm⁻¹ could not be observed because of strong background scattering. Figure 1 shows the microdensitometer reading plotted against the frequency in the range between 430 and 200 cm⁻¹.

METHOD OF CALCULATION

The trigonal bipyramidal XY_5-type molecule shown in Fig. 2 belongs to the point group D_{3h}, and its twelve normal modes of vibration are grouped into the four species A_1', A_2'', E', and E''. The appropriate symmetry coordinates used are given in Table I, and the elements of the G and F matrices are listed in the Appendix. The G matrix elements were calculated using the bond distances reported in the literature [11–14]. The matrix secular equations of the form $|GF - E\lambda| = 0$ were solved for each species using these elements.

In order to express the potential energy, a modified Urey–Bradley force field of the following type was employed:

$$V = \sum_i [K_r' r(\Delta r_i) + \tfrac{1}{2} K_r(\Delta r_i)^2]$$

$$+ \sum_i [K_d' d(\Delta d_i) + \tfrac{1}{2} K_d(\Delta d_i)^2]$$

$$+ \sum_{i<j} [H_\alpha' r_\alpha^2(\Delta a_{ij}) + \tfrac{1}{2} H_\alpha(r_\alpha \Delta a_{ij})^2]$$

$$+ \sum_{i<j} [H_\beta' r_\beta^2(\Delta \beta_{ij}) + \tfrac{1}{2} H_\beta(r_\beta \Delta \beta_{ij})^2]$$

$$+ \sum_{i<j} [F_{rr}' q_{rr}(\Delta q_{ij}) + \tfrac{1}{2} F_{rr}(\Delta q_{ij})^2]$$

$$+ \sum_{k<j} [F_{dr}' q_{dr}(\Delta q_{jk}) + \tfrac{1}{2} F_{dr}(\Delta q_{jk})^2]$$

$$+ \sum_{k \neq i < j} [H_{\alpha\beta}(r_\alpha \Delta a_{ij})(r_\beta \Delta \beta_{jk})]$$

where Δr_i, Δa_{ij}, $\Delta \beta_{ij}$, Δq_{ij}, and Δq_{jk} are, respectively, the change in the bond lengths r and d, the change in the bond angles a and β, and the change in the distances q_{rr} and q_{dr} between nonbonded atoms. The symbols K, H, and F represent stretching, bending, and repulsive force constants, respectively. The primed letters denote first derivatives; the symbol $H_{\alpha\beta}$ represents an angle–angle interaction force constant between the angles a and β, which lie in planes perpendicular to each other and share a common bond; r, d, r_α, r_β, q_{rr}, and q_{dr} are the values of the distances at equilibrium; they are inserted to make the dimensions of the force constants the same. The quantities K' and H' vanish in the final result, since they can be expressed in terms of F' using the equilibrium condition. F' was taken as $-0.1F$ by assuming that the repulsive energy between the nonbonded atoms is proportional to r^{-9}.

Tables III to VI compare the observed frequencies with those calculated using the sets of force constants listed in Table II. In Tables III to VI, the

TABLE I
Symmetry Coordinates for Trigonal Bipyramidal XY_5-Type Molecules

A_1' species
$$\begin{cases} S_1 = (1/\sqrt{3})(\Delta r_1 + \Delta r_2 + \Delta r_3) \\ S_2 = (1/\sqrt{2})(\Delta d_1 + \Delta d_2) \\ S_a = (1/\sqrt{3})(\Delta\alpha_{12} + \Delta\alpha_{23} + \Delta\alpha_{31}) \equiv 0 \\ S_b = (1/\sqrt{6})(\Delta\beta_{14} + \Delta\beta_{24} + \Delta\beta_{34} + \Delta\beta_{15} + \Delta\beta_{25} + \Delta\beta_{35}) \equiv 0 \end{cases}$$

A_2'' species
$$\begin{cases} S_3 = (1/\sqrt{2})(\Delta d_1 - \Delta d_2) \\ S_4 = (1/\sqrt{6})(\Delta\beta_{14} + \Delta\beta_{24} + \Delta\beta_{34} - \Delta\beta_{15} - \Delta\beta_{25} - \Delta\beta_{35}) \end{cases}$$

E' species
$$\begin{cases} S_{5a} = (1/\sqrt{6})(2\Delta r_1 - \Delta r_2 - \Delta r_3) \\ S_{5b} = (1/\sqrt{2})(\Delta r_2 - \Delta r_3) \\ S_{6a} = (1/\sqrt{6})(2\Delta\alpha_{23} - \Delta\alpha_{12} - \Delta\alpha_{31}) \\ S_{6b} = (1/\sqrt{2})(\Delta\alpha_{31} - \Delta\alpha_{12}) \\ S_{7a} = (1/\sqrt{12})(2\Delta\beta_{14} - \Delta\beta_{24} - \Delta\beta_{34} + 2\Delta\beta_{15} - \Delta\beta_{25} - \Delta\beta_{35}) \\ S_{7b} = (1/2)(\Delta\beta_{24} - \Delta\beta_{34} + \Delta\beta_{25} - \Delta\beta_{35}) \end{cases}$$

E'' species
$$\begin{cases} S_{8a} = (1/\sqrt{12})(2\Delta\beta_{14} - \Delta\beta_{24} - \Delta\beta_{34} - 2\Delta\beta_{15} + \Delta\beta_{25} + \Delta\beta_{35}) \\ S_{8b} = (1/2)(\Delta\beta_{24} - \Delta\beta_{34} - \Delta\beta_{25} + \Delta\beta_{35}) \end{cases}$$

TABLE II
Urey–Bradley Force Constants of Various Trigonal Bipyramidal Molecules (mdyn/A)

	K_r	K_d	H_α	H_β	F_{rr}	F_{dr}	$H_{\alpha\beta}$
$SbCl_5$	2.15	1.83	0.05	0.085	0.11	0.13	0.043
PCl_5	2.10	1.00	0.07	0.15	0.10	0.55	0.07
SbF_5	5.20	4.60	0.085	0.10	0.13	0.13	0.07
PF_5	5.20	3.65	0.25	0.53	0.30	0.90	0.10

TABLE III
Fundamental Frequencies of $SbCl_5$ (cm^{-1})

	Siebert [1]	Wilmshurst [7]	Miller[3]	Present work	
				observed	calculated
$A_1'\sigma_1$	356(R)	356	356(R)	356	356
σ_2	307(R)	307	307(R)	307	307
$A_2''\sigma_3$	((390))	387(IR) 380(IR)	371(IR)	387 380	384
σ_4	((131))	(156)	154(IR)	154	156
$E'\sigma_5$	399(R)	398(IR)	395(IR,R)	398	398
σ_6	182(R)	182	172(IR,R)	172	172
σ_7	74(R)	74	—	74	74
$E''\sigma_8$	166(R)	166	165(R)	165	163

following abbreviations were used: frequencies followed by IR (gaseous or matrix-phase infrared spectra) or R (Raman spectra) were actually observed by the investigator reporting the data; frequencies taken from a previous investigator are denoted by numbers only; frequencies calculated from overtones or combination bands are in single parentheses ( ), and those obtained from normal coordinate analysis are in double parentheses (( )). The values calculated in the present investigation are compared in the tables with what we consider the best reported values. Tables III to VI show that the agreement obtained is quite satisfactory for all the molecules studied. The average frequency deviation is ± 1.1% and the maximum deviation is 21 cm^{-1}.

Although only four trigonal bipyramidal molecules were studied in the present work, Table II shows the following trends in force constants: (1) All the force constants decrease as the halogen is changed from fluorine to chlorine; the stretching force constants change most markedly. (2) The stretching force constants increase and the bending force constants decrease as the central atom is changed from phosphorous to antimony. (3) The repulsive force constant F_{dr} is always greater than F_{rr}. This is reasonable since the distance between two halogens in the equatorial plane (e.g., Y_1 and Y_2 in Fig. 1) is greater than the distance between two halogens in the polar plane (e.g., Y_1 and Y_4).

BAND ASSIGNMENTS FOR INDIVIDUAL MOLECULES

$SbCl_5$

Table III compares the calculated frequencies with those observed by Siebert [1] (Raman spectrum), Wilmshurst [7] (infrared spectrum), and Miller [3] (infrared and Raman spectra). All the band assignments made by these investigators are consistent with each other, and the calculated frequencies are in perfect agreement with those observed.

TABLE IV
Fundamental Frequencies of PCl_5 (cm^{-1})

	Siebert [1]	Wilmshurst and Bernstein [2]	Miller [3]	Present work observed	calculated
$A_1' \; \sigma_1$	392(R)	394(R)	393(R)	394(R)	394
σ_2	271(R)	394(R)	?	264(R)	261
$A_2'' \; \sigma_3$	449(R)*	465(IR)	446(IR)	465	470
σ_4	((124))	176(IR)	299(IR)	299	303
$E' \; \sigma_5$	495(R)	592(IR,R)	580(IR,R)	592	588
σ_6	190(R)	335(IR,R)	273(IR,R)	273(R)	278
σ_7	100(R)	100	100(R)	100	100
$E'' \sigma_8$	190(R)	280(R)	289(R)	282(R)	277

*Although this mode is only infrared active, Siebert claims that he observed it in the Raman spectrum, and he attributes the origin of this band to the distortion of the trigonal bipyramidal structure.

PCl$_5$

As is shown in Table IV, the previous band assignments differ seriously among the investigators. Wilmshurst and Bernstein [2] assigned a Raman line at 394 cm^{-1} to both σ_1 and σ_2 on the assumption of accidental degeneracy. On the other hand, Miller [3] assigned a Raman line at 393 cm^{-1} to σ_1, and left the σ_2 unassigned. As is shown in Fig. 1, three Raman lines were clearly observed in the present work. Therefore, the line at 264 cm^{-1}, which is closest to the calculated frequency, was assigned to the σ_2 mode. It should be noted that this Raman line was previously observed and assigned to a combination band by Wilmshurst and Bernstein. The calculated frequency for σ_4 (303 cm^{-1}) is in good agreement with that observed by Miller (299 cm^{-1}) in the infrared region. Thus, the calculated frequencies obtained by Wilmshurst and Bernstein and by Siebert seem to be too low for this mode. The calculated frequency for σ_5 is in good agreement with that observed by Wilmshurst and Bernstein (592 cm^{-1}), and is fairly close to that observed by Miller (580 cm^{-1}). However, the Raman line at 495 cm^{-1} reported by Siebert has not been observed by any of the other investigators. The frequency calculated for σ_6 (278 cm^{-1}) is in good agreement with that observed at 273 cm^{-1} by Miller and by the present investigators. The calculated frequency for σ_8 (277 cm^{-1}) is close to the Raman line observed at 282 cm^{-1} in the present work. This line was also observed by Wilmshurst and Bernstein and by Miller, and assigned by both to the σ_8 mode. The frequency assigned by Siebert (190 cm^{-1}) to this mode is too low.

SbF$_5$

The infrared and Raman spectra of SbF$_5$ were observed by Akers [8] and by Gaunt and Ainscough [13], but their band assignments are in complete disagreement, as is seen in Table V. In order to solve these conflicting band assignments, it is useful to consider the frequency order observed in SbCl$_5$ and PCl$_5$. From Tables III and IV, it is seen that the stretching frequencies (σ_1, σ_2, σ_3, and σ_5) are much higher than the bending frequencies (σ_4, σ_6, σ_7, and σ_8), and the order of the stretching frequencies is always

$$\sigma_{5(\text{asym. in-plane})} > \sigma_{3(\text{asym. out-of-plane})} > \sigma_{1(\text{sym. in-plane})} > \sigma_{2(\text{sym. out-of-plane})}$$

The six fundamentals observed by Akers and the two fundamentals assigned by Gaunt and Ainscough were rearranged with this trend in mind. The four bending frequencies were also rearranged so as to obtain the frequency order $\sigma_6 > \sigma_8 > \sigma_4 > \sigma_7$, which was found to hold for SbCl$_5$. As is seen in Table V, the agreement between the calculated and observed frequencies is quite satisfactory. It should be mentioned that the set of the fundamentals selected by the present investigators can account for all the other observed frequencies on the assumption that they are overtones and combination bands.

PF$_5$

No Raman data are available for PF$_5$ because the compound either strongly reacts with, or is very insoluble in, most solvents. The infrared spectrum of the compound in the gaseous state has been reported by Gutowski and Liehr [15], and also by Pemsler and Planet [16]. These in-

TABLE V
Fundamental Frequencies of SbF$_5$ (cm^{-1})

	Akers [8]		Gaunt and Ainscough[13]	Present work observed	calculated
$A_1' \sigma_1$	397(R)	718(R)	667(R)	718	717
σ_2	265(R)	672(R)	264(R)	672	651
$A_2'' \sigma_3$	758(IR)	758(IR)	(294)	724	744
σ_4	708(IR)	708(IR)	(212)	212	204
$E' \sigma_5$	718(R) 724(IR)	397(R)	710(IR) 716(R)	758	770
σ_6	672(R)	265(R)	498(IR)	265	262
σ_7	224(R)	224(R)	ca.90(R)	93	95
$E'' \sigma_8$	187(R)	187(R)	228(R)	224	232

vestigators agree in their assignments of four of the eight fundamentals (see Table VI). We have assumed the validity of their assignment and have made use of the total observed spectrum to calculate the values of the other four fundamentals, σ_1, σ_2, σ_7, and σ_8. These eight fundamentals form a coherent set which agrees well with the results of the normal coordinate analysis and which can be combined to give the overtones and combination bands whose frequencies also agree closely with those observed (see Table VII).

ACKNOWLEDGMENT

We would like to express our sincere thanks to Professor F. F. Cleveland, of Illinois Institute of Technology, who permitted us to use his spectrograph for the measurement of the Raman spectrum of PCl$_5$. We are also grateful to the U. S. Army Research Office (Durham), which partly supported this investigation.

TABLE VI
Fundamental Frequencies of PF$_5$ (cm^{-1})

	Gutowski and Liehr[15]	Pemsler and Planet[16]	Present work observed	calculated
$A_1' \sigma_1$	—	—	(819)	817
σ_2	—	—	(610)	620
$A_2'' \sigma_3$	948(IR)	948(IR)	948	947
σ_4	576(IR)	576(IR)	576	571
$E' \sigma_5$	1025(IR)	1029(IR) 1026(IR)	1029 1026	1030
σ_6	534(IR)	534(IR)	534	547
σ_7	—	—	(286)	285
$E'' \sigma_8$	—	—	(562)	565

TABLE VII
Band Assignments of the Infrared Spectrum of PF_5 (cm^{-1})

Observed frequency[15, 16]	Calculated frequency	Band assignment
2041	2055	$2\sigma_5$, $A_1' + A_2' + E'$
1842	1846.5	$\sigma_1 + \sigma_5$, E'
1767	1767	$\sigma_1 + \sigma_3$, A_2''
1675	1672	$\sigma_4 + \sigma_6 + \sigma_8$, $A_1' + A_2' + E'$
1582	1589.5	$\sigma_5 + \sigma_8$, $A_1'' + A_2'' + E''$
1458	1472	$\sigma_2 + \sigma_4 + \sigma_7$, E''
1447	1458	$\sigma_2 + \sigma_7 + \sigma_8$, $A_1'' + A_2'' + E''$
1348	1345	POF_3 impurity
1190	1186	$\sigma_2 + \sigma_4$, A_2''
1062	1068	$\sigma_2 + \sigma_6$, $A_1' + A_2' + E'$
1025	1027.5	σ_5, E'
985	990	POF_3 impurity
948	948	σ_3, A_2''
890	896	$\sigma_2 + \sigma_7$, E'
866	873	POF_3 impurity
848	848	$\sigma_7 + \sigma_8$, $A_1'' + A_2'' + E''$
820	820	$\sigma_6 + \sigma_7$, $A_1' + A_2' + E'$
809	810	$\sigma_6 + \sigma_8 - \sigma_7$, $A_1'' + A_2'' + 3E''$
576	576	σ_4, A_2''
534	534	σ_6, E'
486	489	POF_3 impurity
—	286	σ_7, E'

APPENDIX
G Matrices

A_1' species

$$G_{11} = \mu_y$$
$$G_{12} = G_{21} = 0$$
$$G_{22} = \mu_y$$

A_2'' species

$$G_{33} = 2\mu_x + \mu_y$$
$$G_{34} = G_{43} = -2\sqrt{3}\rho_r\mu_x$$
$$G_{44} = 6\rho_r^2\mu_x + 2\rho_r^2\mu_y$$

E' species

$$G_{55} = \mu_y + (3/2)\mu_x$$
$$G_{56} = G_{65} = 3(\sqrt{3}/2)\rho_r\mu_x$$
$$G_{57} = G_{75} = -(3/\sqrt{2})\mu_x\rho_d$$
$$G_{66} = 3\rho_r^2\mu_y + (9/2)\rho_r^2\mu_x$$
$$G_{67} = G_{76} = -3(\sqrt{3}/\sqrt{2})\mu_x\rho_r\rho_d$$
$$G_{77} = 3\rho_d^2\mu_x + (3/2)\rho_d^2\mu_y$$

E'' species

$$G_{88} = 2\rho_r^2\mu_y + (3/2)\rho_d^2\mu_y$$

μ_x and μ_y are the reciprocal masses of X and Y, while ρ_r and ρ_d are the reciprocals of the bond distances, r and d, respectively.

F Matrices

A_1' species

$$F_{11} = Kr + 4 s_{rr}^2 F_{rr} + 2 t_{rd}^2 F_{dr}' + 2 s_{rd}^2 F_{dr}$$
$$F_{12} = F_{21} = \sqrt{6}\,(-t_{dr} t_{rd} F_{dr}' + s_{dr} s_{rd} F_{dr})$$
$$F_{22} = K_d + 3 t_{dr}^2 F_{dr}' + 3 s_{dr}^2 F_{dr}$$

A_2'' species

$$F_{33} = K_d + 3 t_{dr}^2 F_{dr}' + 3 s_{dr}^2 F_{dr}$$
$$F_{34} = F_{43} = \sqrt{3}\, r(t_{dr} s_{rd} F_{dr}' + t_{rd} s_{dr} F_{dr}$$
$$F_{44} = rd\,(H_\beta - s_{rd} s_{dr} F_{dr}' + t_{rd} t_{dr} F_{dr}$$

E' species

$$F_{55} = K_r + 3 t_{rr}^2 F_{rr}' - s_{rr}^2 F_{rr} + 2 t_{rd}^2 F_{dr}' + 2 s_{rd}^2 F_{dr}$$
$$F_{56} = F_{65} = -r(t_{rr} s_{rr} F_{rr}' + t_{rr} s_{rr} F_{rr})$$
$$F_{57} = F_{75} = \sqrt{2}\, d(t_{rd} s_{dr} F_{dr}' + t_{dr} s_{rd} F_{dr})$$
$$F_{66} = r^2(H_\alpha - s_{rr}^2 F_{rr}' + t_{rr}^2 F_{rr})$$
$$F_{67} = F_{76} = r^{3/2} d^{1/2} H_{\alpha\beta},$$
$$F_{77} = rd\,(H_\beta - s_{rd} s_{dr} F_{dr}' + t_{rd} t_{dr} F_{dr})$$

E'' species

$$F_{88} = rd\,(H_\beta - s_{rd} s_{dr} F_{dr}' + t_{rd} t_{dr} F_{dr})$$

Here $s_{dr} = t_{rd} = d/q_{dr}$, $s_{rd} = t_{dr} = r/q_{dr}$, $s_{rr} = \sqrt{3}/2$, and $t_{rr} = 1/2$, where $q_{dr}^2 = r^2 + d^2$.

REFERENCES

1. H. Siebert, Z. anorg. u. allgem. Chem. 265:303 (1951).
2. J. K. Wilmshurst and H. J. Bernstein, J. Chem. Phys. 27:661 (1957).
3. F. Miller, Abstracts of the International Symposium on Molecular Structure and Spectroscopy (Tokyo), A311 (Sept., 1962).
4. A. Maccoll, Proc. Roy. Soc. (New South Wales) 79:133 (1945).
5. J. S. Ziomek and C. B. Mast, J. Chem. Phys. 21:862 (1953); J. S. Ziomek, ibid. 22:1001 (1954).
6. P. C. Haarhoff and C. W. F. T. Pistorius, Z. Naturforsch. 14A:972 (1959).
7. J. K. Wilmshurst, J. Mol. Spectroscopy 5:343 (1960).
8. L. K. Akers, Univ. Microfilms (Ann Arbor, Michigan) Publ. No. 12965 (1955).
9. J. Hiraishi, I. Nakagawa, and T. Shimanouchi, Abstracts of the International Symposium on Molecular Structure and Spectroscopy (Tokyo), A313 (1962).
10. T. Shimanouchi, "Force Constants of Small Molecules," Private Publication, Tokyo (1962).
11. H. Moureau, M. Magat, and G. Wetroff, Proc. Ind. Acad. Sci. 8A:356 (1938).
12. M. Rouault, Ann. phys. 14:78 (1940).
13. J. Gaunt and J. B. Ainscough, Spectrochim. Acta 10:57 (1957).
14. L. O. Brockway and J. Y. Beach, J. Am. Chem. Soc. 60:1836 (1938).
15. H. S. Gutowsky and A. D. Liehr, J. Chem. Phys. 20:1652 (1952).
16. J. P. Pemsler and W. G. Planet, J. Chem. Phys. 24:920 (1956).
Note added in proof: Reference 3 should now be replaced by G. L. Carlson, Spectrochim. Acta. 19, 1291 (1963).

Additional Papers on Infrared – Raman Spectroscopy

● "Gas Maser Spectroscopy in the Infrared," W. L. Faust, R. A. McFarlane, C. K. N. Patel, and C. G. B. Garrett, Bell Telephone Laboratories, Murray Hill, New Jersey.

● "The Vibrational Spectra of XeF_4 and $XeOF_4$," H. H. Classen, J. G. Malm, and C. L. Chernick, Argonne National Laboratories, Argonne, Illinois.

● "Infrared Analysis in Biomedical Research," D. S. Erley, Dow Chemical Company, Midland, Michigan.

● "Infrared Spectra of Pesticides Recorded at Microgram Levels," W. W. Morris, Jr., and E. O. Haenni, Food and Drug Administration, Washington, D. C.

● "Applications of Normal Coordinate Treatments," S. Sundaram, Illinois Institute of Technology, Chicago, Illinois.

● "Normal Coordinate Treatments of 1,2–Dihalo Alkenes," R. Krupp, Illinois Institute of Technology, Chicago, Illinois.

● "Examples of Fermi Resonance and Coupling Involving N-D Vibrations," Brother Columba Curran, University of Notre Dame, Notre Dame, Indiana.

● "Raman Spectral Studies of Aqueous Solutions of Selenic Acid," G. E. Walrafen, Bell Telephone Laboratories, Murray Hill, New Jersey.

● "Raman and Infrared Spectra of the CS_3^{-2} Ion," A. J. Perkins, University of Illinois, Chicago, Illinois.

● "Normal Coordinate Treatments, Potential Constants and Thermodynamic Properties of $PSBr_3$, $PSBrF_2$, $POCl_2F$, and POF_2Cl," J. S. Ziomek, Martin Co., Orlando, Florida; F. Fillwalk, De Paul University, Chicago, Illinois, and E. Piotrowski, Illinois Institute of Technology, Chicago, Illinois.

Emission Spectrography

Atomic Absorption Spectroscopy Using Long Absorption Path Lengths and a Demountable Hollow Cathode Lamp

S. R. Koirtyohann* and C. Feldman

Oak Ridge National Laboratory†
Oak Ridge, Tennessee

Long absorption path lengths are obtained by directing the flame from a Beckman atomizer-burner through a quartz tube as suggested by Fuwa and Vallee. This method and modifications of it are used for the determination of Bi, Cd, Hg, Mg, Mn, Ni, Pb, Sb, Sr, Tl, and Zn in tissues, body fluids, and other samples. Concentrations giving 1% absorption are as low as 0.001 μg/ml for several elements. Solutions of tissue ash are analyzed for Cd, Mg, Mn, Ni, Pb, and Zn without chemical separations. Both oxy-hydrogen and oxy-acetylene flames are used and will be compared. Organic solvents provide greater sensitivity than aqueous solutions for some elements. A demountable hollow cathode lamp is used as a light source. This lamp has a water-cooled cathode and allows rapid changing of elements in a single base. Cathodes for new elements can be prepared quickly and at a very low cost. This source compares favorably in brightness, stability, and sensitivity (line width) with commercial sealed lamps.

INTRODUCTION

Atomic absorption spectroscopy has earned its place among modern analytical methods based on the use of sealed hollow cathode lamps, slot burners, and separate atomizers in essentially the same form as was developed by the early Australian and New Zealand workers [1, 7]. The Beckman total-consumption atomizer-burner, which is almost universally used in this country for emission flame work, suffers from the limitation of having a very short path length available for absorption. Several attempts have been made to overcome this limitation while retaining the basic simplicity of the total-consumption burner.

Robinson [8] directed the flame from this type of burner into the side arm of a T-tube to extend the path length. He obtained no absorption for sodium or nickel but did get a 10-fold increase in platinum sensitivity. Feldman and Dhumwad [3] used a somewhat similar arrangement for the determination of mercury. Here the flame was directed into the side arm of a jacketed stainless steel tube with open ends. Light from a mercury lamp was passed through the tube which was about 25 cm long. Air was passed through the jacket for cooling. This method worked very well for mercury, but was limited by contamination of the tube when it was used for other elements. Tabeling and Devaney [9] used multiple light passes through the flames from three Beckman burners. Here the fuel and sample consumption rates are rather high and optical alignment is more critical.

*Present address: 45 Agriculture Bldg. University of Missouri, Columbia, Missouri.
†Operated by Union Carbide Corporation for the Atomic Energy Commission.

Fuwa and Vallee [4] suggested directing the flame from the Beckman atomizer through a long quartz tube to obtain greatly extended absorption path lengths. Our experiences and adaptations using their method will be given here.

The sealed hollow cathode lamps that were used in nearly all of the published work on atomic absorption spectroscopy represent an investment of several thousand dollars if a large number of elements are to be determined. If long lamp life is expected, the operating current must be low with consequent low intensity. It is also necessary that the need for a given lamp be anticipated 3-6 weeks to ensure that it will be available when it is required. We have found that a simple demountable lamp can give very satisfactory performance and avoid some of the limitations of the sealed type.

INSTRUMENTATION

A flame spectrophotometer designed by the ORNL analytical instrumentation group [6] was modified slightly and used for this work. The instrument uses a Jarrell–Ash 0.5-meter spectrometer mounted on a standard Jarrell–Ash optical bench. Standard bench riders are used to support the external optical components. Fifty-micron entrance and exit slits were used in the monochromator.

Tuned AC amplification is used in conjunction with a 320-cps light chopper. The amplifier from a Jarrell–Ash model 82-350 dual-purpose spectrometer was used in the cases where DC amplification was desired.

The lamp power is supplied from a Jarrell–Ash constant current hollow cathode power supply. A locally built constant current DC power supply was also used part of the time with equally good results.

THE DEMOUNTABLE HOLLOW CATHODE LAMP

A very simple demountable hollow cathode lamp was described by Werner et al. [10] for the determination of lithium isotope ratios. This lamp, which is shown in Fig. 1, uses a brass cathode $\frac{1}{2}$ in. in diameter by $1\frac{1}{4}$ in. long with a $\frac{7}{32}$-in.-diameter by 1-in.-deep cavity. The gas outlet is connected to a vacuum pump which is operated continuously while the lamp is being used. Commercial argon is bled into the system through a needle valve to maintain the desired pressure. No purification or recirculation of the argon is used, but air must be carefully excluded from the system. A water cooling jacket is placed over the protruding portion of the cathode. The tedious clean-up steps needed in the preparation of sealed lamps are not required here because impurities are continuously removed by the flowing argon.

Cathodes for a given element are prepared by placing a small amount (10-50 mg) of the metal or a suitable salt into the brass cathode. The lamp is assembled, evacuated, filled with argon, and operated at 100 mA current for about 2 min to sputter the test element over the inside of the cathode. The current is then reduced to the desired operating level (10-40 mA), and the argon pressure adjusted to give maximum intensity. This is also the pressure that gives the best stability, and is different for each element. After a stabilization period of about 10 min the lamp is ready for use.

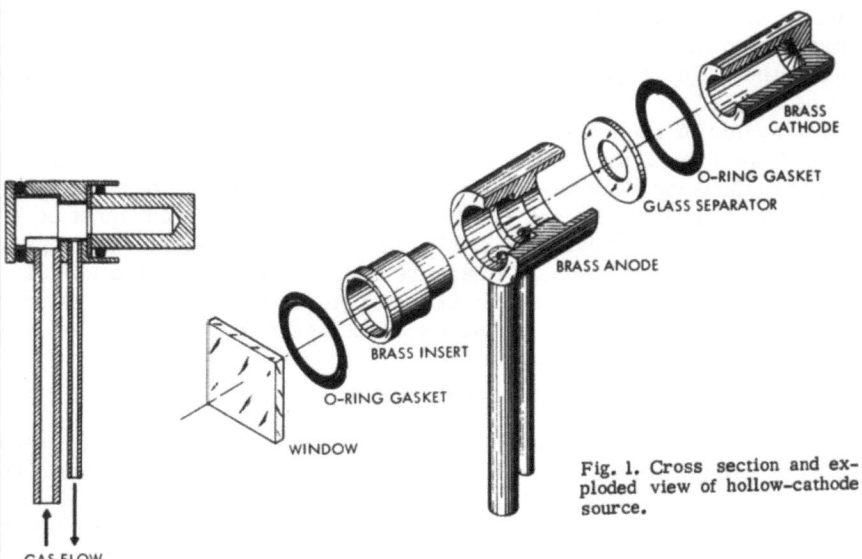

Fig. 1. Cross section and exploded view of hollow-cathode source.

GAS FLOW

A lamp for a new element can be ready for use in about 20 min after deciding that it is needed. The cost for each additional element is only the very inexpensive brass cathode. These are saved and used repeatedly for a given element without further addition of metal.

We have compared the operation of this type of lamp with only one of the sealed type, this being a Micro-Tek lead lamp purchased from the Jarrell–Ash Company. The stability of the two lamps is comparable ($\pm$0.2-0.5%), but the intensity of the sealed one is less than one tenth as great when used at the recommended current (10 mA). The absorption vs. current data for the two lamps are given in Table I. The intensity and the absorption response are much more current sensitive with the sealed lamp. The two lamps have about the same intensity at the higher currents. The increase in lead absorption with increasing lamp current was reported by Elwell and Gidley [2]. They attribute this to improvement in the line-to-background ratio in the lamp. The data in Table I indicate that the sealed lamp is spectrally equal to the demountable one only when operated at about four times the recommended current. The lamp life would probably be considerably shortened under these conditions.

TABLE I

Absorbance from 1 μg/ml Pb at Various Lamp Currents (Pb 2833 25-cm tube)

Current, mA	Demountable lamp	Sealed lamp
10	0.138	—
15	—	0.060
20	0.135	0.090
30	0.127	0.114
40	0.120	0.127

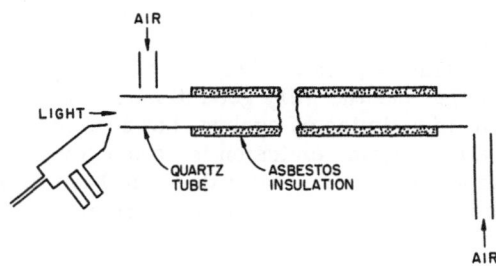

Fig. 2. Burner and tube arrangement for long absorption path lengths.

The demountable lamp can be operated at currents up to 100 mA with no damage to the lamp. Currents up to about 40 mA can be used for most elements with no serious loss in absorption response. Cadmium and zinc cathodes require lower currents (10-15 mA) to prevent excessive line broadening and resultant loss in response.

It should be pointed out, however, that the sealed lamps have the advantage that no assembly, gas pressure adjustment, etc. are necessary, and consequently less operator training is required.

LONG ABSORPTION PATH LENGTHS

Our adaptation of the Fuwa and Vallee method is shown schematically in Fig. 2. The flame from a Beckman burner is directed into the end of the quartz tube from an angle to allow the light to pass. A stream of air is directed onto the outside of the tube at the point where the flame strikes to prevent tube burnout. A second stream of air directed vertically past the end of the tube protects nearby optics. The tube is 1 cm inside diameter and may vary in length from 12 to 75 cm on our instrument. The asbestos insulation raises the temperature of the tube walls and helps to discourage condensation and resulting memory effects. This also increases the sensitivity for some elements. Collimated light from the source is passed through the tube and is then focused on the spectrometer slit.

Fuel-rich oxy-hydrogen flames have been used for most of the work with oxygen-hydrogen ratios ranging from 1:3 to 1:6. The flame usually extends through the entire length of the tube. Fuel-rich oxy-acetylene flames have been used on occasion and have advantages which will be discussed later.

The flame gases absorb appreciably under these conditions. This absorption may be as much as 50% of the incident radiation when working with short-wavelength lines such as Zn 2138 in the longer tubes. This absorption is quite constant and causes no trouble except for the apparent loss in lamp intensity. Hydroxide band absorption in the flame renders Bi 3068 practically useless with these path lengths, but fortunately there are other Bi lines that can be used.

The quartz tubes devitrify and crack after prolonged use, especially when solutions high in alkali salts are aspirated into them. Even under these conditions, a tube will last for several hundred determinations and is easily replaced. Continuous aspiration of solvent into the flame between samples helps prolong tube life.

Surprisingly, blackbody radiation from the hot tube walls has not caused trouble even when an unmodulated source and DC amplification were used.

Wavelengths as long as Sr 4607 have been used with only a small, easily determined correction for continuous radiation needed. No correction is necessary at shorter wavelengths. This is possible because of the high intensity of the hollow cathode lamps and the very good dispersion by the monochromator. The lens system discriminates against, but does not exclude, radiation from the tube walls. Flame emission is so much weaker than the absorption for the elements used here that it does not have to be considered when DC amplification is used. Emission from the flame or the tube had no effect, of course, when the chopper and AC amplification system were used.

RESULTS

We have used these methods for the determination of 13 elements. The lines used and concentrations giving 1% absorptions for aqueous solutions are given in Table II. For most elements the noise level is such that this is about the detection limit. The use of organic solvents modifies the flame conditions required for maximum response, but does not change the sensitivity greatly for most elements on which they were tried. In cases where chemical separations are necessary, the determinations are made directly on the organic phase from an appropriate extraction. This works very well but has the disadvantage that the flame becomes quite noisy when aspirating organics.

We have determined Mg, Zn, Cd, Mn, Ni, and Pb in samples of tissue ash without chemical separations. Standards were prepared in a solution of a synthetic soft tissue ash to avoid errors from matrix effects. Linear working curves were obtained when the $\log (I_0/I)$ was plotted vs. concentration. In most cases the extremely good sensitivity obtained from the long tubes allowed the sample to be diluted until the response was very little different from that obtained from simple aqueous solution of the test element. The values obtained in this way agreed quite well with spectrographic

TABLE II

Lines Used and Concentration for 1% Absorption in Aqueous Solution

Element	Line	μg/ml
Bi	2230	0.1
Cd	2288	0.001
Cu	3247	0.005
Hg	2537	0.5
Mg	2852	0.001
Mn	2795	0.005
Ni	2320	0.01
Pb	2833	0.02
Sb	2176	0.1
Sr	4607	0.1
Te	2143	0.02
Tl	2768	0.05
Zn	2138	0.0005

values on the samples.   A standard deviation of 2.0% was obtained from replicate exposures on a solution containing 0.025 μg/ml of Zn and one of 3.4% from a solution containing 0.010 μg/ml of Mg.   Better precision can be obtained at slightly higher concentrations.

The organic phase from an appropriate extraction was aspirated into the flame to determine Hg, Pb, Tl, and Te in urine and blood.   Some of the commonly used solvents absorb in the flame.   For this reason hexone (methylisobutyl ketone) was used rather than chloroform for the dithizone extractions of Hg, Pb, and Tl.   Tellurium was extracted with benzene and sodiumdiethyldithiocarbamate.

The absorption response obtained from tubes of different lengths has been studied.   Increasing the length of the tube will increase the amount of absorption only if the test element exists as atoms in the added length.   For elements that have little tendency to form stable compounds in the flame one might expect the absorption to be a nearly linear function of the tube length. Figure 3 shows the change in absorption with tube length for lead.   The non-linear relationship is probably caused by increased friction in the longer tubes.   This causes an increase in the pressure inside the tube, which in turn reduces the amount of air entrained with the flame.   Both of these factors lead to increased absorption per unit of tube length.

The same type of curve is obtained with short tubes for elements that form stable compounds in the flame, but the maximum tube length that can be used to advantage is shorter.   Competing reactions deplete the population of atoms before the gases travel through a long tube.   Figure 4 shows the curve that was obtained for magnesium.   The reduced response from longer tube lengths has not been studied, but it may be due to an increased reaction rate toward compound formation, brought about by a slightly greater pressure in the tube.

In general the very long tubes are useful for the determination of Pb, Zn, Cd, and Hg, while 20-25 cm seems to be the maximum useful length for work with Mg, Mn, and Ni.   Longer tubes only contribute to undesirable matrix and memory effects.   Changing tube lengths is an effective way to change the concentration range covered.

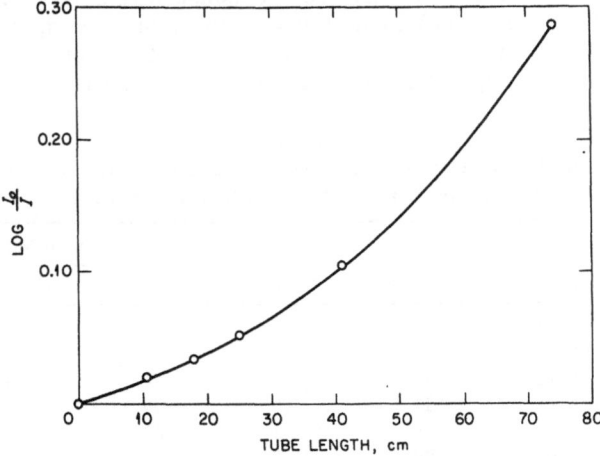

Fig. 3. Effect of tube length on absorption by 0.5 μg/ml Pb.

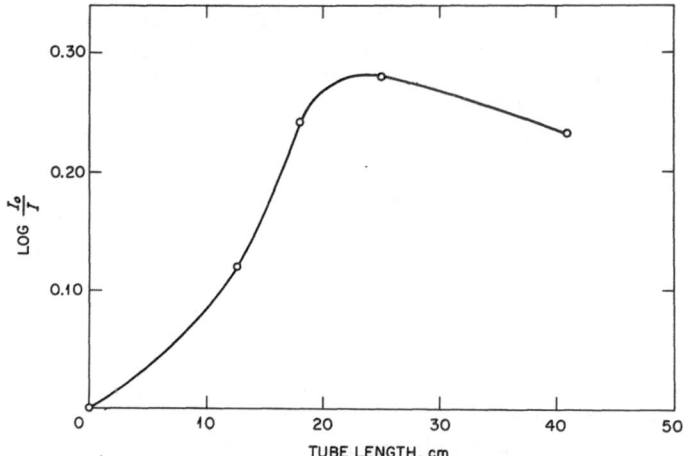

Fig. 4. Effect of tube length on absorption of 0.10 μg/ml Mg.

The effect of interferences on the absorption response for different elements does not appear to be very different for this type of flame than has been reported using slot burners. The effect of interfering ions on three elements are given in Table III. Phosphate gives the only large interference on lead. The formation of refractory oxides in the flame seriously suppresses zinc and magnesium absorption. Magnesium is also affected by the other ions to a greater extent, but all of these suppressions are less serious in the hotter acetylene flame. No attempt has been made to alter these effects by adding releasing agents to the solutions.

While investigating the effects of interferences we noticed that solutions having high salt concentrations frequently showed appreciable absorption at the wavelength of a given element when the element had not been added and was very unlikely to be present as a contaminant. In several cases absorption was detected at the cadmium wavelength. Later determinations on chemically separated portions of these samples established that cadmium was absent. We decided to check for response with solutions high in salts at wavelengths where no elemental absorption would be expected. For this purpose we used argon lines and nonground state copper and lead lines from the cathode lamp. Figure 5 shows the results that were obtained for

TABLE III

Percent Decrease in Absorbance Caused by the Presence
of Various Matrix Materials (25-cm tube)

Solute	Zn, 0.05 μg/ml	Pb, 0.50 μg/ml	0.05, μg/ml Mg	
			H_2 flame	C_2H_2 flame
1 mg/ml Ca	5	1	40	27
1 mg/ml Na	4	−1	46	16
0.1 M H_2SO_4	14	−1	50	4
0.1 M H_3PO_4	11	68	70	29
1 mg/ml Al	−	−1	89	51
1 mg/ml Si	36	7	85	37

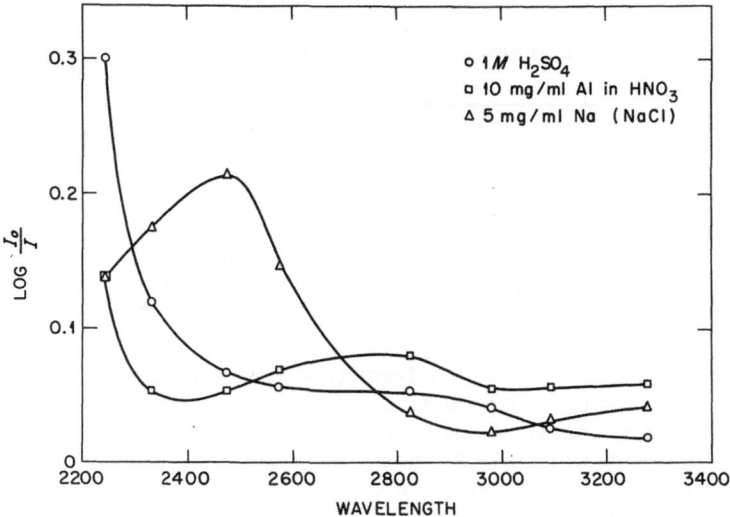

Fig. 5. Apparent absorption from materials in the flame.

three solutions. It should be emphasized that measurements were made only at the indicated wavelengths. The lines connecting the points are intended as an indication of the general change with wavelength, not as a true representation of the total spectrum. We have not studied the detailed causes for these nonspecific absorptions which are obviously not caused by atomic transitions and have no simple relationship with wavelength. Molecular band absorption and scattering by solid particles both undoubtedly contribute. The absorbance increases linearly with concentration of the solute and is generally more serious at short wavelengths. Hydrochloric and nitric acids showed no appreciable absorption.

Using the long tubes, we now have a background of absorption which is not specific for the test element and depends on the concentration and nature of other materials in the solution. The specific elemental absorption is superimposed on this background. The nature of the source prevents scanning wavelengths in the immediate vicinity of the line to correct for this background. In most cases the very high sensitivity of the method will permit the dilution of the sample enough to avoid errors. However, in order to take full advantage of this sensitivity we must find ways to avoid or correct for these nonspecific absorptions.

Attempts were made to minimize the effect by changing tube diameter and length, but they were not successful. Tube diameters both larger and smaller than 1 cm gave ratios of elemental/nonspecific absorptions that were poorer than those for the 1-cm tube. There are small variations in this ratio with tube length, but it appears unlikely that large improvements are to be found here. It may be possible to correct for this effect in some spectral regions by making measurements on an "unabsorbed" line near the analytical line. Detailed information on the nature of these absorptions will be needed before this can be applied generally. If one sample of a given type of material can be obtained that is free of the test element, then a correction can be easily obtained. Table IV gives estimates of the maximum

TABLE IV

Maximum   Errors   from   Nonspecific
Absorptions

Solute	Error, μg/ml			
	Pb	Mg	Zn	Cd
1 mg/ml Na	0.01	0.005	0.005	0.02
1 mg/ml Al	0.07			0.004
1 mg/ml Ca	0.05		0.005	
0.1 M H_2SO_4	0.02		0.010	

error that can be encountered from this effect for several elements and solutes. The actual error might be smaller because we are not sure that the test element concentration was zero. A small amount might have been present as a contaminant. It should be noted that these values are lower than the detection limits using conventional methods [5]. This effect is not restricted to the quartz tube. The apparent cadmium concentration in a solution of NaCl was checked using a quartz tube and also on a Jarrell–Ash Model 82-350 atomic absorption spectrometer. This instrument uses three Beckman burners and five-pass optics [9]. A more concentrated NaCl solution was used in the latter case because of the poorer sensitivity. An apparent cadmium concentration of 20 μg/g of sodium was obtained in the quartz tube and 30 μg/g of sodium from the triple burner flame. These were about the amounts of absorption that would be predicted from the curves in Fig. 5 in the absence of cadmium. This indicates that the effect is not new, but that it becomes more apparent at the higher sensitivities.

CONCLUSIONS

This work has established that a simple demountable hollow cathode lamp can provide a satisfactory light source for atomic absorption work. This lamp has equal stability and superior intensity to the commercial sealed lamp that was tested. Lamp life is no longer a problem because new cathodes are easily and quickly prepared.

Long effective absorption path lengths are easily obtained by passing a flame through a long quartz tube. In this way sensitivities that are 10-100 times greater than those from conventional equipment [5] can be obtained using one burner and single-pass optics. Using the long absorption path lengths, nonspecific absorption from solutes present in sample solutions are encountered. In most cases the sensitivity gain is sufficient to permit dilution of the sample to avoid the effect. The method has proved to be simple and practical and is being applied to routine determinations in our laboratory.

REFERENCES

1. G. F. Box and A. Walsh, Spectrochim. Acta 16:255 (1960).
2. W. T. Elwell and J. Gidley, Anal. Chim. Acta 24:71 (1961).
3. C. Feldman and R. K. Dhumwad, Sixth Conference on Analytical Chemistry in Nuclear Reactor Technology Gatlinburg, Tennessee, October, 1963.

4. K. Fuwa, and B. L. Vallee, International Conference of Spectroscopy, College Park, Maryland, June, 1962.
5. B. M. Gatehouse and J. B. Willis, Spectrochim. Acta 17:710 (1961).
6. H. C. Jones D. J. Fisher, and M. T. Kelley, Fifth Conference on Analytical Chemistry in Nuclear Reactor Technology, Gatlinburg, Tennessee, October, 1961. T. I. D. 7629.
7. W. G. Jones and A. Walsh, Spectrochim. Acta 16:249 (1960).
8. J. W. Robinson Anal. Chim. Acta 27:465 (1962).
9. R. W. Tabeling and J. J. Devaney, Developments in Applied Spectroscopy, W. D. Ashby (ed.) (Plenum Press, New York, 1962).
10. G. K. Werner, D. D. Smith, S. J. Ovenshine, O. B. Rudolph, and J. R. McNally, J. Opt. Soc. Am. 45:203 (1955).

Flame Photometric Investigation of the Factors Affecting the Emissivity of Metal Chelates

Howard C. Eshelman and James Armentor

University of Southwestern Louisiana
Lafayette, Louisiana

The purpose of this project is to investigate some of the factors that affect the emissivity of metals when solutions of their chelates in organic solvents are aspirated into a flame. Solvent extraction methods were used to prepare the solutions of the beryllium, iron, and manganese chelates that were investigated. The following reagents were used to produce the corresponding chelate of each of the metals: 8-quinolinol, 4,4,4-trifluoro-1-(2-thienyl)-1,3-butanedione, sodium diethyldithiocarbamate, salicylaldoxime and the ammonium salt of N-nitrosophenylhydroxylamine. Toluene, chloroform, 4-methyl-2-pentanone, and pentyl acetate were used as extraction solvents. Studies were carried out to determine the optimum combustion conditions for each of the organic extracts. Emission readings were taken as follows: at 470.9 mμ for the BeO bandhead, at 372.0 mμ for the iron line, and at 403.3 mμ for the unresolved triplet of manganese. The effect of the aspiration rate of the sample upon the emission intensity of each of the metals was investigated. The control of the rate of aspiration of the sample independent of the oxygen flow rate was accomplished by using several atomizer burners with capillary tubes of different diameters. These studies made it possible to establish a set of operating conditions for organic aerosols so that a fifty-eight-fold enhancement in the emission intensity of the BeO bandhead was obtained.

INTRODUCTION

The application of organic solvents to flame photometry has been a significant development of the last decade [1, 2]. The introduction of an organic aerosol into a flame in place of an aqueous aerosol increases the emission intensity of metals many times. A large part of the information that is available concerning the use of organic solvents in flame photometry was obtained as a part of investigations to establish methods of analysis [3-6]. Flame photometric methods are influenced by a large number of experimental variables; some of them are difficult to control individually. As a result the literature contains a number of contradictory statements [7].

In 1960 the senior investigator (under the sponsorship of a Petroleum Research Fund Grant) began an investigation into some of the factors that affect the emissivity of metals when solutions of their chelates in organic solvents are aspirated into a flame.

EXPERIMENTAL

A Beckman Model DU spectrophotometer equipped with a photomultiplier attachment was employed in the emission intensity studies. Data were recorded on a 10-mV Bristol recorder having a 1-sec pen response by means of a Beckman spectral energy recording attachment.

Brooks rotameters (Brooks Rotameter Co., Lansdale, Pa.), Sho-Rate Model 1356, were used to regulate the flow of the gases.

Solvent extraction methods were used to prepare the solutions of the beryllium, iron, and manganese chelates that were studied. The following reagents were used to produce the corresponding chelate of each of the metals: 8-quinolinol, 4,4,4-trifluoro-1-(2-thienyl)-1,3-butanedione, sodium diethyldithiocarbamate, salicylaldoxime, and the ammonium salt of N-nitrosophenylhydroxylamine. Toluene, chloroform, 4-methyl-2-pentanone, and pentyl acetate were used as extraction solvents.

Studies were carried out to determine the optimum combustion conditions for each of the organic extracts. The oxygen flow rate was set at 5 cfh and the acetylene flow varied until a maximum in emission intensity was observed. The ratio of the oxygen flow rate to the acetylene flow rate that gave the maximum emission was calculated. This procedure was repeated for oxygen flow rates of 6, 7, and 8 cfh, and was carried out for several burners each having a different capillary diameter. The emission readings were taken as follows: at 470.9 mμ for the BeO bandhead, at 372 mμ for the iron line, and 403.3 mμ for the unresolved triplet of manganese.

The effect of the aspiration rate of the sample upon the emission intensity of each of the metals was investigated. The control of the rate of aspiration of the sample independent of the oxygen flow rate was accomplished by using several atomizer burners with capillary tubes of different diameters. The capillary diameters were measured with a microscope. The following measurements are typical of the Beckman total consumption aspirator burners. For a large-bore burner the capillary diameter was 0.440 mm, for a medium bore 0.323 mm, and for a small bore 0.270 mm.

RESULTS AND DISCUSSION

The data obtained from a study of the emissivity of metals as a function of aspiration rate are shown in a series of plots. The aspiration rates are

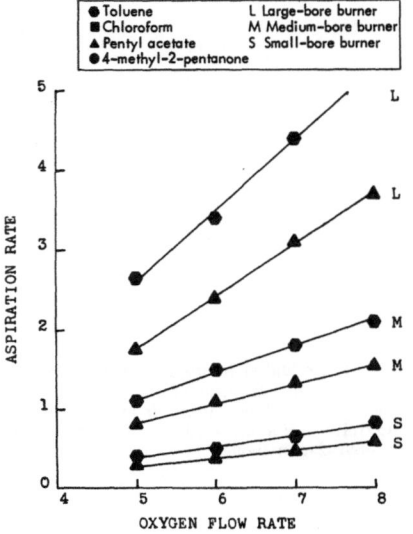

Fig. 1. Aspiration rate of toluene and of pentyl acetate as a function of capillary diameter and oxygen flow rate. Solvents used: pentyl acetate and toluene.

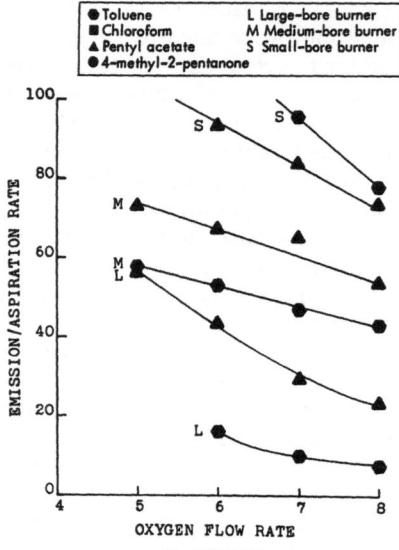

Fig. 2. Emission intensity per unit aspiration rate as a function of capillary diameter and oxygen flow rate. Iron present, 100 μg/ml. Solvents used: pentyl acetate and toluene.

in milliliters per minute, and the oxygen flow rates are in cubic feet per hour.

The effect of oxygen flow rate and capillary diameter upon the aspiration rate of pentyl acetate and toluene is shown in Fig. 1. Chloroform and 4-methyl-2-pentanone are not shown. The curve for chloroform is slightly above pentyl acetate and 4-methyl-2-pentanone is slightly below the toluene curve.

Emission intensity divided by aspiration rate is plotted vs. oxygen flow rate in Figs. 2-6. An examination of the illustrations shows that:

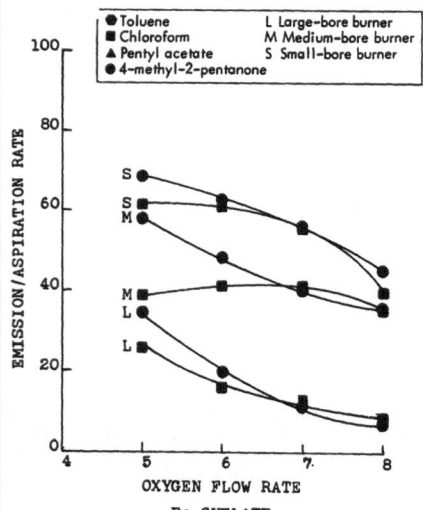

Fig. 3. Emission intensity per unit aspiration rate as a function of capillary diameter and oxygen flow rate. Iron present 100 μg/ml. Solvents used: chloroform and 4-methyl-2-pentanone.

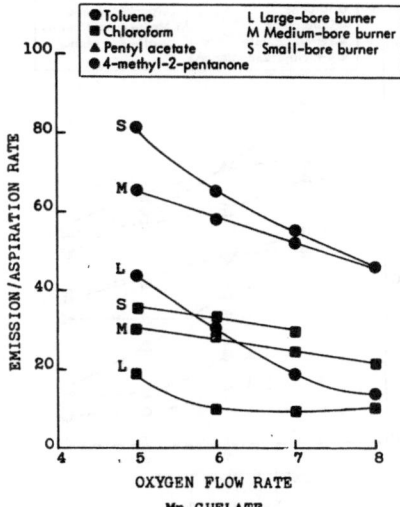

Fig. 4. Emission intensity per unit aspiration rate as a function of capillary diameter and oxygen flow rate. Manganese present, 10 μg/ml. Solvents used: chloroform and 4-methyl-2-pentanone.

1. When a given organic extract is introduced into a given burner flame, the emission intensity per unit aspiration rate diminishes as the oxygen flow rate is increased. In many cases the plot is a straight line.

2. A comparison of the data obtained at a fixed oxygen flow rate when a given organic extract is introduced into the flame of a large-bore burner with the data obtained from the medium-bore and small-bore burners shows that, in general, the emission intensity per unit aspiration rate is greatest for the small-bore burner and least for the large-bore burner.

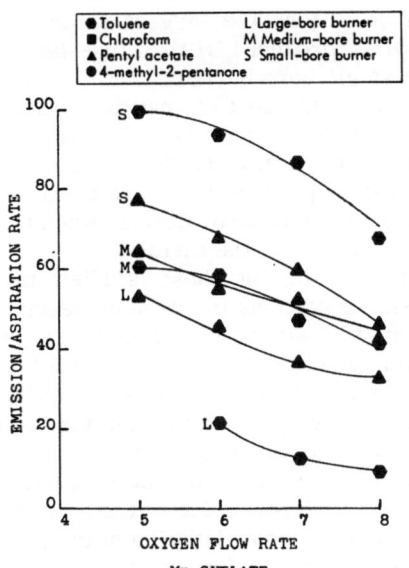

Fig. 5. Emission intensity per unit aspiration rate as a function of capillary diameter and oxygen flow rate. Manganese present, 10 μg/ml. Solvents used: toluene and pentyl acetate.

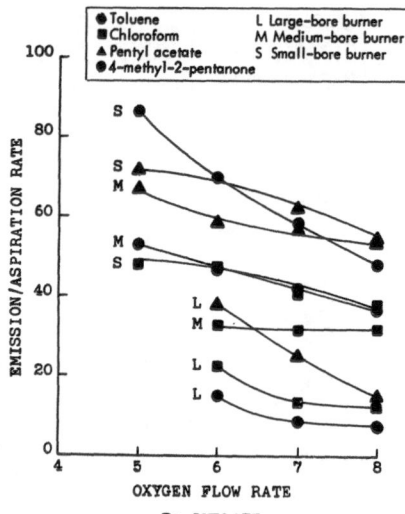

Fig. 6. Emission intensity per unit aspiration rate as a function of a capillary diameter and oxygen flow rate. Beryllium present 200 μg/ml. Solvents used: chloroform, pentyl acetate, and 4-methyl-2-pentanone.

Two factors contribute to produce the effect that was first stated. An increase in the oxygen flow rate increases the velocity of the organic aerosol through the flame; consequently, a larger portion of the solvent passes through the flame without evaporation. The result is that the rate of increase of the concentration of the excited atoms in the flame is not as great as the increase in the rate of introduction of the sample into the flame.

The second effect is influenced by at least two factors. Under a given set of conditions the aspiration rate of the large-bore burner is approximately six times as great as the aspiration rate of the small-bore burner. At a fixed oxygen flow rate the lowering of the flame temperature is greatest for the large-bore burner and least for the small-bore burner. For a fixed set of conditions the largest mean diameter of the droplets of the organic aerosol is produced by the large-bore burner. The finest mist is produced by the small-bore burner.

It was found that the nature of the chelating agent had no effect upon the emissivity of the metals. The use of quinolinol as a chelating agent is not recommended since solutions of its metallic chelates frequently clogged the burner tip when they were aspirated into the flame. Flame background readings were usually erratic when toluene was used as an organic solvent for the chelates of the metals.

The results of these studies made it possible to establish a set of operating conditions for organic aerosols so that a fifty-eight-fold enhancement in the emission intensity of the BeO bandhead was obtained. The beryllium chelate of 4,4,4-trifluoro-1-(2-thienyl)-1,3-butanedione dissolved in pentyl acetate was aspirated into the flame. The ratio of the oxygen flow rate to the acetylene flow rate was 2.0 to 1. The substitution of toluene for pentyl acetate as a solvent seemed to give a greater enhancement. No useful readings could be obtained because of the erratic flame background. Also the interference due to the Swan bands located between 460 and 478 mμ was severe. A consideration of the properties of the two solvents led to the conclusion that a mixture of toluene and pentyl acetate would give a larger usable enhancement than either one of them. A series of experiments was devised and carried out to test the validity of that conclusion. The results showed that emission intensity of the BeO bandhead increased as the concentration of the toluene in the mixed solvent was increased. The emission readings became erratic when the volume of toluene was greater than 30% of the total volume of the solvent. The ratio of flow rates of oxygen to acetylene was increased to 2.33 to 1. The volume percent of toluene could then be raised to 40%. The emission intensity of the BeO bandhead was found to be fifty-eight times the emission intensity obtained when an aqueous solution was used.

ACKNOWLEDGMENT

This work is part of a research program which has been aided by grants from The Petroleum Research Fund of the American Chemical Society. The authors are grateful for this assistance.

REFERENCES

1. J. A. Dean, Flame Photometry, (McGraw-Hill Book Company, Inc., New York, 1960), Chapter 5.
2. J. A. Dean, Special Technical Publication No. 238 (American Society for Testing Materials, Philadelphia 1958), p. 43.
3. J. A. Dean and Carl Cain, Jr., Anal. Chem. 29:530 (1957).
4. H. C. Eshelman, J. A. Dean, Oscar Menis, and T. C. Rains, Anal. Chem. 31:183 (1959).
5. H. C. Eshelman and J. A. Dean, Anal. Chem. 33:1339 (1961).
6. D. E. Fornwalt, Anal. Chim. Acta. 17:597 (1957).
7. W. H. Foster and D. C. Hume, Anal. Chem. 31:2028 (1959).

Mutual Interference Effects of Certain Group III-A Ions with Calcium in Flame Spectroscopy*

M. E. Doty and W. G. Schrenk

Department of Chemistry
Kansas State University
Manhattan, Kansas

The reduction in emission intensity of calcium due to the presence of BO_3^{-3}, Al^{+3}, Ga^{+3} and In^{+3} in the test solutions has been determined. The reverse effect, i.e., the effect of calcium on emissions of Group III-A elements also has been determined. Sample dilution decreased BO_3^{-3} suppression but increased Al^{+3} suppression of calcium radiations. Apparently n-butanol forms a compound with the BO_3^{-3} ion in the evaporating aerosol droplet. This results in an increase in the band emissions owing to an excited state of the boron oxide in the flame, and effects a release of boron suppression of calcium radiations. The presence of methanol (10 to 50% by volume) and sucrose (1%) in the test solution decreases the effect produced by BO_3^{-3} but increases the effect of aluminum on calcium radiations. Difference in the degree of suppression of the Ca (I) line and CaO bands due to the presence of BO_3^{-3} or Al^{+3} in the test solutions cannot be accounted for on the basis of an intermediate refractory solid phase. The effects of Ga^{+3} and In^{+3} on calcium radiations can best be explained in terms of a gas-phase suppression mechanism. In these studies both H_2-O_2 and $C_2H_2-O_2$ fuel systems have been used.

INTRODUCTION

A major factor influencing emission intensities of metal atoms in flames is apparent interaction of the elements being studied with other substances present in the test solution. The literature carries many reports concerning specific interferences, but in most cases studies reported have been limited to a single line or band emission.

In an effort to provide more data useful in explaining interactions of ions in flame photometry this investigation was undertaken (1) to study the extent of mutual interferences of calcium with some Group III-A elements under identical experimental conditions, (2) to compare relative suppression effects of certain ions in the test solution on the calcium resonance line and oxide band emissions under identical experimental conditions, and (3) to further investigate mechanisms of some releasing agents applicable to those systems.

EXPERIMENTAL

A Model DU spectrophotometer equipped with a photomultiplier and associated electrometer amplifier and electronic power supply in conjunction with a Model 9200 flame attachment and S.E.R.A. unit, all manufactured

*Contribution No. 624, Department of Chemistry, Kansas State University. Abstracted from a dissertation submitted by the senior author to the faculty of Kansas State University in partial fulfillment of the requirements for the Ph. D. degree. Supported in part, by National Institute of Health Grant No. GM-08524. Present address: Phillips Petroleum Co., Research Center, Bartlesville, Oklahoma.

by Beckman Instrument Company, were used in most of this work. The output signal of the electrometer amplifier was fed into a Sargent S-72150 recorder. For the remaining portion of the work, a Jarrell–Ash Model 82-200 Ebert scanning monochromator in conjunction with a Model 82-110 recording electronic system was used. A Beckman No. 4030 acetylene atomizer burner and a Beckman No. 4020 hydrogen atomizer burner were used. All relative emission intensity data were obtained at low scan speeds with background corrections made by the base-line method. Data presented within a given figure or table were taken at constant fuel and oxygen flow rates.

Stock solutions of $CaCl_2$ were prepared by dissolving reagent grade $CaCO_3$ in HCl. Primary ammonium orthophosphate solutions were used as the source of phosphate ions. The trichlorides of aluminum, gallium, and indium were prepared by dissolving the respective metals in a minimum amount of HCl.

RESULTS AND DISCUSSION

Calcium–Boron Interaction

The effect of $CaCl_2$ on boron oxide band emission intensities is shown in Fig. 1; also shown is the effect of additions of H_3BO_3 on calcium emissions resulting from aspiration of a $CaCl_2$ solution.

The upper curve shows an abrupt change in slope at a Ca/B concentration ratio slightly greater than one. If these mutual suppression effects are to be attributed solely to the slow release of calcium and boron from the solid phase which formed upon dehydration of the aerosol particles aspira-

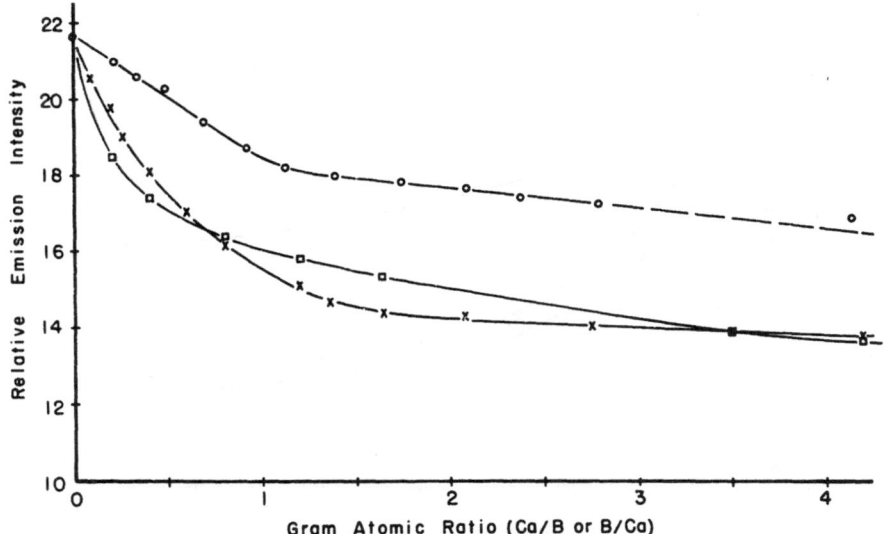

Fig. 1. Calcium and boron mutual suppression effects:
O Boron oxide band head   518 mμ H_3BO_3 = 0.0288 M
□ Calcium oxide band head 554 mμ $CaCl_2$ = 0.00126 M
x Calcium(I) line   422.7 mμ $CaCl_2$ = 0.00126 M
H_2–O_2 flame

ted, then one would expect that adding BO_3^{-3} would decrease calcium emissions in a manner similar to decreases of boron oxide band emission intensities with additions of Ca^{+2} to the BO_3^{-3} test solution.

Calcium–Aluminum Interactions

Since the relative emission intensity of AlO bands resulting from aspiration of pure aqueous solutions of $AlCl_3$ are very low, the mutual effects of Al and Ca could not be tested in that solution.

That the negative aluminum effect resulting from aspiration of pure aqueous solutions may be of a different magnitude for the various calcium emissions is shown in Fig. 2. At a given Al/Ca concentration ratio the aluminum suppression was significantly greater for CaO band emissions than for the Ca(I) line emissions. However, at the same $CaCl_2$ concentration, similar data taken in the H_2-O_2 flame gave nearly superimposable suppression curves. Other investigators have found that the magnitude of suppression may depend on the particular emission wavelength used. Dinnin [5] found aluminum and phosphate suppression of various magnesium emissions to vary considerably. Dean, Rains, and co-workers [4] have pointed out that aluminum suppression of the Ba(I) line was greater than of the Ba(II) line.

Calcium–Gallium Interactions

Decreased calcium emission intensity (422.7 mμ) due to the presence of $GaCl_3$ in the test solution is shown in Fig. 3. It is interesting that calcium intensity was not decreased by the presence of small amounts of $GaCl_3$,

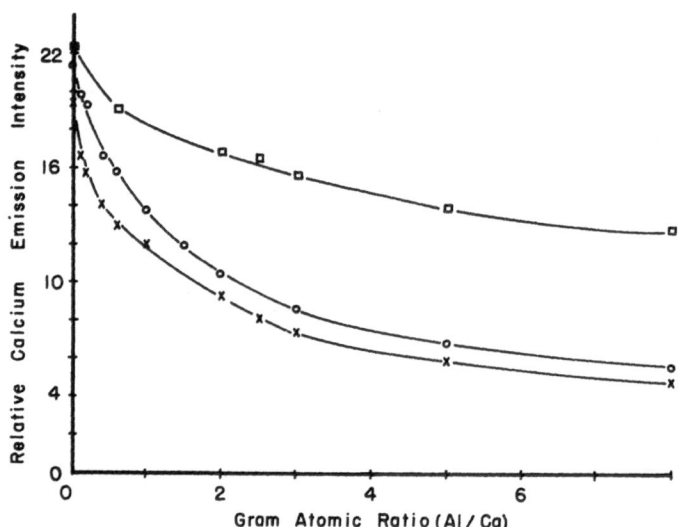

Fig. 2. Aluminum suppression curves:
□ Calcium (I) line   422.7 mμ All solutions 0.0200 M
O Calcium oxide band head   554 mμ $CaCl_2$
× Calcium oxide band head
  $C_2H_2-O_2$ flame

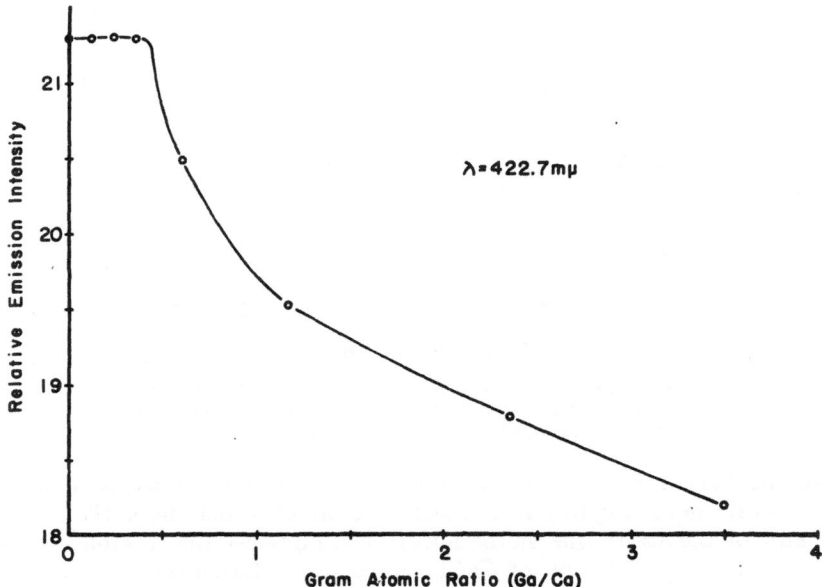

Fig. 3. Gallium suppression curve: $CaCl_2 = 1.25 \cdot 10^{-3}$ M; $C_2H_2-O_2$ flame.

i.e., at Ga/Ca ratios less than 0.4/1. $GaCl_3$ in the test solution also reduced the CaO band emissions at 554 m μ and 622 mμ.

When the effect of calcium on the Ga(I) line emission intensity (417.2 mμ) was investigated under the same flame conditions as those prevailing when the effect of gallium on calcium emissions was determined, it was found that the presence of $CaCl_2$ in the test solution increased the emission intensity of the Ga(I) line. This enhancing effect was small, amounting to from 2 to 3%, and did not vary significantly when the gram atomic ratio of Ca/Ga was changed from 0.3/1 to 1/1. It was found that under different flame conditions gallium suppression of calcium (422.7 mμ) amounted to 30% in the lower flame region and 17% in the higher flame region at a concentration ratio of Ga/Ca = 6/1. Although gallium can exist as the gallate anion in aqueous solution chemistry, it is not known whether a calcium gallate refractory lattice results from dehydration of the aerosol particles when a solution containing $CaCl_2$ and $GaCl_3$ is aspirated into the flame. Since the presence of $CaCl_2$ in the test solution containing $GaCl_3$ did not reduce the Ga(I) line emission intensity, the negative effect of gallium on calcium cannot readily be explained either on basis of stable di- or polyatomic molecules containing both calcium and gallium or on basis of a refractory calcium–gallium lattice.

Calcium–Indium Interactions

By the use of a limited-area technique, the calcium–indium system was studied in the acetylene–oxygen flame. At a $CaCl_2$ concentration of 10^{-3} M and at an In/Ca gram atomic ratio of 2.5/1, the indium suppression of calcium (422.7 mμ) amounted to 2% in the lower flame region (0.4 to 1.0 cm

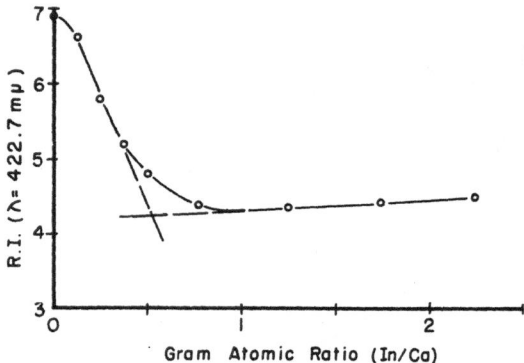

Fig. 4. Relative intensity (R.I.) of calcium (I) line in lower flame region as function of indium to calcium concentration ratio: $CaCl_2 = 0.0200$ M; $C_2H_2-O_2$ flame.

above the burner tip). Near the flame tip, however, suppression amounts to 25% (422.7 mμ). Figures 4 and 5 show the effect of indium on the radiation intensity of the Ca(I) line in the lower- and higher-flame regions, respectively. The concentration of $CaCl_2$ in the test solution was greater in this experiment. Even in the low-flame region a reduction in Ca(I) line emission intensity resulted when $InCl_3$ was present in the test solution. Extrapolation of the linear portions of both curves (Figs. 4 and 5) shows an intersection at an In/Ca ratio of 0.5. Conditions were not found whereby the presence of Ca^{+2} in the $InCl_3$ test solution decreased the In(I) line emission intensity. In Fact, In(I) line emission intensity was increased slightly when $CaCl_2$ was present. These data clearly indicate that indium suppresses calcium radiations by a gas-phase mechanism. The effect of the presence of calcium on indium radiations is not easily measured in the higher-flame regions because the In(I) line emission intensity is very low. The presence of $InCl_3$ in the test solution served to reduce the CaO band emission intensi-

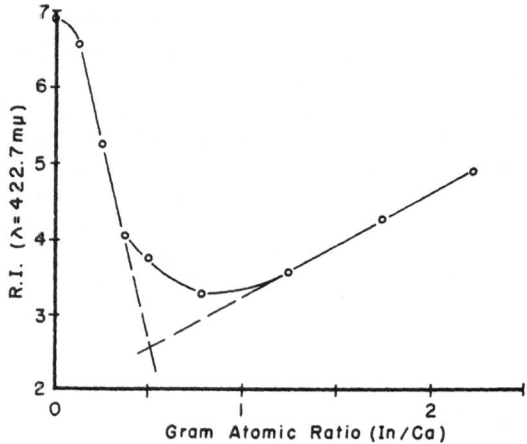

Fig. 5. Relative intensity of calcium (I) line in higher flame region as function of indium to calcium concentration ratio: $CaCl_2 = 0.0200$ M; $C_2H_2-O_2$ flame.

ties as well as the resonance line intensity. The presence of $InCl_3$ in the test solution also decreased strontium radiation intensities, and the presence of $SrCl_2$ increased the In(I) line emission intensity slightly (2 to 3%) at the concentration levels tested.

Data obtained by using the DU monochromator with the aspirator burner and mirror mounted in their normal positions showed a different type of indium suppression curve. The plot (data not shown) showing relative calcium emissions (422.7 mμ) as a function of gram atomic ratio of In/Ca in the test solution (at a constant $CaCl_2$ concentration) was of a form similar to those showing PO_4^{-3} suppression effects on calcium emissions. Calcium emissions decreased as $InCl_3$ concentration in the test solution increased until the In/Ca gram atomic ratio of 0.5/1 was reached, then a sharp break in the curve occurred and further additions of $InCl_3$ gave a small linear decrease in the Ca(I) line emission intensity.

Alcohol Effects

The effect of adding small amounts of n-butanol to the test solution on the boron oxide band emissions resulting from aspiration of the H_3BO_3 solution into the H_2-O_2 flame is shown in Fig. 6. The point of the abrupt change in slope corresponds within 4% with the amount of n-butanol necessary to convert the boron quantitatively into the triester. Possibly the triester formed during the process of solvent evaporation from the aerosol droplets and the ester facilitated more nearly complete conversion of the H_3BO_3 into the boron oxide. Addition of small amounts of n-butanol was accompanied by decreased rate of aspiration of the H_3BO_3 test solution. The aspiration rate was 1.8 g/min for the pure aqueous solution and 1.45 g/min at a 4 v/v % n-butanol level. The continued increase in boron oxide emission intensity with additions of n-butanol in excess of the stoichiometric amount (Fig. 6) probably could be related to the decreasing aspiration rate; hence

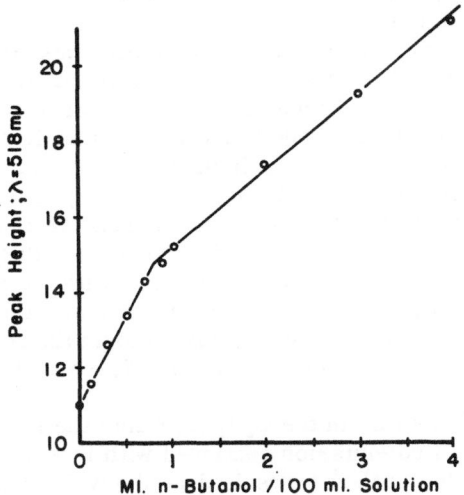

Fig. 6. Boron oxide emissions, 518 mμ, as function of v/v %
n-butanol; $H_3BO_3 = 0.0288$ M; H_2-O_2 flame.

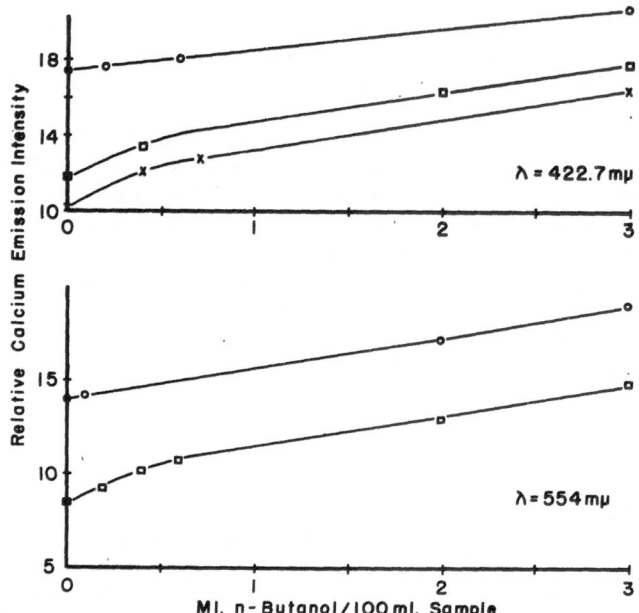

Fig. 7. n–Butanol release of boron suppression:
  O $CaCl_2$ = 0.0100 M, no H_3BO_3 present
  □ $CaCl_2$ = 0.0100 M, H_3BO_3 = 0.0086 M
  ✕ $CaCl_2$ = 0.0100 M, H_3BO_3 = 0.0172 M; $H_2 - O_2$ flame.

the cooling effect of H_2O being added to the flame is less, with resultant higher flame temperatures.

The releasing action of n-butanol on BO_3^{-3} suppression of calcium radiations was subsequently tested, since if n-butanol forms the borate ester, then BO_3^{-3} will not be available to react with Ca^{+2} and release of boron suppression of calcium radiations should result.

Figures 7 and 8 show results of these experiments. Figure 7 shows that when n-butanol was added up to approximately 0.5 ml per 100 ml of BO_3^{-3} containing $CaCl_2$ solutions, partial release of BO_3^{-3} suppression resulted but with larger additions of n-butanol, from 0.5 to 3 v/v %, no significant further release resulted. Similar results were obtained for the CaO band cresting at 622 mμ. That only a partial release resulted may perhaps be explained in terms of two competitive reactions involving BO_3^{-3} ions taking place in the flame within the aerosol droplet. These two reactions are (a) the tendency of calcium to react with the boron which is opposed by (b) the tendency of the n-butanol to react with the boron. The stoichiometric amount of n-butanol necessary to form the triester was 0.23 ml per 100 ml sample for the curves in Fig. 7, where BO_3^{-3} concentration was 0.0086 M.

Figure 8 shows that in the $C_2H_2-O_2$ flame the release of boron suppression of the Ca(I) line emission continued with increased additions of n-butanol, until complete release resulted at 3 v/v % n-butanol. Ethylene glycol was found to release a large fraction of the BO_3^{-3} suppression of calcium radiations when added in small amounts to the test solutions being aspirated into the H_2-O_2 flame.

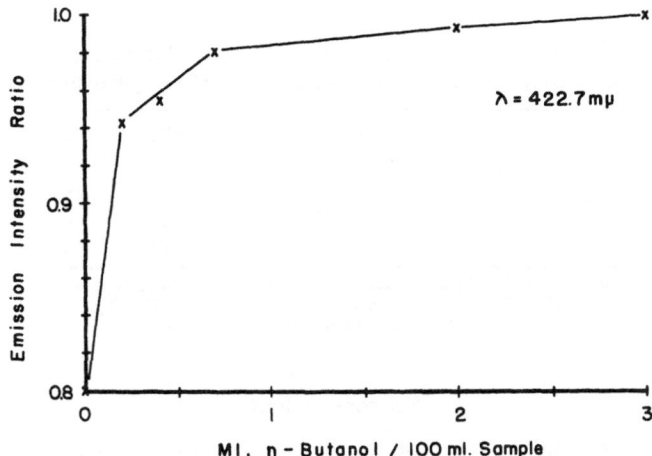

Fig. 8. Ratio of emission intensity of a 0.0100 M CaCl$_2$ solution containing H$_3$BO$_3$ (0.0086 M) to the emission intensity of a pure CaCl$_2$ solution (0.0100 M), as a function of v/v % methanol in the solution; C$_2$H$_2$–O$_2$ flame.

The methanol effect on boron and aluminum suppressions of calcium radiations is shown in Table I. Table I data are ratios of the emission intensity (of the CaCl$_2$ solution containing the indicated suppressor ion compared to that with the pure CaCl$_2$ solution) at the indicated calcium line or band head and at the indicated v/v % methanol level. For all calcium emissions tested and in both flames, methanol reduced the BO$_3^{-3}$ effect but increased the aluminum effect on calcium. Methanol in the test solution also was found to reduce the effect of phosphate on calcium emissions. Methanol also reduced BO$_3^{-3}$ and PO$_4^{-3}$ suppression of barium radiations and increased the aluminum suppression of barium radiations.

TABLE I
Effect of Methanol on Aluminum and Boron Suppression of Calcium Radiations

	Hydrogen-Oxygen Flame									
	$\lambda = 422.7$ mμ				$\lambda = 622$ mμ				$\lambda = 554$ mμ	
Suppressor Ion	0%	10%	30%	50%	0%	10%	30%	50%	0%	50%
Al^{+3}	0.86	0.85	0.85	0.80	0.83	0.83	0.78	0.73	0.82	0.70
BO$_3^{-3}$	0.70	0.70	0.76	0.84	0.47	0.58	0.63	0.66	0.48	0.69

	Acetylene-Oxygen Flame									
	$\lambda = 422.7$ mμ				$\lambda = 622$ mμ			$\lambda = 554$ mμ		
	0%	10%	30%	50%	0%	10%	30%	0%	10%	30%
Al^{+3}	0.89	0.85	0.84	0.85	0.75	0.73	0.67	0.76	0.78	0.70
BO$_3^{-3}$	0.85	0.96	0.98	0.96	—	—	—	0.62	0.89	0.86

CaCl$_2$ = 0.0100 M throughout; AlCl$_3$ (where present) = 0.0200 M; H$_3$BO$_3$ (where present) = 0.0144 M.

TABLE II

Effect of Sucrose on Aluminum and Boron Suppression of Calcium Radiation: C_2H_2—O_2 Flame

(mμ)	$\dfrac{CaS}{Ca}$	$\dfrac{CaAlS}{CaS}$	$\dfrac{CaAl}{Ca}$	$\dfrac{CaBS}{CaS}$	$\dfrac{CaB}{Ca}$
422.7	0.86	0.23	0.71	1.05	0.73
554	0.75	0.17	0.63	1.04	0.59
622	0.71	0.16	0.64	1.04	0.57

$CaCl_2 = 0.0200$ M throughout; Al/Ca = 1/1 for solutions containing $AlCl_3$; B/Ca = 1.44/1 for solutions containing H_3BO_3; S = indicates sucrose present; sucrose (where present) = 1%.

Sucrose Effect

The effect of adding sucrose to the test solution, on the boron and aluminum suppression of calcium radiations is shown in Table II. Column one shows that at all wavelengths tested sucrose decreased the calcium emission intensity in the absence of suppressor ions, i.e., solution designated as CaS, containing $CaCl_2$ plus sucrose, gave less calcium emission than the solution designated as Ca, which was the pure $CaCl_2$ solution. The ratio was 0.86 at 422.7 mμ. This general decrease in emission intensity cannot be attributed to the change in sample aspiration rate since the aspiration rate of pure water was (within 1%) the same as the aspiration rate of a 1% sucrose solution.

Comparison of columns two and three shows that aluminum suppression increased greatly in the presence of sucrose. However, the effect of sucrose on boron suppression differed. Column four shows that the emission intensity of the solution containing $CaCl_2$, H_3BO_3, and sucrose was in every instance greater than the emission intensity resulting from the solution containing only the $CaCl_2$ and sucrose. Substantial release of phosphate suppression was also observed in the presence of sucrose in the test solution. When $CaCl_2$ was replaced by $SrCl_2$ in the test solution, sucrose again decreased boron and phosphorus suppression but significantly increased aluminum suppression of strontium radiations.

Dilution Effect

Figure 9 shows the change in emission intensity ratio (of a solution containing $CaCl_2$ plus a suppressor ion, to a $CaCl_2$ solution at the same $CaCl_2$ concentration level in the absence of the suppressor ion), as a function of $CaCl_2$ concentration in the solution being aspirated. Using $SrCl_2$ instead of $CaCl_2$ and measuring strontium emission intensities at the Sr(I) line, Sr(II) line, and at the SrOH bandhead cresting approximately at 665 mμ showed that dilution affected Al^{+3}, BO_3^{-3}, and PO_4^{-3} suppressions of strontium in a manner completely analogous to the calcium systems. The data indicate that perhaps the slow vaporization of calcium, from the calcium-boron- and/or calcium-phosphorus-containing particles, resulting from dehydration of the aerosol droplets aspirated into the flame, reduced the calcium emission intensities relative to emission intensities resulting from aspiration of a pure $CaCl_2$ solution.

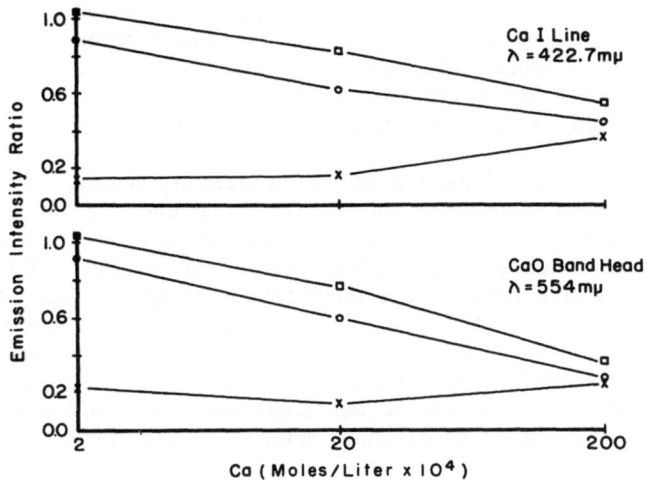

Fig. 9. Boron, aluminum, and phosphate suppression as a function of calcium concentration:
  □ $H_3BO_3/CaCl_2$ = 5.75/1 throughout
  O $NH_4H_2PO_4/CaCl_2$ = 0.5/1 throughout
  x $AlCl_3/CaCl_2$ = 4.0/1 throughout
    $C_2H_2-O_2$ flame.

The average volume of the refractory crystals resulting after dehydration of the aerosol particles would be expected to increase with increased Ca^{+2} concentration in the test solution at constant calcium to suppressor ion concentration. The vaporization rate of calcium from the lattice, being a surface phenomenon, would be a linear function of the crystal surface area. Increasing the volume, by increasing test solution concentration, by a factor of x, would (assuming spherical crystals for the purpose of a simple model) increase the surface area only by a factor of $x^{2/3}$. This very rapid increase of suppression as a function of increased calcium concentration in the test solution at constant BO_3^{-3}/Ca^{+2} ratio was not observed, which was not surprising since the mass of the secondary fragments formed would not be expected to be a linear function of the test solution concentration. The initial crystal formed must be very imperfect because of the rapid rate of dehydration of the aspirated aerosol particles. Any increase in the average crystal fragment volume, regardless of the geometric shape of the crystal, would, however, result in a decreased area-to-volume ratio. Gibson, Grossman, and Cooke [6] have recently reported that SO_4^{-2} and PO_4^{-3} suppression of calcium radiations, decreases at low calcium concentrations.

CONCLUSIONS

Since aluminum suppression did not decrease with additions of methanol or sucrose to the test solutions, and also since it did not decrease with decreased test solution concentration, while boron and phosphorus suppression decreased significantly under all three conditions, it is likely that the aluminum suppression mechanism is basically different from that involved in the case of boron and phosphorus suppression.

Alcohol effects on emission intensities in general have been discussed in terms of decreased aerosol droplet size [3]. This phenomenon, in addition to the higher flame temperature realized when alcoholic solutions are used, seems applicable to the observed decreased boron and phosphorus suppression in alcoholic solutions. Baker and Garton [2] discussed their observed sucrose effect in terms of the added sucrose acting as a dispersing agent and resulting in increased fragmentation of the crystals in the calcium phosphate system.

They reported that Ca(I) line emission intensity resulting from aspiration of a solution containing Ca^{+2}, PO_4^{-3}, and sucrose was greater than Ca(I) line emission intensity resulting from aspiration of a pure calcium solution in the absence of both phosphate ion and sucrose. Calcium concentrations in both solutions were identical. Although they did not compare phosphate suppression of calcium in the presence of sucrose with that in the absence of sucrose, and hence obtained no concrete evidence for their proposed sucrose releasing mechanism, their suggested mechanism of release is tenable, and seems to be substantiated by decreased boron and phosphorus suppressions of calcium radiations in the presence of sucrose indicated in our results.

The mechanism of aluminum suppression has not been elucidated. It now appears, however, that it is unlikely that occlusion of calcium in the refractory aluminum oxide lattice is the mechanism by which aluminum decreases Group II-A element emissions because aluminum suppression of calcium radiations was found to intensify when sucrose or ethanol was added to the test solution and also to increase with decreased test solution concentration at constant Al/Ca concentration ratio.

Differences in the degree of suppression of the Ca(I) line and CaO bands in the presence of BO_3^{-3} or Al^{+3} in the test solutions cannot be accounted for solely on the basis of an intermediate refractory solid phase. The releasing action of small amounts of n-butanol on boron suppression and the large increase in B_xO_y emission intensities when small amounts of n-butanol are present in the test solution are most readily explained on the basis of compound formation between BO_3^{-3} and n-butanol in the evaporating aerosol droplet. That the releasing action of small amounts of n-butanol on BO_3^{-3} suppression of calcium radiations was complete only in the $C_2H_2-O_2$ flame indicates that if releasing agents are to be used, they must be first tested for effectiveness in the particular system. As a further example of this we found that in the $C_2H_2-O_2$ flame PO_4^{-3} suppression of strontium radiations was increased significantly when small amounts of $CaCl_2$ were added to the test solution and that the well-documented releasing action resulted only at high Ca/Sr concentration ratios.

The effects of Ga^{+3} and In^{+3} on calcium radiations can best be explained in terms of gas-phase suppression mechanisms in the light of the data reported here.

REFERENCES

1. C. T. J. Alkemade and M. H. Voorhuis, Zur Frage des Phosphor Einflusses auf die Calcium Emission in der Flamme. Z. anal. Chem. 163:91 (1958).
2. C. A. Baker and F. W. J. Garton, A Study of Interferences in Emission and Absorption Flame Photometry. U.K.A.E.R.E. R 3490.
3. J. A. Dean, Flame Photometry (McGraw-Hill, New York, 1960).
4. J. A. Dean, J. C. Rains, et al., Flame Spectrophotometric Study of Barium. Anal. Chem. 33:1722 (1961).
5. Joseph I. Dinnin, Use of Releasing Agents in the Flame Photometric Determination of Magnesium and Barium. U.S.G.S. Prof. Pap. 424–D, 1961.
6. James H. Gibson, William E. L. Grossman, and W. D. Cooke, Excitation Processes in Flame Photometry. Anal. Chem. 35:266 (1963).

New Developments in Practice and Technique of Flame Spectrometry

John A. Dean

Department of Chemistry
University of Tennessee
Knoxville, Tennessee

A review of the major trends and achievements in flame emission spectroscopy for the last three years will be broken down into these several categories: improvements in equipment components, flames and their composition, combustion pattern, feed rate, and interference effects. The use of organic solvents has continued to increase, and, in conjunction with fuel-rich flames or with observation of spectra from restricted areas of the combustion zone, has extended the utility of flame photometry to many elements whose emission lines are now observed in unusual strength. Gradually some of the factors which influence the intensity of emission spectra are being unraveled.

INTRODUCTION

Forty-five years ago there appeared the first volume of the classic studies on flame spectral analysis by Lundegardh. Since then the utility of flame spectrometry for the analysis of sodium and potassium has been well established. Gradually the method is displacing more troublesome methods for other elements. In this selective review we shall pause and examine the trends in flame spectrometry, the progress made in understanding the interwoven chemical and physical interferences that often arise in practice, and the changes in technique which have extended the application of flame spectrometry to a number of elements hitherto considered unexcitable in flames with adequate sensitivity. Major innovations in recent years have been the increasing use of solvent extraction techniques followed by direct aspiration of the organic phase into the flame, the use of fuel-rich, reducing flames, the exploration of emission lines emanating from the reaction zone of a flame, and a recognition of the role played by chemiexcitation processes in the flame gases.

FACTORS AFFECTING THE TOTAL ATOM POPULATION

Wavelengths of emission lines are fixed by the discrete levels of internal energy within each atomic (or molecular) configuration. The problems that face the analyst are (1) how to bring about sufficiently energetic excitation conditions within the flame gases, (2) how to minimize inevitable variations in physical factors among analyte and standard solutions, and (3) how to circumvent spectral and radiation interferences.

The integrated intensity of the emission of a line is given by

$$\int I\, d\nu = C N_j f \qquad (1)$$

where N_j is the number of atoms in an excited state involved in the transition responsible for the line, f is the oscillator strength of the line, and the proportionality constant C depends on the dispersing and detecting systems. The occupational number of atoms in an excited energy level is related to the number of atoms present in the gound state (proportional to the number introduced into the flame gases), N_0, by the Boltzmann equation, when thermal equilibrium prevails,

$$N_j = N_0 (P_j/P_0) \exp (- E_j/kT) \qquad (2)$$

where P_j and P_0 are the statistical weights for the particular energy levels, and E_j is the excitation energy of the excited state.

Feed Rate

When the intensity of the emission spectrum is plotted against sample feed rate, it usually goes through a maximum; that is, at low feed rates the intensity is low, increasing to a maximum as feed rate increases, and then decreasing as the feed rate further increases [3, 4, 25, 28]. Lady [37] pointed out that increasing the aspiration rate does not have the same effect as increasing the number of atoms by increasing the concentration. With direct sprayers the supply of liquid is known to depress the flame temperature by its cooling effect. The depression is more severe with aqueous solutions than with nonaqueous solutions. With higher-temperature flames whose gases do not involve the dissociative equilibria of the solvent molecules, the feed rate becomes very critical (e.g., oxygen—cyanogen flames). Fortunately, for the usual flame mixtures the feed rate is not too critical and, provided it is kept constant, does not present a problem in obtaining reproducible results.

Feed rate has been interrelated with viscosity [52], surface tension, and pressure of the aspirating gas. Drop size of the aerosol is also affected. Viscosity plays a major role in determining the intensity of turbulence of the issuing jet of liquid from the atomizer. An increase in surface tension results in a decreased rate of disintegration since formation of ligaments and deformation of drops is counteracted. Drop size decreases as the orifice diameter decreases on the aspirator. The atomization becomes finer and more uniform when smaller orifices are used. Pressure is important by virtue of the velocity imparted to the liquid. Varying the pressure does not greatly change the size of the smallest drops, but as the pressure is increased the number of smaller drops is increased so that the mean size is decreased.

For heterodisperse sprays the Sauter mean-drop diameter can be taken to be the main spray parameter of importance. In a study of aerosol drop size, Dean and Carnes [14] showed that methyl isobutyl ketone and chloroform produced finer droplets, in comparison with water, and that these would vaporize rather completely before reaching the plume of the flame. One advantage of organic solvent extraction is that it provides a homogeneous system for atomization and enhances the emission intensity through an increase in effective feed rate. Larger droplets from aqueous aerosols often fail to evaporate completely before reaching the plume of the flame.

A satisfactory burner has yet to be designed. It should enable a constant feed rate to be obtained, irrespective of changes in the physical properties of the solvent. It should also provide that the solvent be aspirated in very small droplet size to enable easy evaporation and a high efficiency of production of atoms. Forced feed pumps are a partial solution [25, 29, 30]. Until the "ideal" burner becomes available, the rates of flow of the gases and the pressure of the aspirating gas must be specified in order to reproduce conditions on other instruments.

A slurry of powdered material in isopropanol-glycerine solvent can be sprayed directly into a flame if one uses a wide-bore integral-aspirator burner [32]. This should prove convenient in certain soil and mineral analyses and, in general, for samples which can be reduced readily to a fine mesh size but whose dissolution is lengthy and fraught with difficulties.

Temperature

The main parameter controlling the flame emission is the "effective" temperature of the flame gases. Quite naturally workers sought special combustion mixtures to yield higher temperatures in order to excite additional elements and to increase the emission of others. Mavrodineanu [39] has provided an excellent summary of these efforts. Expectations were seldom realized. One contributory factor was the cooling effect of the solvent mentioned in the previous section.

Surrounding the normal flame with a protective sheath of gases [31] offers several advantages. Entrainment of air at the base of the flame is avoided thereby reducing flame turbulence and introduction of nitrogen molecules which prevent the flame from attaining its maximum temperature. A special premixed burner assembly [13] utilizing a regular integral-atomizer burner offers advantages also. Detection sensitivity in the sheathed or premixed flames is improved 2 to 10 times as compared with unsheathed flames.

Chemiluminescence reactions, which may give rise to suprathermal excitation, will be discussed later.

Solvent

The emission intensity is enhanced significantly when elements are aspirated into the flame from an organic solution instead of an aqueous solution [10]. Also minimized are variations of the physical properties of the aspirated solution, such as viscosity, droplet size, and volatility. Use of water-organic-solvent mixtures and of organic solvents alone has become a standard feature of many recent publications.

For water-organic-solvent mixtures, in the interval from 20 to 70 volume percent of organic solvent (often a C_1 to C_4 alcohol, or acetone), the sensitivity is increased from two- to sevenfold, depending on the solvent and the

TABLE I

Relation Between Solvent Composition and Intensification of Flame
Emissions*

Element	Wavelength, mμ	Solvent	Volume percent of organic solvent present with water					Reference
			10	30	50	70	90	
Barium	455.4	Methanol	1.6	3.1	4.8	7.8	14.7	[13]
	553.6	Methanol	1.5	2.2	2.7	4.8	10.0	[13]
	488	Methanol	1.5	1.8	2.0	2.9	8.6	[13]
Chromium	425.4	Methanol	1.1	2	4	6	8	[42]
Gallium	417.2	Acetone	3	11	25	39	—	[6]
Indium	451.1	Acetone	1.5	5	12	24	—	[6]
Thallium	377.6	Acetone	2	4	7.5	15	—	[6]

*The enhancement is expressed as a ratio of emission readings in the water–organic–solvent mix-
ture to the reading in water. An integral-aspirator burner was used in each case.

atomizer. Significantly larger increases are reported for Group III-B elements, gallium, indium, and thallium, continuing the trend observed for boron
and aluminum. Results for these and other elements are included in Table I.
For barium, data are available for the ionic and atomic lines plus the molecular band system peaking at 488 mμ. Although enhancements are observed for
all three types of emission, the greatest increase occurred with the ionic
line and probably arises from the thermal shift upon the overlapping molecular dissociation and ionization equilibria.

More spectacular enhancements (10- to 1000-fold) are reported for elements aspirated into a flame from essentially a nonaqueous solvent (Table II).
In a series of reports Goto and Sudo [34, 50] have extracted the oxinates of
calcium, cobalt, copper, gallium, indium, magnesium, manganese, nickel, and
titanium into methyl isobutyl ketone or amyl acetate. The rare earths, except
for cerium and promethium, when extracted with 2-thenoyltrifluoroacetone in
methyl isobutyl ketone, give a number of strongly emitting band systems [43].
Bands were found for lanthanum, yttrium, and neodymium which were free
from spectral interference by other rare earths. Several other metals
suffered only limited interference. Other methods include the extraction of
boron into methyl isobutyl ketone [20], the extraction of tellurium from steel
samples with diethyldithiocarbamate [17], the extraction of thallium in 1 M
HBr from the ash of urine samples with 2-octanone [48], and extraction of
vanadium from dilute H_2SO_4 solution with cupferron and ethyl acetate following a mercury cathode electrolysis [47]. Additional reports concerning the
use of organic solvents will be discussed in subsequent sections.

Numerous possibilities exist for the simultaneous extraction and successive determination of several elements in a sample matrix. Unfortunately,
this attractive approach remains largely unexplored although the author has
called attention to this oversight before [11]. One paper has described a
method for iron, cobalt, nickel, and vanadium using diethyldithiocarbamate
as extractant [46]. Many extractive colorimetric methods, including those
dismissed formerly as unsatisfactory, are directly applicable for flame

TABLE II
Emissions from Fuel-Rich Hydrocarbon Flames

Element	Wavelength, mμ	Emission sensitivity, μg/ml/div	Excitation potential, eV	Dissociation energy of MO, eV	Reference
Antimony	252.8	1.0	6.12	3.8	[15]
Arsenic	235.0	2.2	6.6	4.9	[15]
Beryllium	234.9	0.1	5.3	–	[33]
Bismuth	223.1	6.4	5.5	3.5	[15]
Boron	249.8	7	5.0	>7.6	[12]
Cadmium	228.8	4.2	5.4	<3.8	[15]
Cobalt	242.5	1.7	5.1	–	[12]
Mercury	253.6	2.5	4.88	–	[15]
Molybdenum	379.8	0.5	3.26	5.0	[23]
Niobium	405.9	12	3.18	–	[23]
Nitrogen (NO)	236*	14	–	–	[33]
Phosphorus (PO)	253*	1.0	–	5.4	[33]
Platinum	265.9	15	4.66	–	[15]
Rhenium	346.1	3	3.58	–	[23]
Silicon	251.6	4.5	4.9	8.0	[33]
Sulfur (CS)	258*	24	–	–	[33]
Tellurium	238.6	2.0	5.8	4.0	[17]
Tin	243.0	1.6	5.1	5.7	[15]
Titanium	399.9	5	3.15	6.9	[23]
Vanadium	437.9	3	3.13	6.4	[23]
Tungsten	400.9	90 (4)†	3.09	–	[23]
Zinc	213.8	77	5.8	<4.0	[15]

*Band head.
†Premixed rich acetylene-oxygen-alcohol flame (21).

spectrometry. Often the only change required is to substitute an oxygenated solvent, ketone or ester, for the halogen-containing solvents. Cleaner's naphtha, alone or as a 1:1 mixture with isopropyl alcohol, is an excellent solvent for metal naphthenates and petroleum products in general. Methyl isobutyl ketone or amyl acetate is ideal for many metal complex extractions. It is not necessary that completely specific methods be used as long as other metals coextracting do not enhance or inhibit the signal emitted by the test element or elements, or emit radiation of their own at the same wavelength as the test element(s). Successive scanning of pertinent emission lines can be done rapidly. There is less likelihood of interferences in the relatively dilute extractant and in view of the chelated condition of the metal atoms which is analogous to protective chelation employed sometimes to circumvent condensed-phase interference. The enhancements achieved with organic solvents permit more dilute solutions to be analyzed which, in turn, often circumvents the effects of self-absorption on the linearity of working curves, and thereby enables the standard-addition method to be employed for evaluation of samples. Solvent extraction also affords a means for concentrating the test element, invaluable in trace analysis. Lastly, as will be shown in subsequent sections, the presence of an organic solvent is essential if one wishes to elicit chemi-ionization or chemiluminescence.

The gain factor obtained with the substitution of organic solvents for water depends in a complicated and subtle way on the combined action of many factors [2, 10, 45]. Although the flame temperature is often higher (or at least the cooling effect of the solvent is less), this fact cannot explain the high degree of signal enhancement. An increased amount of solution reaches the flame (see the section on Feed Rate). Smaller mean-drop diameters and more rapid and complete evaporation of spray droplets are directionally helpful, particularly with integral-aspirator burners. The organic ligand in a metal-organic complex will burn and leave the free metal atom unfettered by a shell of hydrated water (or OH or O radicals). The increase in absorption signals can be attributed largely to these factors. The usual enhancements in flame emission spectroscopy can also be explained by the foregoing factors, but not the unusually large enhancements observed for the elements of Group III-B or the appearance of many high-energy lines in the spectra from reaction zones of flames fed by organic aerosols.

In an interesting study with dual atomizers, Dvorak and Novobilsky (19) and Pungor et al. [41] found that the metal emission showed the usual increase on changing from water to 50 vol. % alcohol when the same atomizer was used. However, on aspirating the alcohol into the flame with a separate burner, the metal reading remained unchanged or even decreased slightly. The amount of water in the droplets remained relatively constant as the alcohol content increased, which implies more fine spray.

EMISSIONS FROM FUEL-RICH HYDROCARBON FLAMES

Lines of a number of elements that are absent or very weak in the plume of an oxygen-hydrogen or oxygen-acetylene flame appear in unusual strength in the spectrum of a rich acetylene-oxygen or an air-hydrogen or oxygen-hydrogen flame, when the analyte is dissolved in a hydrocarbon. These flames are saturated with carbon vapor and provide an environment favorable for existence of free atoms in the flame gases, owing to the strong chemical reducing action of carbon as expressed by the reaction

$$C + MO \rightarrow CO + M^0 \text{ (or } M^*) \tag{3}$$

Since ΔH_f of CO is 11.1 eV, and the dissociation energy D_{MO} for many monoxides seldom exceeds 7 eV, the reaction expressed by equation (1) will always be exothermic by the amount $\Delta H_{CO} - D_{MO}$, the energy available for excitation of metal atoms.

Once conditions conducive to the existence of free atoms exist in the flame, Fassel and co-workers [22, 23] found the excitation to be principally thermal in character for the line spectra of the rare earths, Ti, Nb, V, W, Re, and Mo. Excitation potentials of these elements are not particularly high, and are usually within 3.5 eV of the ground state of the atom. The high dissociation energy of their monoxides had prevented the appearance of their line spectra in ordinary flames. In fact, for the rare earths, the monoxide band spectra show no diminution in intensity but rather an enhancement as compared with their intensity in oxygen–hydrogen flames.

The spectrum of the inner cone and reaction zone of hydrocarbon flames provides another rich source of lines whose intensity far exceeds that expected from purely thermal excitation. Lines are primarily from elements of high excitation potential and high ionization potential. Dean and co-workers [12, 15, 17] have studied the line spectra of Sb, As, Bi, B, Cd, Co, Hg, Pt, Te, Sn, and Zn from oxygen–acetylene–ketone flames. Buell [8, 9] has made a similar study with an oxygen–hydrogen–naphtha flame. Except for the manner of introducing the hydrocarbon component, and perhaps a slight difference in emission profile, the two types of flames are apparently comparable. Maximum emission emanates from a region extending from 1 to 4 mm above the tip of the inner blue cone. Also observed in these flames are the spectra of the OH, CH, and C_2 radicals, the fourth positive system of CO, and the atomic line of carbon at 247.8 mμ. By contrast, lines of lower excitation potential, or elements whose oxides have lower dissociation energies, usually appear higher in the plume of the flame. Thus, the emission profile becomes a critical factor in selecting optimum working conditions, and some technique [8, 15] is necessary for observing limited areas of the flame.

Elicitation of atomic lines from so many elements hitherto considered unexcitable with normal oxygen-fuel flames constitutes the most significant advance in operating technique since the introduction of organic solvents into flame spectrometry. The emission sensitivity of a number of elements will be found in Table II. Whenever available in adequate strength, discrete emission lines are usually preferable to the use of molecular band spectra and the resulting problems of spectral resolution. In the case of boron, its emission line at 248.7 mμ, which appears in the reaction zone, offers virtually the same sensitivity as do the fluctuation bands in the green (now assigned to the radical BO_2 [35], although either type of spectra is superior to the indirect method employing precipitation as barium borotartrate and subsequent determination of the barium content of the precipitate [7]. Indirect methods have also been suggested for a number of other elements, but methods of this type, which depend upon a depressant effect on the emission of calcium, strontium, or magnesium lack specificity.

Some band spectra are enhanced and useful in addition to those of the rare earths mentioned earlier. Band spectra for NO, CN, CS, and PO are sharp and distinctive [33], and eliminate dependence on a continuum or a depressant action for the determination of nitrogen, sulfur, and phosphorus.

Excitation Mechanism

For the elements studied by Buell [9] and Dean and co-workers [12, 15, 17], the relative emission intensity, as a function of the flame region viewed, parallels closely the pattern observed for the atomic line of carbon, but occurs slightly higher in the flame than the band emission from the CH and C_2 radicals. Undoubtedly energetic conditions prevail in these flames. Electronic excitation is probably produced directly by chemical reaction (i.e., chemiluminescence) with, perhaps, a concomitant reduction of the monoxides of these elements or, at least, preventing the formation of monoxides, caused by the high vapor pressure of carbon in the reaction zone. The fourth positive system of CO forms much of the background emission of the far-ultraviolet portion of the spectrum.

Quite definitely the abnormal strength of the emission spectra from the reaction zone does not arise because of any significant change in the adiabatic flame temperature. Alkemade [1] has pointed out that if all processes leading to excitation of metal atoms were perfectly equilibrated at the apparent temperatures of the gases, there would be no possibility of enhancing the metal excitation. Deviations from equilibrium exist in the reaction zone, and also in the burned-gas mixture several millimeters beyond the reaction zone. Chemiexcitation will be found mainly in these regions where radicals are formed in excess and the chemical reaction involving the combustion of the flame gases is not equilibrated. Equipartition of energy over the vibrational degrees of freedom of molecules is a noticeably slower process than equipartition of translational and rotational energies.

Although Padley and Sugden [40] found chemiluminescent radiation at a short distance above the reaction zone with Na, Tl, and Fe in a cool air–hydrogen flame, chemiluminescent radiation is more pronounced with elements having higher excitation energies and high ionization potential. Addition of a hydrocarbon compound to the flame, either by spraying some organic solvent or by additionally supplying acetylene gas, seems to be a prerequisite for most chemiluminescent radiation. If the hydrogen content of a system is gradually replaced by acetylene, keeping the stoichiometry of the mixture the same, Dean and Kuper [16] found that the emission intensity is increased. While the atomic lines of aluminum are scarcely visible in an oxygen–hydrogen–ketone flame, addition of as little as a few mole percent acetylene elicits the atomic lines in surprising strength. In fact, the emission spectra of all the neutral atoms and molecular band systems studied exhibited an increase in intensity as the acetylene content of the flame gases was increased. This was true whether aqueous solutions or organic solutions were aspirated. The sharp rise in emission intensity as a small amount of acetylene was introduced into the fuel stream hints that perhaps the acetylene plays a direct role. In fact, incandescent rich-acetylene–oxygen flames were found necessary to break down the more stable molecular oxides [33].

A chemiluminescent transfer of excess vibrational energy from one or more flame radicals or molecules to the analyte is plausible. Hypotheses have been advanced by Gilbert [33] and Sternberg [49] in which the principal reaction responsible for the chemiluminescence is variously

$$CH + O = CO^* + H \tag{4}$$

and

$$CO^* + M = M^{+} \text{ (or } M^*) + e^- + CO \tag{5}$$

or

$$CO^* + MO = M^* + CO_2 \tag{6}$$

In another sequence, the H_3O^+, the most abundant positive ion in flames of hydrocarbons, is considered to be the vector of the high excitation energy for chemi-ionization:

$$H_3O^+ + MO = M^+ + H_2O + OH \qquad (7)$$

or

$$CO + MO = M^+ + e^- + CO_2 \qquad (8)$$

If one reviews the optimum excitation conditions that have been reported for a variety of elements over the past years that have shown unusual enhancements, several facts become apparent. (1) Fuel-rich flames were found to be optimum in many cases. For fuel-rich conditions to be apparent, the oxygen requirements of the organic solvent in combustion must be considered. (2) The ratio of oxygen to fuel (acetylene plus combustible solvent) for maximum emission intensity occurred at values ranging from 1.5 to 1.1 which correspond to $\lambda = 0.6$ to $\lambda = 0.44$ (a stoichiometric flame of oxygen—hydrocarbon, by comparison, corresponds to a ratio of 2.5, or $\lambda = 1.0$, and a burned gas composition of $CO_2 + H_2O$). At $\lambda < 0.43$ smoke formation is evident. Is it surprising then that optimum excitation conditions seem to correspond to a burned-gas composition largely $CO + H_2O$ in light of the proposed chemiluminescent mechanisms? Ratios of oxygen to fuel less than 1.5 imply intermediate products of combustion such as carbon atoms, C_2H, CH, etc.; hydrocarbon radicals are significant. Probably no singular route to chemiluminescent excitation should be expected to exist, although it is simpler to assume one reaction mechanism is dominant. Gilbert [33] and Buell [9] have compared the energy available with the excitations observed and conclude that equation (6), in which

$$\Delta H = -13.1 + E(M) + D(MO) \qquad (9)$$

and equation (8), in which

$$\Delta H = -13.1 + I(M) + D(MO) \qquad (10)$$

fit the data rather well.

INTERFERENCES

The twin problems of ionization and molecular dissociation pose a unique problem for rubidium, cesium, and barium [24]. For these elements the dissociation energy is 0.3 to 0.6 eV higher than the ionization energy. Accordingly, there is no temperature and concentration range in which complete dissociation and negligible ionization take place. Both emission and absorption will be affected.

Ionization occurs to a measurable extent with elements having ionization potentials of about 6 eV, and lower, particularly when their concentration is low. It becomes a serious problem with the higher-temperature flames. Incorporation of a second easily ionizable element into sample and standards (radiation buffer) is mandatory. Ionization is minimized when the emission, or absorption, is observed close to the outer boundary of the blue zone of the inner cone where there is a relatively high concentration of flame ions [26].

Chemical interference arises from formation of condensed phases in the flame, composed generally of involatile compounds which are difficult to volatilize and dissociate into free metal atoms. Flames containing hydrogen are advantageous when interference due to formation of halides is feared [40]. Change of oxidation state eliminates interference of pentavalent vanadium with calcium. The use of releasing agents or protective chelating agents has been recommended by a number of workers [18, 51]. Two routes are feasible: (1) Addition of a competing cation which preferentially combines with the interferent or simply by mass action denies the interferent access to the test element. An example is the addition of a large concentration of strontium or lanthanum to free calcium from the depressant action of phosphate. (2) Addition of a competing anion, such as EDTA, which preferentially complexes calcium in the presence of phosphate, and yet which is promptly decomposed by the flame. One might suppose that a series of releasing cations could be arranged in the relative order of their releasing ability. Such does not seem to be the case, at least with the limited information available. Although atomic absorption spectroscopy is plagued by the same type of chemical interference, the absence of spectral interference from the releasing agent when it is a metal ion in high concentration, is an advantage compared with difficulties sometimes encountered in emission spectroscopy.

Rains et al. [44] recommended 10% by volume glycerol as a releasing agent to overcome the inhibition effect of phosphate and sulfate on strontium. Control of hydrogen ion concentration was important, and a complete release was not accomplished until the pH was adjusted between 1 and 2 (probable formation of $H_2PO_4^-$ from the $HPO_4^=$ added originally).

Konopicky and Schmidt [36] observed a specific enhancement of aluminum in the presence of HF; enhancement was increased further by addition of butyl alcohol.

In some flames, there are departures from chemical equilibrium near the reaction zone, these taking time to disappear. Thus, sodium does not form appreciable concentrations of hydroxide, so the strength of the lines at $589.0/589.6$ mμ are fairly constant throughout the flame, but the lithium concentration is affected by the formation of LiOH. Consequently, the strength of the lithium red lines is much increased at the base of the flame by an excess concentration of free hydrogen atoms which shifts the equilibrium

$$LiOH + H \leftrightarrows Li + H_2O \tag{11}$$

to the right so that the lithium atomic lines are abnormally strong [40].

Spectral interference is minimum when an instrument with high resolution, such as the Jarrell–Ash Ebert monochromator, is used. Lithium present in high concentration exerts no interference from the proximity of its 460.3-mμ line upon the strontium line at 460.7 mμ [44]. Multiplier phototubes whose response curves extend into the near infrared are becoming increasingly available. A surface with a S-20 response (E.M.I. phototube) possesses a useful range which extends from approximately 800 mμ down to 300 mμ (or 220 mμ if a fused silica window is incorporated).

EVALUATION METHODS

The method of standard addition is well established as an evaluation method. By its use a number of radiation interferences can be overcome; it is particularly good for residual matrix effects and in trace analysis. However, limitations exist which have not always been recognized. The method requires that the working curve be linear and pass through the origin, although methods have been proposed for nonlinear working curves [5, 27], which are somewhat involved. Proper correction for background radiation must be made on both the sample and the sample plus added standard solution. Ionization effects must be stabilized by means of a radiation buffer. Relatively dilute solutions of ionizable materials present a problem if the working curve exhibits an upward concavity from the origin.

Except for dual-beam instruments designed specifically for the determination of sodium and potassium, the use of an internal standard element is rarely encountered anymore in flame spectrometry.

Alkemade [1] has pointed out that one should be careful in interpreting the working curve of chemiluminescent resonance lines. Slopes of the linear segments of a double logarithmic plot of signal against concentration do not possess the expected values.

REFERENCES

1. C. Th. J. Alkemade, Proceedings of the Xth Colloquium Spectroscopicum Internationale (Spartan Books, Washington, 1963), p. 143.
2. R. Avni and C. Th. J. Alkemade, Mikrochim. Acta 460 (1960).
3. M. R. Baker and B. L. Vallee, Anal. Chem. 31:2036 (1959).
4. B. L. Beck, Ph.D. Dissertation, University of Wisconsin (1957).
5. T. E. Beukelman and S. S. Lord, Appl. Spectroscopy 14:12 (1960).
6. H. Bode and H. Fabian, Z. anal. Chem. 170:387 (1959).
7. E. Bovalini, L. Pucini, and M. L. Conti, Ann. Chim. Roma 49:1051 (1959).
8. B. E. Buell, Anal. Chem. 34:635 (1962).
9. B. E. Buell, Anal. Chem. 35:372 (1963).
10. J. A. Dean, Flame Photometry (McGraw-Hill Book Co., New York, 1960), Chap. 5.
11. J. A. Dean, Analyst 85:621 (1960).
12. J. A. Dean and J. E. Adkins, Jr., unpublished studies.
13. J. A. Dean, J. C. Burger, Jr., T. C. Rains, and H. E. Zittel, Anal. Chem. 33:1722 (1961).
14. J. A. Dean and W. J. Carnes, Anal. Chem. 34:192 (1962).
15. J. A. Dean and W. J. Carnes, Analyst 87:743 (1962).
16. J. A. Dean and H. S. Kuper, unpublished studies.
17. J. A. Dean and J. C. Simms, Anal. Chem. 35:699 (1963).
18. J. I. Dinnin, Anal. Chem. 32:1475 (1960).
19. J. Dvorak and V. Novobilsky, Acta Chem. Hung. 30:355 (1962).
20. H. C. Eshelman and J. A. Dean, Anal. Chem. 33:1339 (1961).
21. V. A. Fassel, personal communication.
22. V. A. Fassel, R. H. Curry, and R. N. Kniseley, Spectrochim. Acta 18:1127 (1962).
23. V. A. Fassel, R. B. Myers, and R. N. Kniseley, Spectrochim. Acta 19:1194 (1963).
24. J. Fisher and A. Doiwa, Spectrochim. Acta 11:28 (1957).
25. W. H. Foster and D. N. Hume, Anal. Chem. 31:2028 (1959).
26. S. Fukushima, Mikrochim. Acta 332 (1960).
27. S. Fukushima, K. Takahashi, S. Terasaka, and K. Otozai, Mikrochim. Acta, (1960), p. 533.
28. K. Fuwa, R. E. Thiers, B. L. Vallee, and M. R. Baker, Anal. Chem. 31:2039 (1959).

29. J. H. Gibson, W. E. L. Grossman, and W. D. Cooke, Anal. Chem. 35:266 (1963).
30. P. T. Gilbert, Jr., Am. Soc. Testing Materials, Spec. Tech. Publ. 269:88 (1960).
31. P. T. Gilbert, Jr., Analyzer (Beckman Instruments, Inc.) 2, No. 4:3 (1961).
32. P. T. Gilbert, Jr., Anal. Chem. 34:1025 (1962).
33. P. T. Gilbert, Jr., Proceedings of the Xth Colloquium Spectroscopicum Internationale (Spartan Books, Washington, 1963), p. 171.
34. H. Goto and E. Sudo, Sci. Repts. Research Insts., Tohoku Univ., Ser. A 11:355 (1959); 13:282 (1961); 13:290 (1961).
35. W. E. Kaskan and R. C. Millikan, Eighth Symposium on Combustion (Williams and Wilkens, Baltimore, 1962), p. 262.
36. K. Konopicky and W. Schmidt, Z. anal. Chem. 174:262 (1960).
37. J. H. Lady, Ph. D. Dissertation, University of Tennessee (1955).
38. W. J. Maeck, M. E. Kussy, B. E. Ginther, G. V. Wheeler, and J. E. Rein, Anal. Chem. 35:62 (1963).
39. R. Mavrodineanu, Spectrochim. Acta 17:1016 (1961).
40. P. J. Padley and T. M. Sugden, Seventh Symposium on Combustion (Butterworths, London, 1959), p. 235.
41. E. Pungor, B. Weszpremy, and M. Palyi, Mikrochim. Acta 436 (1961).
42. A. Puschel and S. Eckhard, Arch. Eisenhuttenw. 30:731 (1959).
43. T. C. Rains, H. P. House, and O. Menis, Anal. Chim. Acta 22:315 (1960).
44. T. C. Rains, H. E. Zittel, and M. Ferguson, Anal. Chem. 34:778 (1962).
45. J. W. Robinson, Anal. Chim. Acta 23:479 (1960).
46. E. Schöffmann and H. Malissa, Mikrochim. Acta 319 (1961).
47. C. M. Stander, Anal. Chem. 32:1296 (1960).
48. W. B. Stavinoha and J. B. Nash, Anal. Chem. 32:1695 (1960).
49. J. C. Sternberg, private communication.
50. E. Sudo and H. Goto, Sci. Repts. Research Insts., Tohoku Univ., Ser. A 14:220 (1962); 14:231 (1962).
51. A. C. West and W. D. Cooke, Anal. Chem. 32:1471 (1960).
52. J. D. Winefordner and H. W. Latz, Anal. Chem. 33:1727 (1961).

A New Look at Light

J. W. Robinson

Ethyl Corporation
Baton Rouge, Louisiana

Light rays exhibit properties which are characteristic of waves and particles. This dualistic nature has been a major stumbling block in understanding the real nature of light. As a possible explanation it is proposed that the photon is composed of two centers of electrical energy—one positive, one negative. These charges rotate around each other giving the photon rotational energy $E = h\nu$. Movement of the whole photon gives the photon translational energy. A combination of these two energies can explain many of the dualistic properties of light. Some illustrations are given.

INTRODUCTION

The object of this paper is to describe a model of light which can be interpreted physically. It is intended that this model will fit the mathematical models described in wave mechanicals, and will also have properties that can be interpreted as wave-like and particle-like. This model explains to some extent the dualistic nature of light.

IS LIGHT A WAVE OR A PARTICLE?

A number of classical experiments have been described which give contradictory results. Some of these indicate that light is a wave form of energy. Such experiments include the observation of interference fringes, the polarization of light, and diffraction of X-rays and light. Simultaneously, equally conclusive experiments indicate that light is particulate. These experiments include Compton scattering, Einstein's photoelectric effect, and Raleigh's detection of weak gamma emission over long distances.

The conclusions drawn from the results of these experiments seem to be completely contradictory. It requires severe mental strain to accommodate these conclusions as a single physical phenomenon. However, it is equally inescapable that there is a physical explanation for all these results. The following model allows a mental picture to be drawn which will permit simultaneous wave-like and particle motion.

THE PROPOSED PHYSICAL MODEL OF A PHOTON

It has long been assumed that the photon has zero electrical charge and zero mass. However certain considerations, such as the electromagnetic character and the Compton effect, indicate that it has both charge and mass.

The model may therefore be envisioned as in Fig. 1. The representation of Fig. 1 is not intended to portray the actual physical shape of the photon, only the arrangement of the component parts.

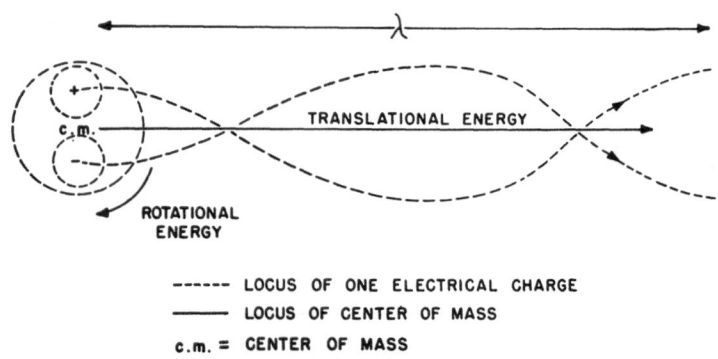

----- LOCUS OF ONE ELECTRICAL CHARGE
——— LOCUS OF CENTER OF MASS
c.m. = CENTER OF MASS

Fig. 1. Proposed model of a moving photon.

It is made up of one negative and one positive charge, each charge equal to half the charge of an electron and a positron, respectively. The charges rotate around the locus of the center of mass, each producing a magnetic field. The locus of either of these charges will describe a distorted sine wave. This produces a "wavelength" which is equivalent to the distance traveled by the center of mass during a complete revolution of the particle. The energy of this particle has two forms, i.e., translational and rotational.

Translational Energy

The velocity of the particle is approximately the speed of light. Collisions, between photons and other particles, are laestic, causing transfer of translational energy to be a fairly uncommon phenomenon. The momentum of the particle is therefore rarely involved in reactions in which this energy is transferred. However, it is probably noted in the absorption and emission of light if a change in the velocity of the particle takes place, but not in reflection or in simple collisions. The elastic collisions enable the particle to maintain a constant velocity in spite of collisions with atoms, etc. However, its electrical charges should allow some interaction with its environment causing a change in velocity in different transparent materials. We can conclude therefore that its velocity is a function of its environment. Also, unless other interaction takes place, it should be controlled by its velocity at the point of origin.

Rotational Energy

This is the energy involved in the electrical charges and gravitational forces rotating around the center of mass. The quantity of energy involved is defined by quantum mechanics as $E = h\nu$. It will be noted that when $\nu = 1$ then $E = h$, and Planck's constant is the rotational energy of the photon at unit wave number. This is of course a fundamental physical constant and would explain the importance and widespread application of Planck's constant. Transfer of rotational energy is common between the photon and other interacting bodies. The energy involved is governed by the standard equation $E = h(\nu_1 - \nu_2)$, where ν_1 and ν_2 are the wave numbers before and after absorption. After absorption, the photon may be re-emitted as radiant energy giving rise to fluorescence and phosphorescence, or infrared radiation. The transferred rotational energy probably emerges ultimately as heat. If the photon falls into the nucleus of the absorbing atom, its translational energy may also be transferred to the absorbing atom and emerge as heat.

It is further proposed that IR, visible, UV, gamma rays, and X-rays are all similar particles differing only in their rotational energy. Their physical properties, such as penetrating power, are functions of how easily this energy is transferred (absorbed) to other bodies. This would be a function of the wavelength of the photon which must be sympathetic to the absorbing material before absorption can take place. As can be seen from the model, the wavelength and the rotational energy of the photon are inversely related. Hence these energy forms IR, UV, etc., which differ only in wavelength can be interpreted as being the same particle with different rotational energies.

WAVE MOTION OF THE MODEL

There have been observed in light studies properties ascribed to wave motion. These phenomena include X-ray diffraction, interference fringes, and polarization. The behavior of the proposed model in these phenomena is described below.

X-ray Diffraction

The standard proof of the Bragg formula $n\lambda = 2d \sin \theta$ is indicated below.

A coherent wave front occurs when the photons reinforce each other, i.e., when their charges are synchronized or in phase. Using the standard Bragg proof, this occurs when $n\lambda = 2d \sin \theta$, where λ is the wavelength of the X-rays, but is interpreted here as the distance traveled during one complete revolution of the photon. When $n\lambda \neq 2d \sin \theta$, then the photons which would constitute the wave front would be out of phase. They would repel each other, and no reinforcement would take place, i.e., we would observe the absence of photons—or darkness.

Interference Fringes

It has been stated [5] that "each photon then interferes only with itself. Interference between different photons never occurs." Although this conception fits the mathematics of the phenomenon, it is hard to conceive the physical interpretation.

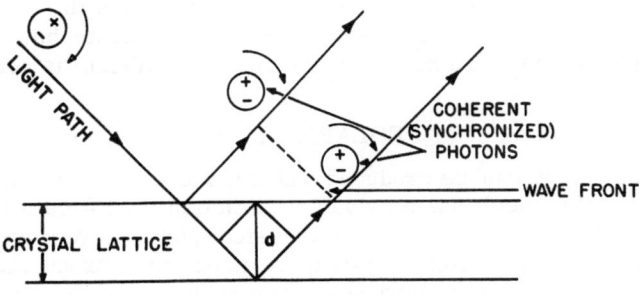

Fig. 2. X-ray diffraction.

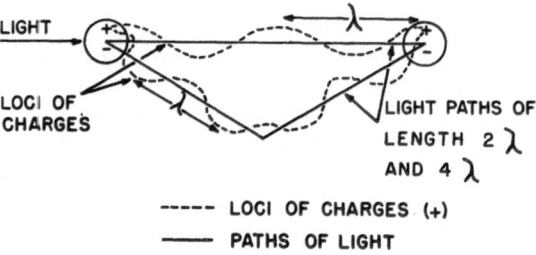

Fig. 3. Interference fringes.

Recently a significant contribution was made to the understanding of this phenomenon. Interference fringes were produced by superimposing two independent maser light beams, polarized to be parallel to each other [1]. This result supports a classical interpretation of interference fringes, i.e., by the interaction of the independent light beams, but seems to refute the idea of such fringes being caused by the interaction of photons with themselves since this type of interaction should be independent of light beams present. In short, what takes place to produce interference fringes is not yet clearly understood but the data can be interpreted in terms of the model.

In order to explain their occurrence it is proposed that a light beam is composed of groups of photons, coherent with themselves, but incoherent with other groups—even of the same wavelength. When such a group of photons reaches a beam splitter, the group is divided into two paths in the conventional way and later recombined. With light in wave motion, interference fringes are observed under these conditions; reinforcement takes place when the path length difference is $n\lambda$, but interference (or dark rings) occurs when the path difference is $(n + 1/2)\lambda$. In the model, it can be seen that when the path length difference is $n\lambda$, then the photons come together in phase and attract each other producing a light fringe. When the path length difference is $(n + 1/2)\lambda$ the photons are out of phase and repel each other. The absence of photons produces a dark ring. This is illustrated in Fig. 3.

It will be noted in the production of interference fringes using superimposed laser beams that these beams were polarized parallel to each other. The necessity of this step could have been predicted from the model described in this article, since this model requires that the charges be correctly oriented in three dimensions before interaction takes place. Further, information could be obtained on the degree of interaction between the photons by decreasing the synonymity of the polarization between the two beams until no further interference was noted. This would indicate how much out of phase the photons must be before interference (repulsion) is noted.

POLARIZATION

Polarized light can be produced under numerous conditions, e.g., using various crystals, nicol prisms, double refraction, light-scattering, and reflection. For this phenomenon the classical pictorial representation is of light waves vibrating at right angles to the direction of propagation. Polarization takes place when the light is separated into two beams, the vibrations of which are in planes which are mutually perpendicular.

With the proposed model we can conceive a similar situation. The direction of propagation of the light is the direction of translational motion of the center of mass. The vibration noted in the classical model is reproduced by the rotating charges of the photon. As with classical polarization vibrations, the direction of the rotation of the electrical charges on the proposed model is not specified, but can be resolved into two directions mutually perpendicular to each other and to the direction of translational motion of the center of mass. The properties of polarized light can be interpreted in terms of the proposed model by substituting this model for the classical vibrating light wave used in current interpretations.

PARTICLE MOTION OF THE PROPOSED MODEL

The most accepted proofs of the existence of light as a particle include Compton scattering, Einstein's photoelectric effect, and Raleigh's work with γ rays. The behavior of the proposed model under these conditions is given below.

Compton Scattering

It was observed by Compton that collisions took place between light particles and electrons.

From the angles of scatter of the photon and the electron, Compton calculated that the energy of the photon was $h\nu + h\nu/c$, where $h\nu/c$ was the momentum of the photon. It is important to note that Compton had to include a term for the momentum of the photon in his calculations. Substituted in the proposed model for the photon, it can be seen that $h\nu$ is the rotational energy of the photon and $h\nu/c$ is its momentum.

Further, if the momentum is considered in classical terms, it is $m \times c$ where m is the mass of the photon and c is its speed.

Comparing this with Compton's calculations,

$$h\nu/c = mc$$

or

$$h\nu = mc^2$$

or

$$E = mc^2$$

As indicated, this relationship develops into Einstein's mass and energy relation. It also indicated that the mass is proportional to ν, the wave number. This shows the particle has variable mass. However, it may only indicate that the forces used to measure mass vary under different electrical

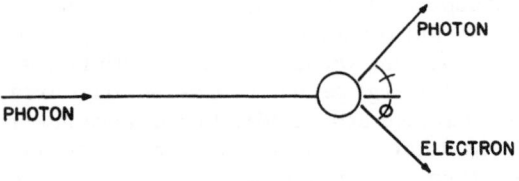

Fig. 4. Compton scattering.

environments. This difficulty points up our lack of understanding of the real significance of mass and energy.

Photoelectric Effect

It was observed by Einstein that if a metal surface is negatively charged, then irradiated with UV light, that electrons are emitted. The required wavelength of the light is a function of the metal used for the surface. Einstein proved the relationship $h\nu_0 = \phi_e = Kb_0$ where ν_0 is the minimum wavelength of the light and ϕ is the work function to release the electron. According to wave theory the excess energy of the emitted electrons should be a function of I_0, the intensity of the UV light. It was found, however, that the rate of emission of electrons was proportional to I_0, but that their K.E. was a function of the wavelength of the light, as $1/2\, mv^2 = Ve = h\,(\nu - \nu_0)$.

If $h\nu > h\nu_0$ emission takes place. This was considered proof of the particulate form of light, since if it was a wave form it would dissipate its energy with increasing distance from the source and would be dependent on I_0. This photoelectric effect has also been observed with X rays and γ rays.

If we substitute our model for the photon, we see that its rotational energy is involved. When this energy is in excess of the threshold energy of the electrons, emission takes place. Unless prior absorption takes place, it is independent of the distance traveled by the photon from its source. Also we see that the excess rotational energy of the photon is translated into kinetic energy of the electron, and the equation $1/2\, mv^2 = Ve = h(\nu - \nu_0)$ still applies.

Further, when I_0 is increased, the number of photons making up the light beam is increased. This in turn would liberate more electrons from the metal surface per unit time but would not change the energy of each electron. This interpretation fits in with the observed experimental evidence.

Rutherford's Observations

Further amplification of the photoelectric effect was made by Rutherford. He calculated that the energy emitted as γ rays from 1 g of radium is about 4.7×10^4 ergs. At a distance of 100 m, it should require 10^{15} sec to build up sufficient energy to be detected, according to wave theory. However these rays were in fact observed immediately. His evidence strongly supports the contention that light is particulate and not a wave form, since light in a wave form would have greatly dissipated its energy over this distance.

Interpretation in terms of the proposed model is as follows. Its energy would remain constant until some interaction with another body took place. This would be independent of the distance traveled. Hence the energy of individual photons would remain constant, however, it must be remembered that a beam of such photons would spread out with increasing distance from the source. The total energy per unit cross-sectional area of the light would diminish with this distance and would follow the laws governing wave motion. Hence, although the total energy per unit cross-sectional area diminishes according to wave form energy, the energy in each particle remains constant until interaction takes place.

RELATIONSHIP BETWEEN PHOTON, ELECTRONS AND POSITRONS

It is known that electrons and positrons react to form two γ rays. This suggests a relationship between electrons, positrons, and photons. It is proposed that this relationship is as follows:

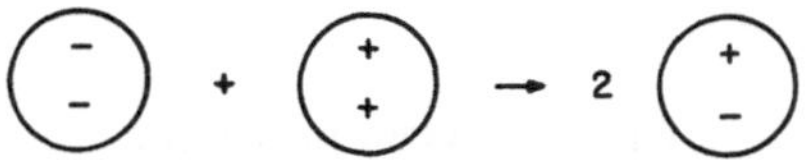

ELECTRON + POSITRON → 2 γ RAYS

Fig. 5. γ rays from electrons and positrons.

Since two γ rays are produced from one electron and one positron, we have produced two $\oplus$ and $\ominus$ charges from the same source. In order to maintain constancy of electrical charge, it is suggested that the electron and positron are each composed of two charges rotating in the same fashion as the photon. It must be pointed out, however, that the charge and size of the electron are such as to produce tremendous repulsive forces at this close proximity. This would normally preclude the existence of such a particle.

However, it may be that the two halves of the electron have different magnetic properties towards each other, than the combined pair have towards other bodies. It is also conceivable that these halves are not actually separated from each other in the electron but provide two centers of negative charge rotating about the center of mass. In this case no strong mutual repulsion would take place within the electron—only a distortion of its shape and the formation of a "duo-pole"—a phenomenon similar to dipole but with equal, similar charges.

Under these conditions the electron should exhibit wave motion similar to the photon. Some experimental evidence supporting this view is listed below. According to Bohr theory, $2\pi r m v = nh$. De Broglie suggested that the electron had wave motion, and further stated that it was inconceivable that the orbital path was anything but a whole number of wavelengths, i.e.,

$$n\lambda = 2\pi r$$

hence

$$m v n \lambda = 2\pi r m v = nh$$

or

$$\lambda = h/mv = h/p$$

where p is the momentum of the electron. It will be remembered that Compton's value for the momentum of a photon was $h\nu/c$, i.e.,

$$p \text{ (photon)} = h/\lambda \text{ (Compton effect)}$$

$$p \text{ (electron)} = h/\lambda \text{ (de Broglie wavelength)}$$

This indicates a strong relationship between the mass and energy characteristics of photons and electrons.

The model proposed for an electron would exhibit a de Broglie wavelength. The locus of the center of mass would be the electronic orbit, the locus of either of the rotating charges would have wave-like properties, with a wavelength equal to the distance traveled during one complete revolution of the electron.

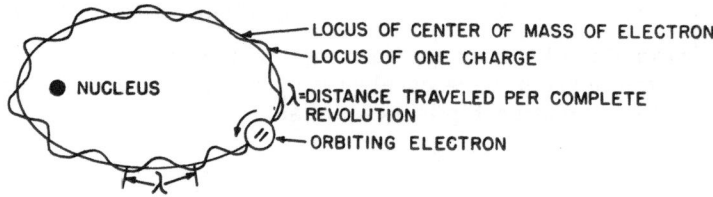

Fig. 6. De Broglie wavelength.

This suggests atomic structure with a pulsating electronic field set up by the electrons; for stability this pulsing must be physically and mathematically reproduced during each revolution of the electron. This would require the de Broglie relationship $n\lambda = 2\pi r$. With the proposed model, the rotating electron would provide a pulsating magnetic field, which would be a function of the "duo-pole" charge of the electron, its wave number, and the velocity of the center of mass. The rotation of the "duo-pole" would produce a pulse with wave-like characteristics and would be in sympathy with this classical work.

The Schrodinger wave equation is based on harmonic or wave motion of the electron—as before this can be considered to be based on the properties of the rotating charge of the electron and a physical interpretation of the process can be obtained. In contrast, the Bohr theory was based on the locus of the center of mass of the electron, since only the energy, or locus, of the electronic orbits were considered.

Other supporting evidence for the wave-like character of electrons is found in electron microscopy where many of the properties of light are found, including diffraction phenomena.

PROPERTIES OF THE PROPOSED MODEL OF THE PHOTON

The proposed model of the photon is basically a Newtonian particle, composed of two rotating electrical charges. From this it should be possible to deduce its properties. Examples of some of these deductions are given below.

It should be mentioned at this point that the mathematical interpretations of wave mechanics should be strictly adhered to. This thesis is intended only to present a physical picture of light particles which may have particulate and wave-like properties. The sustaining of wave mechanical mathematics however is necessary if the model is valid. Happily no contradictions have yet been noted.

The presence of the rotating charges would set up intense local magnetic fields. This may be a possible explanation for the intense absorption exerted by very low concentrations of atoms and molecules, the basis of many analytical techniques. An attractive magnetic field is suggested between orbiting electrons and nearby photons. Of course the photons and electrons would have to be in phase and in the same frequency or some order of frequency, in order to interact and be absorbed by the atom. The absorption itself may be due to the temporary formation of a combined electron-photon particle, which moves to a higher energy state giving an "excited" atom. Emission spectrography suggests that the life time of this combined particle is between 10^{-6} and 10^{-8} sec. After this time the photon is re-emitted. If, however, the photon is transferred to another electron, the original electron would descend to the ground state without emission, but the photon would be emitted later from another part of the molecule. This is the basis of fluorescence and phosphorescence. The intense local field of the photon would also cause mutual attraction between synchronized photons and repulsion between unsynchronized photons. Bunches of photons would result, coherent in themselves, but not coherent with other bunches. Such bunches may explain the blinking of stars, where the great distance traveled may enable substantial separation of these bunches from each other. Also, the size of the bunches may increase by picking up stray photons in space.

It is observed with interference fringes that the rings are not sharp and that the total energy involved remains constant. This suggests that some degree of unsynchronization between bunches of photons and approaching photons can be tolerated. If it is small enough, the phases of the photons concerned may be mutually adjusted to become compatible.

Information on this point may be obtained by superimposing two parallel maser beams to produce interference fringes. The two beams should then be made increasingly nonparallel until no interference fringes are obtained. The angle between the beams should be related to the lack of synchronization of the photons.

Although there are intense local magnetic fields in the photon, the effective radius would only be of the same order of magnitude as atoms. External magnetic fields would have little effect because both the positive and the negative charges involved are oscillating around each other and would produce a net zero charge to a remote superimposed magnetic field. This would explain the apparent lack of response of light to magnetic and electrical fields, although the Zeeman effect indicates that some interaction takes place, possibly between bunches of photons and the superimposed magnetic field.

Some proof of the existence of charged photons should be forthcoming using continuous coherent light—as emitted from a laser. Coherent light should be light in which the photons are rotating in phase. If such a light source was trained on a small metal surface, as in Fig. 7, light from the laser should arrive at the metal surface in a steady state of orientation between the charges on the photon and the metal surface. If the distance between the metal surface and the laser source is kept constant and the coherency of the light maintained, the metal surface would be struck continuously by one charge, e.g., the negative charge of the photon. If now the metal surface is charged positively, the photon should be attracted and absorption should take place. If the surface is negatively charged repulsion, or reflection, should occur. Similarly alternate absorption and reflection should occur

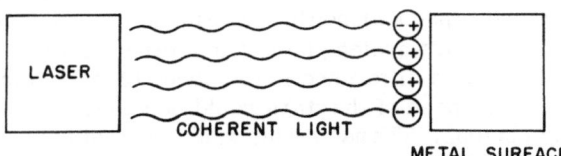

Fig. 7. Coherent light focused on a metal surface.

if the charge on the metal surface is kept constant, but the distance from the coherent source is varied. These alternate reflections and absorptions should be a function of the wavelength of the light. One problem may arise with this experiment. The effective radius of each charge on the photon is limited, since it is counterbalanced by the opposite charge at remote points; hence, interaction is only over a small distance. A negative charge on the metal surface is in effect an excess of electrons. In this case the reaction should be as predicted. However a positive charge may be only a depletion of electrons on the surface. This reaction may not be as predicted. The photons may interact with the residual electrons rather than the shielded positive nuclei of the metal and a result other than that predicted would result.

It will be noted that the model of the photon has mass. This conclusion was also implicit in calculations derived from Compton scattering. It will be noted that these calculations lead to the momentum $h\nu/c$ being a function of the wavelength of the particle. As mentioned earlier, when $h\nu/c$ is equated to mc, Einstein's equation $E = mc^2$ is derived. An alternate proof of Einstein's equation may be presented at this point.

Supposing an unstable atom is about to emit a γ ray from its nucleus. The γ ray is at rest before emission, but travels at the speed of light after emission. Its gain in kinetic energy is therefore $1/2\, mv^2$ or $1/2\, mc^2$. Simultaneously the recoil experienced by the residual atom is also $1/2\, mc^2$. The total energy produced is therefore mc^2. Hence the amount of energy produced on the loss of mass m from the nucleus is E and is equal to mc^2, i.e., $E = mc^2$. This calculation of course assumes that the photon has virtually no kinetic energy when in the nucleus of the atom. There is of course no proof for this assumption.

THE DOPPLER EFFECT

If light is a particle as proposed in this thesis, it is necessary to provide an alternate explanation for the "Doppler" effect. This effect—originally observed with sound—states that the velocity of the wave through the transmitting medium is constant, the waves are emitted at constant frequency, but to an observer in the path of the approaching emitter, the waves appear to be bunched closer together and hence arrive at increased regularity, i.e., at greater frequency. With the proposed particle the same apparent change in frequency will be observed if the velocity of the emitter is superimposed on the velocity of the photon. If the frequency at the point of emission remains constant, but the velocity relative to the receiver is increased, then the number of pulses passing the observer per unit time will increase. This again results in an apparent increase in frequency, a result similar in all respects to the Doppler effect, but caused by a change in velocity rather than wavelength. Use of this phenomenon has been made in measuring the relative speed of the stars by the wavelength shift in their spectrum. These results could be interpreted as a difference in the speed of light relative to the observer. The relative speed of the stars would be superimposed on the speed c of the emitted photons. The net result would be a shift in the spectrum of the stars from which the relative velocity of the stars could be calculated as before.

RELATIONSHIP WITH RELATIVITY THEORY

The development of the Relativity Theory and in particular the Special Relativity Theory is intimately concerned with the speed of light. It is based on the observation that the speed of light is always c and is independent of the relative motion of the emitter and the receiver. The current views of light, i.e., as waves or photons, give no reason why this should be so. The model proposed here would also give no reason to expect this observation. The particle is Newtonian in nature and should travel at a constant velocity unless acted upon by a superimpressed force. It is conceivable that there is an interaction between the magnetic field of the photon and its environment, giving rise to an interaction akin to Fresnel drag. However, movement of the observer should still make a difference to the relative velocity of the photon contrary to observed accepted data. This problem has frustrated physicists for a long period of time. Although Einstein's treatment mathematically resolves the problem, it is difficult to accept as a physical reality.

Einstein's relativity theory and the special relativity theory are based on two axioms, i.e., if two observers are moving in a straight line relative to each other, they cannot tell who is moving; and, if both measure the speed of light, the same answer is obtained. From this is evolved the clock paradox that time is different for these observers, and that $\tau = t\sqrt{1 - v^2/c^2}$.

The proposed model says nothing about relativity. Being Newtonian in nature, it should obey classical laws. Its velocity should not necessarily be c but $c + V$, where V is the relative velocity of the emitting particle and c is the velocity of emission relative to the source.

To suppose that the speed of light is always c defines an absolute velocity. By difference (c – c) is absolute rest. Since it is not possible to identify absolute rest it must be equally impossible to identify an absolute velocity.

Recently Dingle [2] pointed out an apparent inconsistency in the special theory of relativity. This of course met with a storm of protest, but no concrete evidence that he was wrong. Other attempts to confirm the inequivocability of c have been made [3,4]. It was concluded from each of these that the speed of light was confirmed as c. However, there was a weakness in each case. In the former [3], two γ sources were used, one recoiling and one not recoiling. It was hoped to demonstrate that the difference in velocity of the source was superimposed on the velocity of the emitted γ ray. However, it was assumed that one source was in a state of recoil and the other was not. This is tantamount to assuming the result, since it can be argued equally successfully that the results indicate no differences in velocity between the sources.

The second experiment uses a source of white light which passes through a rotating semicircular glass disc. It is claimed that this gives rise to light from a source with varying velocity. However, since the actual source is not moved, only the varying Fresnel drag of the rotating semicircular disc is superimposed on the light rays. As indicated in the article, this difference would not be detectable. However, we would not expect any other difference to be obtained since the actual source itself is not moved.

CONCLUSION

This model of a photon presented in this thesis provides a tangible understanding of many of the contradictory properties of light. The usefulness of any model is to provide a starting point which may be successively tested and improved until an ultimate truth is developed. It is hoped that in some way, this model will be useful in this respect.

ACKNOWLEDGMENT

The author wishes to thank Ethyl Corporation for permission to present and publish this article.

REFERENCES

1. G. Magyar and L. Mandel, Nature 198, 225 (1963).
2. Dingle, Nature 195, 985 (1962).
3. T. Alvager, A. Nilsson, and J. Kjellman, Nature 197, 1191 (1963).
4. J. F. James and R. S. Sternberg, Nature 197, 1192 (1963).
5. P. A. M. Dirac, Quantum Mechanics, (Clarendon Press, Oxford, 1958), 4th edition.

Spectrophotometric Observations of Hydrogen Arc-Jets

R. L. Fox

Research Associate
McDonnell Aircraft Corp.
St. Louis, Missouri

The object of the work reported herein was to study the applicability of spectrophotometric methods for determining the physical and chemical characteristics of hydrogen arc-jets operating in the power range of 20 to 42 kW. The spectrum was observed from 2500 to 7000 A. No molecular spectra were detected in these spectral scans, nor were there any measurable impurities present in the plume. The visible plasma radiation was emitted at the wavelengths of the hydrogen Balmer lines, but the continuum beyond the series limit was not detectable. The excitation temperatures at various points ranged from 2000 to 15,000 K, with the hottest temperature being in the arc region of the engine. The experimental temperatures were higher than the calculated bulk average temperatures. Electron densities in the plume were less than 10^{13} electrons/cm^3 and in the arc about 10^{16} electrons/cm^3. The results of this investigation indicate that (1) the population of excited states obeys Boltzmann's law; (2) the plasma is optically thin in the visible region; and (3) apparently the plasma is not in chemical equilibrium.

INTRODUCTION

An arc-jet is an engine for converting electrical energy into kinetic energy. This is done by passing a gas through an electric discharge, thus causing a great rise in the temperature. The heated gas is then passed out the exit nozzle. An expansion of the hot gas while it is moving through the exit nozzle is accompanied by a large increase in the velocity. This increase in the velocity causes a corresponding increase in momentum, which in turn exerts a force on the engine, the magnitude of which is given by Newton's second and third laws of motion. Hydrogen is commonly used as a propellant in these engines because of the greater specific impulse attainable due to its smaller atomic mass.

Various diagnostic techniques are being used at McDonnell Aircraft Corporation to analyze the physical and chemical properties of these engines in order to provide further understanding of the arc phenomenon and to improve performance characteristics. The spectrometric method is one of these techniques.

This report describes an investigation of the applicability of spectrophotometric methods for determining the physical and chemical characteristics of hydrogen arc-jets operating in the power range of 20 to 42 kW.

THEORETICAL CONSIDERATIONS

Teller–Inglis Electron Density Method

This method is based on the broadening of spectral lines due to the electric microfields of the free electron. The number density of the free electrons is found by determining the last member of the Balmer series which can be resolved; the quantum number of the last distinguishable line is related to the electron density by

$$\log_{10} N_e = 23.3 - 7.5 \log_{10} n_{max} \tag{1}$$

To use this method, the electron densities must be at least 10^{13} electrons/cm^3. If it is assumed that the gas is in thermodynamic equilibrium, then the electron density is related to the unique plasma temperature by the Saha equation [1]

$$\frac{N_{(r+1)}}{N_{(r)}} = \frac{Z_{(r+1)}}{P_e Z_{(r)}} \frac{2\sqrt{8\pi^3 m_e^3 k^5 T^5}}{h} \exp(-x_{(r)}/kT) \tag{2}$$

Plume Region Temperature Measurement

The spectroscopic temperature measurement method used for the plume region was the method of relative line intensities. This method was chosen not only because it gives a numerical temperature, but also because it provides a partial check of the validity of that temperature.

This method, in common with most spectrographic temperature measurement methods, begins with the equation for the time-averaged spectral line intensity of an optically thin plasma [2],

$$I_{nm} = \mathcal{N}_n A_{nm} E_{nm} \tag{3}$$

E_{nm} can be found using Planck's law, and $\mathcal{N}_n$ can be found (with assumption of equilibrium) by Boltzmann's law [1]:

$$\mathcal{N}_n = \mathcal{N}_0 [g_n/U(T)] \exp(-E_n/kT) \tag{4}$$

It should be noted that fulfillment of Boltzmann's law is not proof of thermodynamic equilibrium—but if thermal equilibrium is obtained, then Boltzmann's law is certainly obeyed.

Instead of the Einstein A-value, the oscillator strength for emission is often used. These quantities in cgs units are related by

$$A_{nm} = (8\pi^2 e^2/m_e c\lambda_{nm}^2)f_{nm} \tag{5}$$

Substituting this into the original equation gives

$$I_{nm} = \left(\frac{8\mathcal{N}_0 h\pi^2 e^2}{U(T)m_e}\right)\frac{g_n f_{nm}}{\lambda_{nm}^3} \exp(-E_n/kT) \tag{6}$$

For the Balmer series, it is more convenient to use absorption oscillator strength rather than emission oscillator strength (in order to make the statistical weight constant). Doing this, and taking the logarithm to the base 10, gives:

$$\log_{10}(I\lambda_{nm}^3/f_{mn}) = \text{constant} - (5040/T)E_n \tag{7}$$

Thus a plot of $\log_{10}(I\lambda_{nm}^3/f_{mn})$ should be a straight line, the slope of which is inversely proportional to the temperature. As can be seen, this provides a partial check on the original assumptions, which are:

1. The population of excited states obeys Boltzmann's law, with a unique temperature for the plasma.
2. The plasma is optically thin.
3. The plasma environment does not affect the oscillator strength, as calculated by nonrelativistic quantum mechanics.

It should be noted that even if the plasma were not in thermodynamic equilibrium, it would still be possible for a combination of effects to give a straight line on a plot of the type discussed above. However, the temperature then given by the relative line intensity method would not necessarily be the only temperature of the plasma.

Arc-Region Temperature Measurement

If it has been established by such temperature plots that the population of the excited states is given by Boltzmann's law, then another temperature method can also be used in hotter arc regions of the flame. This other method required that the plasma obey Saha's equation with an ionization temperature equal to the excitation (Boltzmann) temperature. We now consider this Fowler–Milne method.

Figure 1 represents a cross section of the plasma, and shows the usefulness of the concept of the plasma as concentric layers of gas. Each concentric layer is in thermodynamic equilibrium within itself, but the temperature changes from layer to layer. This condition is a type of mild nonequilibrium, which is called local thermodynamic equilibrium.

If the plasma is thin and radially symmetric about its axis, then the radiation as seen by the spectrometer is the sum of the radiation emitted by all the plasma layers which lie along the line of sight and is given by

$$F(x) = 2 \int_0^Y I(r)\,dy \tag{8}$$

In order to find the actual intensity emitted per unit volume at radius r, it is necessary to solve equation (9) for $I(r)$. This transformation gives

$$I(r) = -\frac{1}{\pi} \int_r^R \frac{F'(x)}{\sqrt{x^2 - r^2}}\,dx \tag{9}$$

The second step in the Fowler–Milne method is to combine Saha's equation with the Boltzmann law and then use this combination to give the $\mathcal{N}_n$ value in the formula for the spectral line intensity [equation (2)]. Thus the intensity of a line emitted by a volume of plasma is given as a function of temperature $I(T)$.

These two functions, $I(r)$ and $I(T)$, can be combined to give the temperature as a function of plasma radius [3].

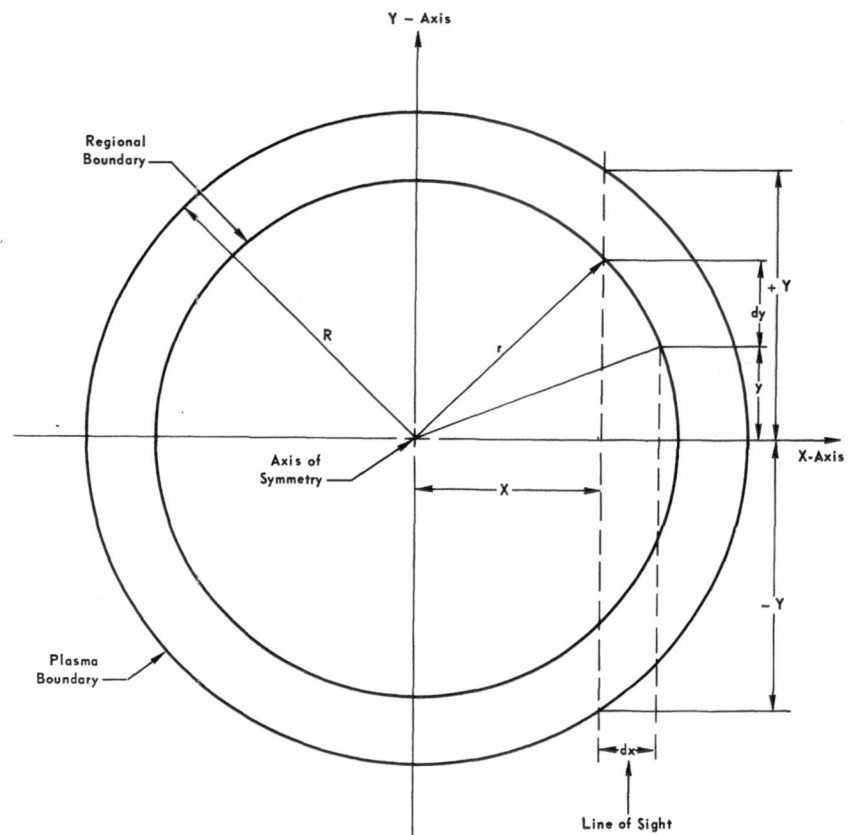

Fig. 1. Geometry of the plasma cross section.

Model—Equilibrium Assumed

If it has been established that the method of relative line intensities will give a temperature, then it is necessary to establish that this is the plasma temperature. Perhaps the best place to start this analysis is to assume a very simple mathematical model and then to test the model by applying the conservation laws. It is assumed that:

1. The arc region is in local thermodynamic equilibrium.
2. The net energy given to the gas results in an increase of the translational energy of the monatomic gas (where net energy is the energy after dissociation into monatomic hydrogen).

The temperature at the nozzle exit is then related to the arc temperature by assuming an adiabatic expansion from the arc to the nozzle exit. Then

$$\frac{T_A}{T_P} = \left(\frac{R_P^2 \, V_P}{R_A^2 \, V_A}\right)^{(C_p/C_v)-1} \tag{10}$$

The expansion as a whole is not adiabatic, because energy is lost at the nozzle walls; however, equation (10) should still be applicable to the temperature along a single stream line which is near the plasma center.

Model—Nonequilibrium Consideration

If the plasma is not in local thermodynamic equilibrium it might still be possible to have a Boltzmann distribution of bound electrons dominated by the temperature of the electrons.

Also, if the plasma is not in local thermodynamic equilibrium but if a species of plasma particles were to have a Maxwellian velocity distribution, then the plasma could be described as having an individual specie temperature.

In the case of a slightly ionized hydrogen plasma, the principal particles are electrons and neutral hydrogen atoms. If it is assumed that the electrons and atoms each have a Maxwellian velocity distribution, but are not at the same temperature, it is of interest to see under what conditions the electrons dominate the population of excited states of the atoms.

A mathematical model of such a plasma will be developed using the following assumptions:

1. The principal means of energy transfer is by two-body collisions.
2. The probability of an inelastic collision is approximately equal for electron–atom collisions as for atom–atom collisions.

The ratio of atom–atom collisions to electron–atom collisions which have enough energy to cause excitation will provide a criterion for the conditions under which the electron translational temperature can dominate the Boltzmann temperature.

The velocity distribution of the individual specie is given by [5]

$$f(v)dv = \left(\frac{m}{2\pi kT}\right)^{3/2} \exp(-mv^2/2kT)v^2 dv \qquad (11)$$

The only collisions which could shuffle the excited atomic states are collisions with electrons or other atoms which have kinetic energy great enough to cause excitation. The collision frequency for electron-atom collisions is

$$\nu_{e,a} = N_a \hat{N}_e (\bar{v}_e^{\,2} + \bar{v}_a^{\,2})^{1/2} \left(\frac{d_e + d_a}{2}\right)^{1/2} \qquad (12)$$

Let A be the minimum amount of energy which could cause a change in the internal energy of the atom. Then the minimum speed which an electron must have in order to collide inelastically with an atom is

$$v_0 \geq \sqrt{\frac{2A}{m_e}} \qquad (13)$$

Therefore, combining equations (11) and (13) will give the number of electrons which are in a unit volume of the plasma and which could, upon collision with an atom, cause an increase in the atom's internal energy:

$$\widehat{N}_e = 4\pi N_e \left(\frac{m_e}{2\pi kT}\right)^{3/2} \int_0^\infty v_e^2 \exp(-m_e v_e^2/2kT)\,dv_e \tag{14}$$

$$= 2N_e \left\{\frac{v_0}{a_e} \frac{\exp[-\frac{1}{2}(v_0/a_e)^2]}{\sqrt{2\pi}} + \operatorname{erfc}\left(\frac{v_0}{a_e}\right)\right\}$$

where $a_e = \sqrt{m_e/kT}$.

The average speed of the electrons which have energy greater than A is

$$\overline{v}_e = \frac{\displaystyle\int_{|\vec{v}|\,\geq\,v_0} |\vec{v}_e|\,f_e(\vec{v})\,d\vec{v}}{N_e} \tag{15}$$

Therefore, combining equations 11, 14 and 15 gives the desired electron-atomic collision frequency:

$$\nu_{e,a} = 2N_a N_e \pi \left\{\frac{v_0}{a_e} \frac{\exp[-\frac{1}{2}(v_0/a_e)^2]}{\sqrt{2\pi}} + \operatorname{erfc}\frac{v_0}{a_e}\right\} \times$$

$$\times \left\{\frac{8kT}{\pi m_a} + \frac{4a_e^2[\frac{1}{2}(v_0/a_e)^2 + 1]\exp[-(v_0/a_e)^2]/2\pi}{(v_0/a_e)\exp[-\frac{1}{2}(v_0/a_e)^2](2\pi)^{1/2}\operatorname{erfc}(v_0/a_e)}\right\} \tag{16}$$

The atomic-atomic collision which causes excitation could be calculated in a similar manner. The ratio of these collision frequencies for hydrogen is then given by

$$\frac{\nu_{H,H}}{\nu_{e,H}} = \frac{N_H \Phi_H^{1/2} d_H^2 \Psi_H^2}{N_e \Phi_e^{1/2} \left(\dfrac{d_H + d_e}{2}\right)^2 \Psi_e^2} \tag{17}$$

where

$$\Phi = \Omega\,\frac{\exp(-\Omega/2)}{\sqrt{2\pi}} + \operatorname{erfc}\Omega$$

$$\Psi = \frac{8kT}{\pi m} + \frac{4a^2(\frac{1}{2}\Omega^2 + 1)^2\exp(-\Omega^2)}{2\pi\Phi^2}$$

and $\Omega = v_0/a$, where a has the subscript e or H, depending on the subscript on Φ and Ψ.

A graph of this ratio (holding everything constant except the electron temperature) is given in Fig. 2.

EXPERIMENTAL APPARATUS AND PROCEDURE

The spectrometer used in the spectrophotometric observations was a Baird Atomic three-meter concave-grating spectrometer, having a modified

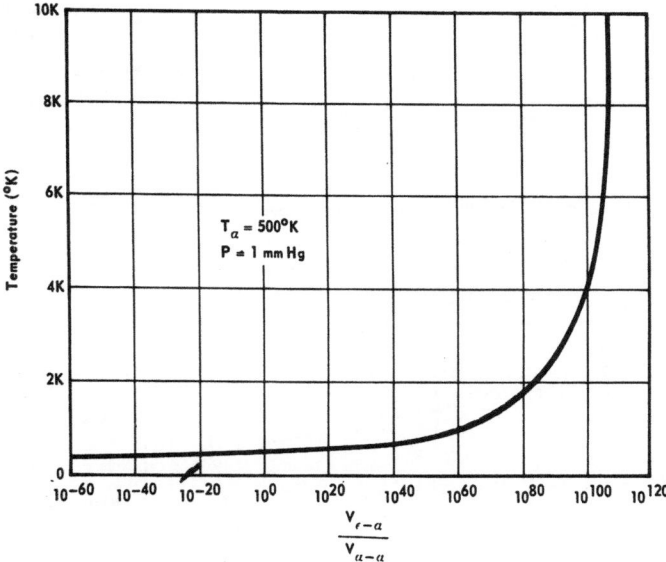

Fig. 2. Ratio of atom–atom to electron–atom collision which could cause
excitation.

Eagle-mount design. The grating has 15,000 lines/in., which gives a dispersion of 5.6 A/mm in the first order. The spectrum was scanned from 2500 to 7000 A, using RCA 1P21 or 1P28 photomultipliers. The phototube response was recorded on a strip-chart recorder.

The arc-jet exhausts into a vacuum chamber which is maintained at the pressures of 0.3 to 0.8 mm Hg. A diagram of the experimental system is shown in Fig. 3.

There were two principal regions of the engine which were studied: the arc region, where the electric discharge transfers energy to operating gas, and the plume region, which is the flame resulting as the hot gases leave the exit nozzle. Figure 4 is a spectrogram of part of the Balmer series for these observed conditions; the top seven exposures were taken in the plume region and the bottom seven exposures were taken in the arc region. Figure 5 shows an arc-jet with windows in both the arc and the plume.

The plume spectrum was scanned with a photomultiplier and the signal was recorded on a strip-chart recorder. The photomultiplier was calibrated by similarly observing the spectrum from a tungsten-strip lamp. The effective temperature (at 6500 A) of the tungsten-strip lamp was found with an optical pyrometer, and the surface temperature of the tungsten filament was found by correcting the observed temperature for the emissivity of tungsten by Planck's law [4]

$$T = \frac{C_2/\lambda}{\ln\{\epsilon(\lambda T)\,[\exp(C_2/\lambda\,T_{pyro}) - 1] + 1\}} \tag{18}$$

where $C_2 = hc/k$.

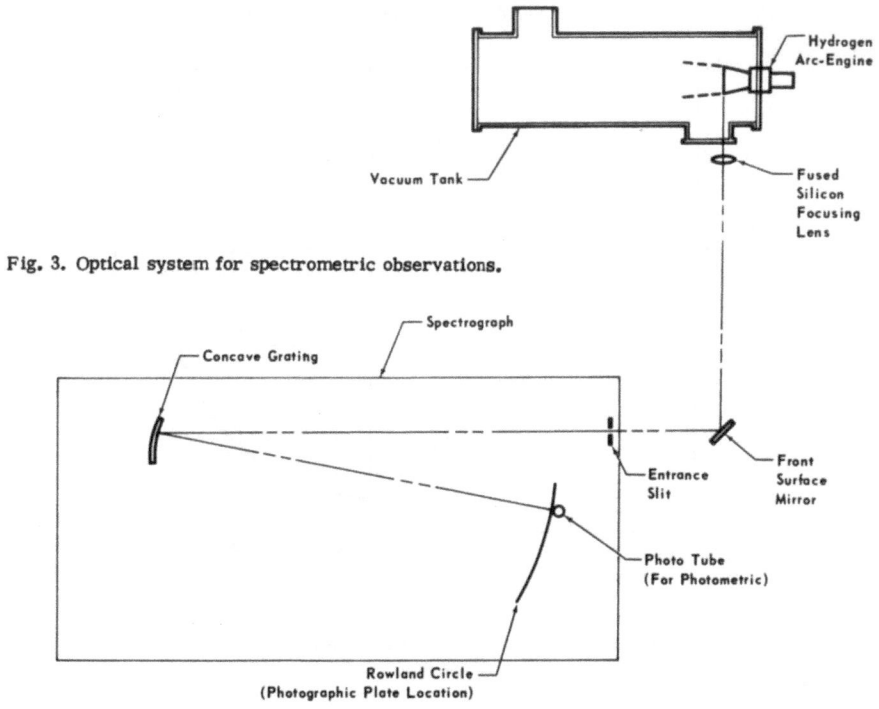

Fig. 3. Optical system for spectrometric observations.

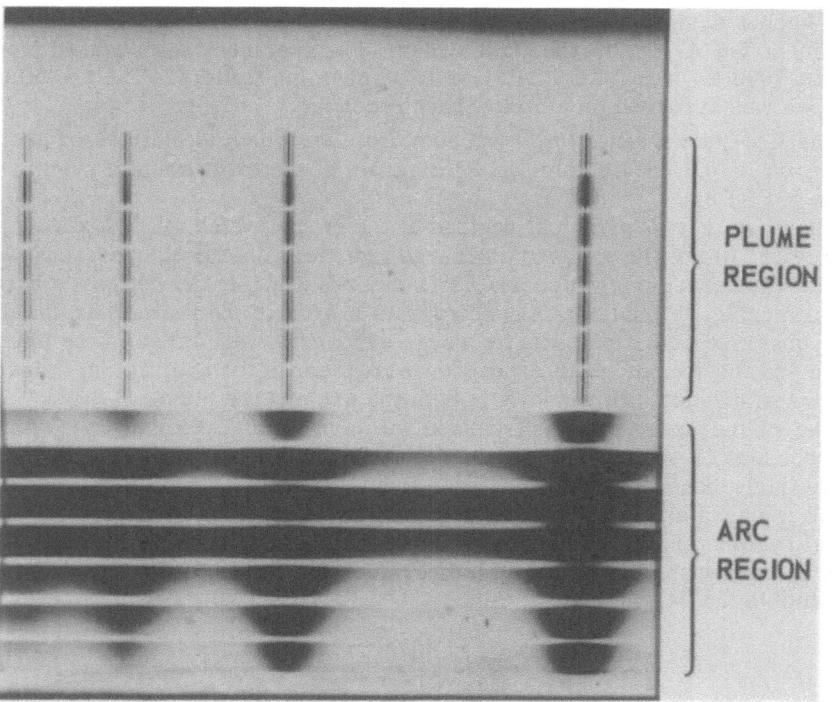

Fig. 4. Spectrogram of part of the Balmer Series.

Fig. 5. Arc-jet exhausting into vacuum chamber.

The strip-chart recordings of the spectrum were then planimetered. The relative intensity of the spectrum line was then found by

$$I_{line} = \frac{\epsilon(\lambda_1 T) B_\lambda(T)}{R_{lamp}} R_{line} \qquad (19)$$

DISCUSSION OF RESULTS

Electron densities, as estimated by the Teller–Inglis formula, equation (1), were below 10^{13} electrons/cm³ in the plume and were about 10^{16} electrons/cm³ in the arc.

Figures 6 and 7 show typical results of temperature plots described earlier from photoelectric data for the hydrogen jet plume. As can be seen, the experimental results shown on these graphs fit straight lines. The excitation temperatures ranged from 1600 to 5000 K in the plume at the nozzle exit, and ranged from 10,000 to 16,000 K in the arc.

There was no observable molecular spectrum nor was there any continuous spectrum beyond the series limit. This implies that the plasma is in that state of chemical nonequilibrium known as "frozen flow." A stagnation probe was inserted into the plasma in order to bring it to equilibrium in the shock region at its tip. This resulted in causing the molecular

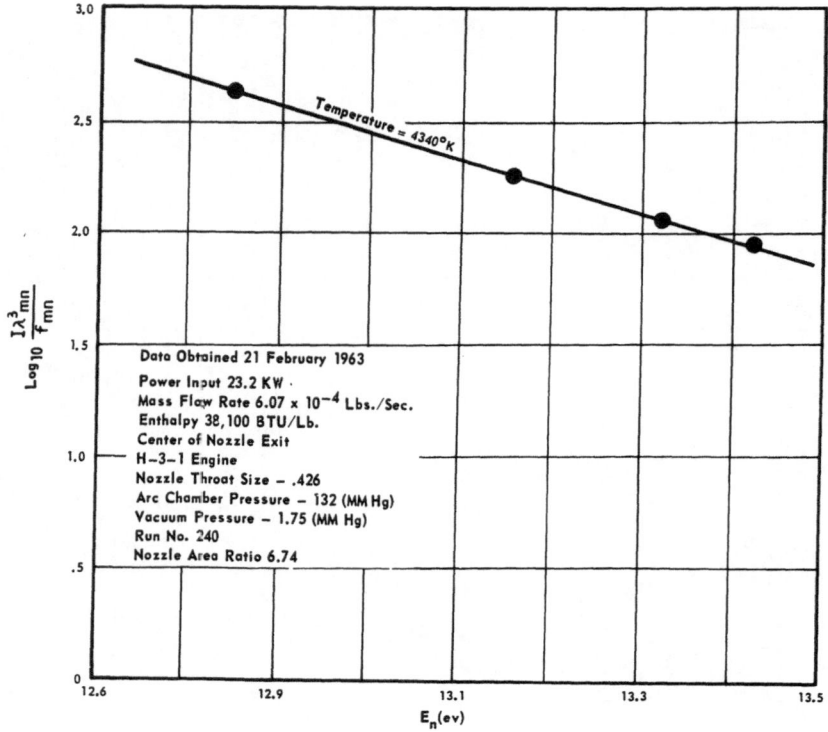

Fig. 6. Boltzmann temperature graph taken on the center line of the plasma.

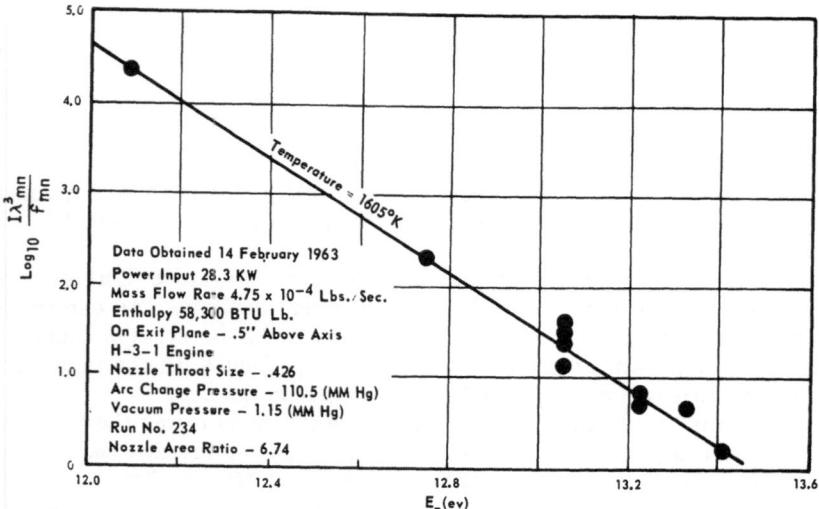

Fig. 7. Boltzmann temperature graph taken near the plasma boundary.

spectrum to appear. It is interesting to observe that even under conditions of frozen flow the Boltzmann law is obeyed.

The spectrometer charts did not exhibit the existence of any impurities in the plume. This fact implies that there was very little erosion of the electrodes in the arc. The immeasurable amount of electrode erosion is important because it implies a long engine life.

As concerns the model constructed to test the validity of the arc and plume temperatures obtained experimentally, the spectroscopically measured temperatures are higher than the calculated plume temperatures for the given arc temperature. It has been shown theoretically that it is possible for the electrons to dominate the population of excited states. This possibility and others which could affect the spectroscopic temperature are being investigated.

CONCLUSIONS

Analysis of the results of spectrometric observations implies that:

1. The population of the excited states in the plasma obeys Boltzmann's law.
2. The plasma is probably in "frozen flow."
3. Spectrometric plume temperatures exceed calculated bulk average plume temperatures.

In addition, the following assumptions have been shown highly plausible:

1. The plasma is optically thin in the visible wavelength region.
2. The plasma environment does not affect the oscillator strengths as calculated from nonrelativistic quantum mechanics.

In order to increase the efficiency of these engines, it is imperative to understand the methods of energy transfer which takes place in the arc. Therefore, engines with visible arc regions are being developed for further spectroscopic studies.

NOTATION

A_{nm} = Einstein probability for pure emission
$B_\lambda(T)$ = Blackbody function
C_p = Specific heat at constant pressure
C_v = Specific heat at constant volume
d = Diameter of particle
e = Electronic charge
E_{nm} = Energy of proton emitted in $n \rightarrow m$ transition
$f_e(\vec{v})$ = Non-normalized Maxwellian distribution of the electrons
$f(v)dv$ = Probability of a particle having a velocity between v and $v + dv$
f_{nm} = Oscillator strength of $n \rightarrow m$ transition
$F(x)$ = Spectrometer observation at position x
g_n = Statistical weight of the n th quantum state
h = Planck's constant
I_{line} = Intensity of spectrum line
I_{nm} = Spectral intensity in an $n \rightarrow m$ transition
$I(r)$ = Actual radiation from plasma cyclinder of radius r

k = Boltzmann constant
m = Mass of particle
N = Number density of particles
$\hat{N}$ = Number density of particles which could cause excitation
$\mathscr{N}_0$ = Electron population of ground state
$\mathscr{N}_n$ = Electron population of the nth quantum state
P = Pressure
R_{lamp} = Phototube responses to standard lamp
R_{line} = Phototube responses to spectrum line
R_A = Radius of nozzle in arc region
R_P = Radius of the plume exit
T = Thermodynamic temperature
V_A = Plasma velocity in the arc region
V_P = Plasma velocity at the plume exit
v = Velocity of particle
$Z_{(r)}, Z_{(r+1)}$ = Partition function of neutral and ionized hydrogen, respectively
$\epsilon(\lambda, T)$ = Emissivity of tungsten
λ_{mn} = Wavelength of radiation resulting from $n \rightarrow m$ transition
$\nu_{e,a}$ = Collision frequency between and electron and an atom having energy greater than the excitation energy
$\chi_{(r)}$ = Ionization potential

Subscripts

a = Atom
A = Arc
e = Electron
H = Hydrogen atom
P = Plume

REFERENCES

1. L. H. Aller, The Atmospheres of the Stars and Sun, Ronald Press, New York, (1953).
2. Van Camp, W. M., McVey, F. D., Merrifield, S. E., Painter, J. H., Brock, F. J., Fox, R. L., "Hydrogen Arc Jet Exhaust Diagnostics," McDonnell Report No. 9621, (1963).
3. Dickerman, P. J., Optical Spectrometric Measurements of High Temperatures, University of Chicago Press, (1961).
4. Poland, D. E., Green, J. W., and Margrave, J. L., "Corrected Optical Pyrometer Readings," NBS Monograph 30, (1961).
5. Present, R. D., Kinetic Theory of Gases, McGraw-Hill Book Co., Inc., (1958).

Health Physics Application for Spectrography Monitoring for Beryllium in Air

J. T. Rozsa, J. Stone, and J. D. Golland

National Spectrographic Laboratories, Inc.
Cleveland, Ohio

Application of direct-reading spectrographs to the health physics field provides rapid information. By use of a coupon-type filter, sensitivities of 0.1 μg/m of air are readily obtained at acceptable precision levels. Alternative utilization of a Plasma Arc involves spectrochemical preparation, but permits a wider latitude of chemical form and physical size at increased accuracy.

INTRODUCTION

The health physics aspects of beryllium as a toxic material have been well established in recent years. However, in the missile, nuclear, and metallurgical fields the demonstrated advantageous properties of beryllium and its compounds have warranted development of various methods for protection of personnel.

To facilitate the various analytical control problems of beryllium [1-8] several spectrochemical methods of sample handling and excitation are described for application to direct-reading spectrometers.

Typical analytical practices [9, 10] for exposure evaluation are reviewed as follows:

Air Sampling

The average exposure per day shall not exceed 2 μg of beryllium per cubic meter of air. The maximum exposure at any one time shall not exceed 25 μg of beryllium per cubic meter of air. The usual analytical range is 0.4 to 25 μg of beryllium per cubic meter of air.

Urine Sampling

Exposure values, as measured by this biological fluid, vary somewhat but 0.001 μg of beryllium per milliliter of urine is often considered normal. An exposure value of 0.020 μg per milliliter of urine will call for the exposed person to be transferred from the contaminated area and medically treated. The usual analytical range is 0.001 to 0.2 μg of beryllium per milliliter of urine.

Smear Sampling

Smear samples are taken to ascertain the degree of beryllium contamination of a suspect area. Essentially, an area of 3 in. by 3 in. is swiped when contamination is known. An area of 12 in. by 12 in. is swiped when little contamination is suspected.

A contaminated area may have up to 1000 μg of beryllium per square foot of area, while a clean or decontaminated area is usually below 10 μg of beryllium per square foot of area. The usual analytical range is 0.1 to 100 μg of beryllium per whole smear sample.

APPARATUS

Direct-Reading Spectrograph

The direct-reading spectrograph, or Monitor, developed for this analytical problem and successfully applied to the similar problem of analysis for lead in air is shown as Fig. 1 [16].

The spectrometer is a 50-cm Ebert type unit with a 55,000-groove/in. grating, which affords a reciprocal linear dispersion of 4.1 A per millimeter in the second order. A vertical mounting of the spectrometer is used with the slits horizontally positioned to minimize the effect from wandering of the arc column. The total light entering the spectrograph is sampled by a quartz beam splitter located behind the entrance slit. An interference filter is utilized at the exit of this sample of total light prior to the photomultiplier for selection of specific wavelength ranges and control of the exposure. The entrance slit is 40 μ and the exit slit 80 μ in width. Wavelength setting and temperature compensation is achieved with a bimetal sine bar drive. The speed rating as calculated by the effective aperture ratio is f/8.

The detection system employs RCA 1P28 photomultipliers. The output of both photomultipliers is fed into electrometer input circuits and then into the respective summing direct-current amplifiers. The whole light circuit controls the preset exposure by an electronic trip circuit. The output of the signal amplifier is read out by a strip chart, or digital recorder. Stability of the photomultipliers is assisted by an electrical bucking circuit.

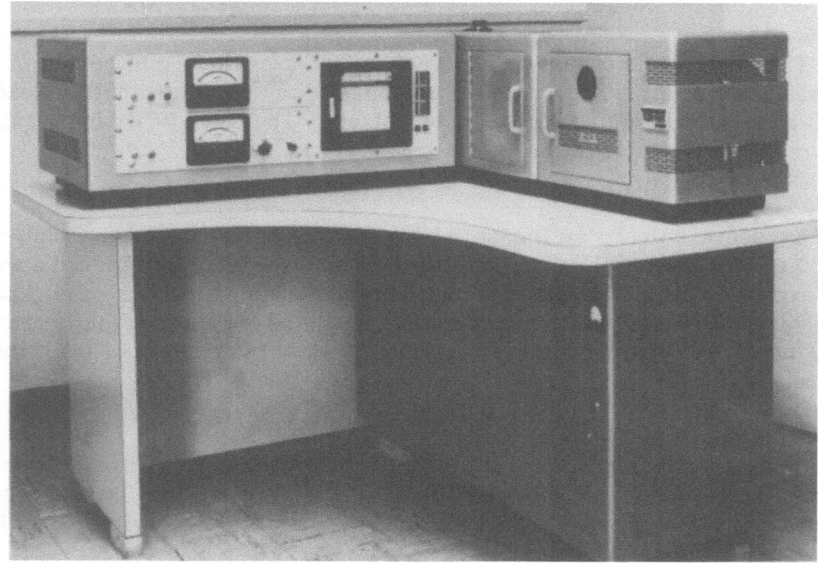

Fig. 1. Direct reading spectrometer—monitor.

TABLE I
Evaluation for Precision by Coupon and Plasma Arc Methods

Parameter	Coupon method	Plasma Arc method
Excitation		
Type	AC spark	DC arc
Amperage	3.2 A	17 A
Capacitance	0.005 μF	High-frequency ignitor
Inductance	Residual	
Discharges per half-cycle	5	
Exposure	About 20 sec	About 25 sec
Sample		
Type	Powder	Solution
Dilution		50% ethyl alcohol
Aspiration rate		0.8 ml per minute
Electrodes		
Anode	Platinum tipped 0.250 in. diameter	Graphite disc, 0.500 in. diameter, 0.125 annular hole
Cathode	Platinum tipped 0.312 in. diameter	Graphite cylinder, 0.500 in. diameter, 0.275 annular hole
Gas flow		
Tangential		Argon at 14.2 liters/min
Siphon		Argon at 6.1 liters/min
Spectra		
Be wavelength	3131.04 A	2348.61 A
Internal standard	Undispersed light	Undispersed light

The excitation is provided by an air interrupted AC spark and a DC arc with parameters as listed in Table I.

Sample Holders

The coupons are supported in a modified Petrey-type stand. The coupon* is held rigidly and centrally within the 4-mm analytical gap produced by two flat-end platinum capped electrodes.

The solutions are excited by a Plasma Arc which has been modified from the units of Margoshes [11], Owen [12], and Korolev [13]. The cross section is depicted in Fig. 2. Because of the very low concentrations being handled, all usual external plumbing has been consigned to the rear. Since the unit is not turned off between exposures, extensive water cooling of the unit including the tungsten rod is provided. Transfer of the arc from cathode to tungsten rod is readily made by electrical high-frequency discharge ignition. The tangential flow of the gases has been modified to obtain greater efficiency and to permit higher flow rates of solution to be introduced axially. Variations of the Beckman medium-bore atomizers require calibration of each unit

*Patent pending.

separately. To overcome the hydrolyzing of solutions at the atomizer tip, cooling of this section has proven effective.

AIR SAMPLING—COUPON METHOD

An air sample of 200 liters is drawn by a positive displacement pump through a Micretain filter of 0.312 in. diameter. This filter will retain all particulate matter above 0.3 μ diameter and was selected primarily because of the known extremely fine size of beryllium oxide particles. The filter paper is physically supported by a U-shaped cardboard mounting as depicted in Fig. 3a. Volume of air is measured by a rotameter and precision timer on the output side. The 200-liter sample size is often varied to accommodate various degrees of concentration of beryllium in air contamination. Once the sample is collected, it is covered with an adhesive paper to ensure against mechanical loss in transport or handling.

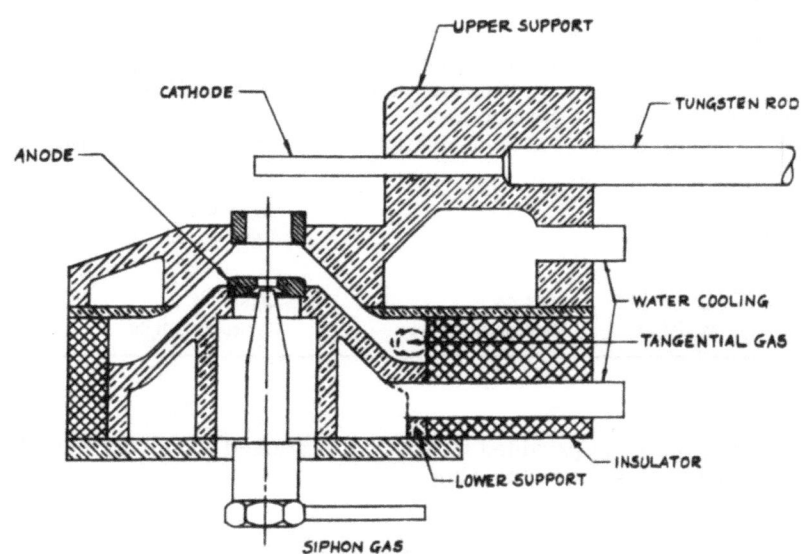

Fig. 2. Cross-sectional view of Plasma Arc.

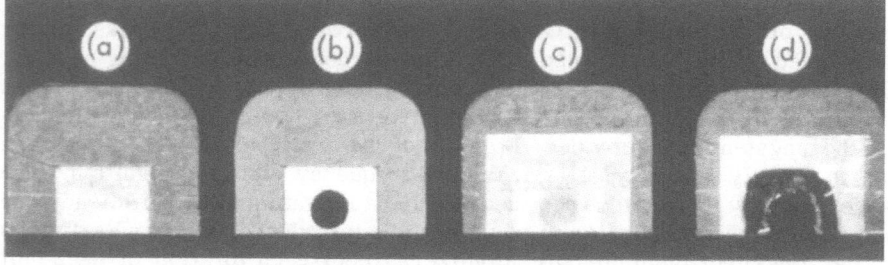

Fig. 3. Typical coupon for air sampling illustrating (a) coupon, (b) coupon with sample, (c) coupon with sample covered, and (d) coupon after sparking.

The coupon samples are inserted into the stand and the entire area of collection consumed (Fig. 3d) in the spark. Blank samples are run systematically and subtracted from all readings. Operating parameters are listed in Table I.

The provision of calibration standards proved to be the major problem but is overcome by two methods. Beryllium oxide when mixed with graphite in the desired concentrations can be deposited in 1-mg portions on blank coupons. These standards have correlated very well with field-obtained samples. For more accuracy, dilute solutions of beryllium nitrate in desired concentrations are deposited on 0.250-in.-diameter Whatman 41 filter paper discs and dried carefully below 100 C under a heat lamp. Deposition of 0.05-ml volumes is readily accomplished by a standard chromatography-type precision pipette. These discs are then inserted into the sandwich, and when kept desiccated, provide a readily available source of standards to be used as needed.

URINE SAMPLING—SOLVENT EXTRACTION

For the biological fluid, urine, a modification of the rotating platform methods [14] has demonstrated repetitively that the beryllium contamination is not uniformly dispersed throughout the solution. To offset this heterogeneity and overcome the high incidence of solids, a liquid-to-liquid solvent extraction method was developed as modified from Morrison and Freiser [15]. A 100-ml volume of urine is complexed at a pH 0.5-1.5 with 3 ml of 10% EDTA. The pH is readjusted to 7.5 with 0.1 M sodium hydroxide. Aqueous acetylacetone solution, 8 ml at 5% concentration, is then added. Extraction is made with successive 9-, 4-, and 2-ml portions of benzene.

With an effective concentration of 6.7×, direct introduction into the Plasma Arc is now feasible. No alcohol dilution is utilized. Otherwise, the operating parameters are the same as listed in Table I.

SMEAR AND AIR SAMPLING—PLASMA ARC METHOD

For reasonable accuracy in smear sampling, a solution-type method was necessary. When taken into solution, the irregular distribution of beryllium on the smear paper could be readily handled. All chemical forms of beryllium—chloride, fluoride, sulfate, oxide, and metal—when taken into solution assume a common matrix. Particularly desirable was the ability of a solution method to handle various heat-treated beryllium materials, such as high-fired beryllium oxide. The effect of physical sizes could also be minimized.

Essentially, the solution method was designed for batch analysis since it was compatible with current protective practices. While an elapsed time of 1 hr is required for solution, a batch of 15 samples can be completely analyzed in 1.5 hr. The method of solution is as follows:

1. Compact filter into a 15-ml platinum crucible.
2. Add carefully five drops of 1 : 1 sulfuric acid.
3. Fume to dryness.
4. Char and ignite at 600 C.
5. Add 5 ml of concentrated hydrofluoric acid.

6. Evaporate to dryness.
7. Take up with 5% hydrochloric acid and dilute to 10 ml with water.

The 10 ml can be adjusted up or down, dependent upon the suspected concentration, and thus provides flexibility for use of a single calibration curve.

Air samples, particularly those of higher beryllium contamination, may be handled in a similar fashion. For high-volume air samplers capable of screening 1 m^3 of air, the solution method has proven particularly attractive.

The parameters for Plasma Arc operation are given in Table I. The use of an ethyl alcohol addition to the aqueous solution provides about a threefold increase in beryllium sensitivity. The actual percentage of alcohol is noncritical. Variations of ±15% from the recommended 50% dilution have produced no observable variations. Spectral-line selection is necessarily a compromise. The beryllium line at 3130.42 A had interference from a hydroxyl band head, which limited its effective detection limit to 1 ppm. The beryllium line 3131.07 lay between banding and would have required a reciprocal linear dispersion of 2 A/mm for satisfactory analysis. A mixed gas system, helium for tangential and argon for siphoning, is required for the 3130 beryllium doublet. Beryllium 2348.61 A has the advantage of permitting the use of an all-argon system with its optimum sensitivity characteristics. Virtually no banding or background exists in this wavelength region. An all-argon gas system, however, lowers the normal precision associated with Plasma Arc analyses. No matrix effect is found for magnesium, copper, or sodium up to 400 μg/ml. Iron exhibited no apparent effect up to 1000 μg/ml. No improvement in detection limit is found using a large-bore atomizer in preference to the medium-bore atomizer.

For ppb detection, a high degree of caution is required to avoid contamination. The greatest opportunity for contamination, from our experience, exists within the atomizer itself. To overcome this problem, a wetting agent is added to all solutions, and the atomizer is flushed with the blank for 5 sec under operating conditions.

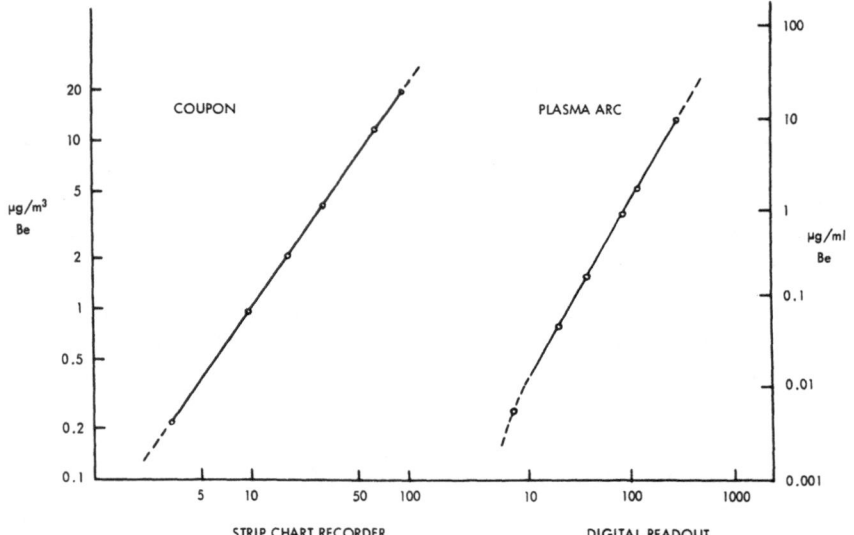

Fig. 4. Typical working curves for coupon and smear methods.

TABLE II
Operating Parameters

Coupon method		Plasma Arc method	
Concentration of Be per m^3, μg	Coefficient of variation (95%)	Concentration of Be per ml, μg	Coefficient of variation (95%)
0.1	12.3%	0.005	9.8%
0.2	7.0%	0.050	11.0%
0.4	11.2%	0.200	9.8%
1.0	10.2%	2.5	7.8%
2.0	11.5%	10.0	5.8%
10.0	8.5%	100.0	5.2%

WORKING CURVES AND PRECISION

The working curves are shown in Fig. 4. Precision values (Table II) as obtained from synthetic solutions and cross-checked field samples repetitively analyzed photographically and electronically are evaluated by the percent coefficient of variation formula:

$$\text{Percent Coefficient of Variation at 95\% Confidence Level} = \frac{100\%}{C} \, 2\sqrt{\frac{\Sigma d^2}{n-1}}$$

SUMMARY

The use of new approaches of sample handling and excitation are found feasible and permissive for application of direct-reading spectrographs into the health physics field of analytical determination of beryllium and its compounds.

REFERENCES

1. J. Cholak and D. M. Hubbard, Anal. Chem. 20:73 (1948).
2. F. P. Landis and M. C. Coons, Appl. Spectroscopy 8:2, 71 (1954).
3. P. W. Kehres and W. J. Poehlman, Appl. Spectroscopy 8:1, 36 (1954).
4. R. H. A. Crawley, Anal. Chem. Acta 22:413 (1960).
5. J. T. Rozsa, U.S. Patent No. 3,088,364.
6. J. Norris, Unpublished.
7. W. A. Van Sandt, V. C. Santomassino, R. P. Rumble, and O. M. Barlow, Ind. Hygiene J. 204 (May 1962).
8. W. L. Churchill and A. H. C. P. Gillieson, Spectrochim. Acta 5:238 (1952).
9. "Threshold Limit Values," A.M.A. Archives of Industrial Health 18:178 (1958).
10. N. I. Sax, Dangerous Properties of Industrial Materials (Reinhold Publishing Co., New York, 1957).
11. M. Margoshes and B. F. Scribner, Spectrochim. Acta 15:138 (1959).
12. L. E. Owen, Appl. Spectroscopy 15:150 (1961).
13. V. V. Korolev and E. E. Vainshtein, Zhur. Anal. Khim. 14:658 (1959).
14. J. T. Rozsa and L. E. Zeeb, Petroleum Processing 11:1708-12 (1953).
15. G. H. Morrison and H. Freiser, Solvent Extraction in Analytical Chemistry (John Wiley and Son, New York, 1957).
16. J. T. Rozsa and J. Stone, Ind. Hygiene Association Symposium, Washington, D.C. (1960).

Identification of Chemical Species in a Plasma by Emission Spectroscopy

Elliott Raisen, Richard A. Carrigan,[†] Victor Raziunas, W. A. Loseke, and E. L. Grove

IIT Research Institute
Chicago, Illinois

Optical spectroscopy was used to identify the chemical species and the state of excitation of these species in argon, air, and nitrogen plasmas. The 2400 to 8800 A wavelength region was studied. Copper–tungsten and graphite–graphite electrodes were used. The species observed were Ar, Ar^+, O, N, N_2, N_2^+, C, C_2, and CN, the latter three being due to graphite electrodes. The major species found correlated with those species predicted from thermodynamic calculations in air. However, certain species as NO and O_2 were not observed.

INTRODUCTION

The objective of this work was to obtain a better understanding of the chemical processes occurring in the plasmas, prior to making basic studies using the plasma jet to produce the high-temperature gas. Optical spectroscopic methods were used to identify the observed chemical species and the states of excitation in plasmas consisting primarily of argon, air, and nitrogen. Other species expected to be present either were not excited, were obscured by other dense band structures, or failed to yield spectra within the spectral range observed.

EXPERIMENTAL EQUIPMENT AND MATERIALS

Plasma Jet

A versatile plasma similar to a previous design [15] was developed for the purposes of this study. The essential components (Fig. 1) were a cylindrical cathode, an annular anode (both water cooled), a vortex chamber, and an environmental chamber. The stabilizing gas was injected tangentially into the vortex chamber, and after passing through the electrical arc between the electrodes, emerged through the orifice in the annular electrode. As it passes through the arc, large amounts of kinetic energy are imparted to the gas and it emerges from the orifice as a high-temperature plasma. When a gas other than air was used, the environmental chamber was flushed continuously with the stabilizing gas to avoid air contamination.

Graphite electrodes were used with each of the stabilizing gases. These electrodes erode and contaminate the plasma, but are necessary in an oxidizing atmosphere.[‡] When argon or nitrogen is used, the cylindrical

*This study was supported by the Aeronautical Systems Division of the U.S.A.F., Wright-Patterson A. F. B., Ohio, under contract AF 33(616)6686. Accepted for publication in the Journal of Applied Spectroscopy.

†Present address, National Science Foundation, Washington, D. C.

‡Studies are in progress to design a plasma jet using electrodes with oxidizing gases which will not contaminate the plasma.

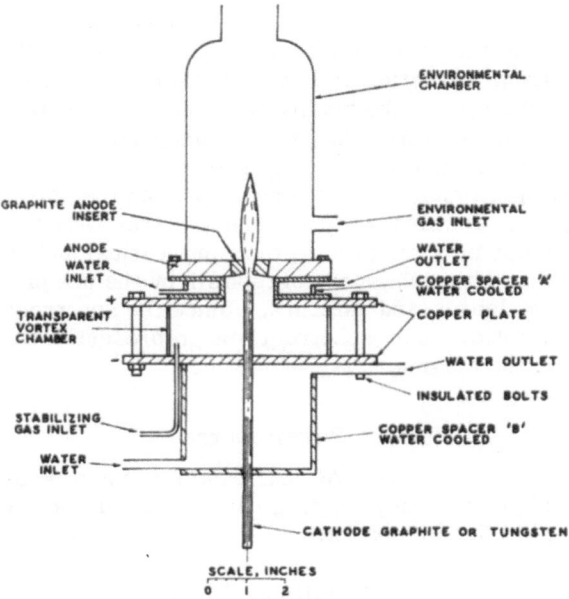

Fig. 1. Schematic diagram of plasma jet.

graphite cathode may be replaced by a tungsten cathode, and the water-cooled copper spacer "A" and the annular graphite anode may be replaced by the copper anode shown in Fig. 2. Under these conditions an essentially uncontaminated plasma is produced.

Electrodes

Spectroscopically pure graphite was used for the electrodes and inserts. The inserts (Fig. 1) were fabricated from graphite discs 1 in. in diameter and $\frac{1}{2}$ in. thick. The tungsten electrodes were 2% thoriated tungsten $\frac{1}{4}$-in. rods used in commercial welding.

Spectroscopy

The spectrograph was the Jarrell–Ash 3.4-m Ebert mount instrument with a 15,000-line/in. plane grating blazed for 4000 A in the first order.

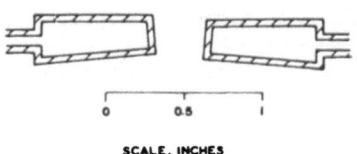

SCALE, INCHES

Fig. 2. Copper annular electrode.

The spectra were recorded on standard photographic plates in the first and higher orders in the wavelength region from 2400 to 8800 A. Exposure times varied from a fraction of a second to 30 sec, depending on the wavelength region, the order, the plasma jet variables, and the type of spectra. Most of the spectrograms represented integrated light from the whole plasma. These were made by focusing an image of the plasma on an external limiting diaphragm which, in turn, was imaged on the collimating mirror of the spectrograph. Detailed studies of the spectral distribution for different portions of the plasma were made by focusing vertical and horizontal images of the plasma directly on the spectrograph slit by the use of suitable external arrangements of lenses and mirrors. Since the spectrograph used in this work was stigmatic, such spectrograms represented the distribution of emission within the plasma.

Power Source

The Miller Model 360-P AC-DC welding unit was used as the power source. The high-frequency starting unit in this source was convenient for starting the arc in the plasma jet.

Stabilizing Gases

Argon, air, and nitrogen were used as stabilizing gases. Argon, supplied by Linde, had a reported analysis of 99.995% argon, 3-5 ppm nitrogen, 5 ppm hydrogen, 5 ppm carbon monoxide and hydrocarbons, and a maximum of 5 grains/1000 ft^3 for water.

EXPERIMENTAL RESULTS AND DISCUSSION

Plasma Jet Characteristics

The plasma jet operated smoothly with argon as the stabilizing gas. However, when nitrogen or air was used, the arc was generally initiated with argon, then gradually converted to the other gas. A summary of the different conditions at which the plasma jet was operated is shown in Table I.

The electrical characteristics of the plasma jet were examined as a function of the electrode gap between graphite electrodes. The results for

TABLE I
Plasma Jet Operating Conditions

Stabilizing gas	Flow rate, ft^3/hr	Voltage	Amperage	Cathode material (cylindrical)	Anode material (annular)
Argon	15 to 60	12 to 27	250 to 800	Tungsten	Copper
	30	22 to 29	380 to 480	Tungsten	Graphite
	10 to 60	26 to 39	240 to 800	Graphite	Graphite
Nitrogen	2 to 15	44 to 46	160 to 225	Graphite	Graphite
Air	7 to 35	40 to 46	165 to 340	Graphite	Graphite
	28 to 59	30 to 42	245 to 400	Graphite	Copper

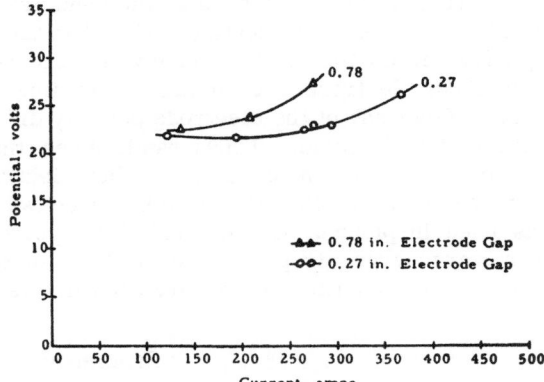

Fig. 3. Electrical characteristics of the argon stabilized plasma jet with graphite electrodes.

the argon stabilized jet are shown in Fig. 3. With a constant electrode gap the jet operation was in the positive but nonohmic region. At a constant current the potential drop across the arc and the power input increased when the electrode gap was increased.

Plasma Contamination

Graphite electrodes erode quite rapidly and contaminate the plasma with various carbon species. Barely detectable traces of Mg, Ca, Al, Si, Ti, and Fe were observed from spectroscopically pure graphite electrodes. Erosion also was observed on the metal electrodes and the metal species were observed in the spectra.

The gross contamination of the argon plasma is illustrated in Table II. These results were obtained by recording the weight loss of the electrode

TABLE II
Plasma Contamination Due to Graphite Electrodes in an Argon-Stabilized Arc

	Run number					
	1	2	3	4	5	6*
Gas flow rate, g/sec	0.209	0.418	0.418	0.418	0.835	0.835
Normalized arc power, kW/g-sec^{-1}	11.5	15.5	16.5	18.2	13.2	11.8
Electrode gap, in.	—	0.27	0.28	0.78	0.38	0.47
Contamination due to annular anode, atom %†	1.50	1.73	2.14	1.81	2.23	1.2‡
Contamination due to cylindrical cathode, atom %	0.29	0.32	0.30	0.27	9.7	4.9§
Contamination due to both electrodes, atom %	1.79	2.05	2.44	2.08	11.9	6.1

*Reverse polarity.
†The atom % refers to the g-atoms of carbon per g-atom of argon in the plasma. Two g-atom % represents 0.1 weight %.
‡Annular cathode.
§Cylindrical anode.

after a run. At the low flow rates the annular electrode contributed approximately six times more contamination than the cylindrical electrode. At the high flow rates the total erosion was increased by a factor of five, and the cylindrical electrode eroded about four times faster than the annular electrode.   Reversal of the electrode polarity did not alter the relative annular and cylindrical erosion. Thus it would appear that the predominant mechanism of erosion was mechanical rather than electrical or chemical. Qualitative evidence indicates that the contamination was even more severe when air was used in place of argon. Bonin [3] reported 3%, by mass, contamination of a very large air plasma jet with graphite electrodes, operating under more severe conditions than were used in this work.

Efficiency

The efficiency of the argon plasma jet, based on the relative amount of electrical power transferred to the gas, varied from 40 to 57%, with an average of 50%. The efficiency was much lower when air was used, varying from 12 to 47%, with an average of 30%. However, it appears that under proper conditions the latter efficiency could be increased. Stokes et al. [26], working with a similar argon plasma jet, reported an efficiency of approximately 56%.

TABLE III
Summary of Optical Spectroscopic Observations of the Plasma Jet

Gas	Flow rate, ft^3/ hr	Potential, v	Current, A	Species observed	Species sought but not found		Comments
Ar	10 to 40	12 to 30	430 to 800	Ar, Ar^+, N, N_2, N_2^+, O, C, C_2, CN	Ar^{++}, N^+, O^+ O_2^+, C^+, C^{++}, C_3, CO_2, NO, NH, NH^+, OH, OH^+, CN^+	(1)	Four spectra were made with tungsten cathode and copper anode. The remainder were made with graphite electrodes.
						(2)	All spectra were made with environmental chamber in place.
Air	7.6 to 20	42 to 56	160 to 210	N, O, C, C_2, CN	C^+, N^+, O^+ CN^+, C_3, N_2		Spectroscopic grade graphite electrodes.
Air	7.6 to 20	42 to 56	160 to 210	C, N, O, C_2, CN, N_2, N_2^+, Ar, Cu, Cu^+	C^+, C_3, CN^+, N^+, O^+, O_2, O_2^+, CO, CO^+, NO, Ar^+		Rod cathode of spectroscopic grade graphite. Copper plug anode supported in copper plate.
N_2	6 to 10	44 to 46	160 to 190	C, N, O, C_2, CN, N_2, N_2^+,	C^+, C_3, CN^+, N^+, O^+, O_2^+, CO^+, Ar, Ar^+		Spectroscopic grade graphite electrodes.

TABLE IV
Temperature Measurement of Plasmas

No.	Stabilizing gas and electrode materials	Potential drop, V	Current, A	Gas flow rate, g/sec	Enthalpy,[†] cal/g	Point of measurement	Spectrographic* temperature, K	Enthalpy temperature, K
Argon Plasma								
1	Tungsten rod cathode and copper plate annular anode	25	330	0.835	1310	On axis of jet, about 1.2 cm from its base	4200	10,500
2	The same	25	330	0.835	1350	The same	n.m.[§]	10,800
3	The same	25	250	0.835	718	The same	n.m.	5,750
4	The same	26	280	0.835	1110	The same	n.m.	8,880
5	The same	26	280	0.835	1090	The same	n.m.	8,700
6	The same	27	270	0.835	1080	The same	n.m.[‡]	8,650
7	Graphite rod cathode and graphite plug annular anode	29	500	0.418	n.m.	Integrated light from whole plasma	5820[‡]	n.m.
							5670[‡]	n.m.
Air Plasma								
1	Graphite rod cathode and graphite plug annular anode	26	480	0.418	n.m.	Integrated light from whole plasma	5300	n.m.
2	The same	42–45	160–210	0.107–0.166	n.m.	The same	4800	n.m.
3	The same	46–48	130–160	0.129	n.m.	The same	4650[‡] 4200[‡]	n.m.
4	Graphite rod cathode and copper plate annular anode	33	400	0.359	2750	Stigmatic spectra See Figs. 4 and 5	See Figs. 4 and 5	5200
5	The same	30	380	0.288	1070	Stigmatic spectra on axis of jet, about 1.2 cm from its base	5450	3000
6	The same	41	340	0.359	3420		6200	5700
		41	330	0.359	2920	1.2 cm from its base		5500
7	The same	41	280	0.497	2110	The same	6500	4300
		41	280	0.497	3040			5600
8	The same	40	400	0.359	4130	The same	n.m.	6000
		40	370	0.359	4000		n.m.	6000
9	The same	41	340	0.334	4680	The same	n.m.	6400
10	The same	34	350	0.359	1560	The same	n.m.	3500
11	The same	40	245	0.359	855	The same	n.m.	2700

*Temperatures are not corrected for radial intensity distribution.
†Based on constant heat capacity of 0.125 cal/g for argon being constant to 10,000 K.
‡Different spectra exposures during same run.
§Not measured.

Enthalpy Measurements

The enthalpy of the plasma was determined in a few experiments from temperature difference between the inlet and outlet water after stabilization, from the water flow rate, from gas flow rate, and from the electrical power input.   The difference in temperature between the inlet and outlet water and the water flow rate were used to calculate the energy loss to the water. The gas enthalpy was calculated assuming that the electrical energy that did not enter the water was abstracted by the gas. The results are approximate because corrections for radiative and conductive losses were not considered.   These enthalpies, therefore, represent maximum values, the actual values being somewhat lower.   Results of these measurements are listed in Table IV.

TABLE V
Summary of Species in the Plasmas

Species	Argon plasma		Air plasma		Nitrogen plasma
	W–Cu electrodes	C–C* electrodes	C–C electrodes	C–Cu electrodes	C–C electrodes
Ar	+	+		+	O
Ar^+	+	+		O	O
Ar^{++}	O	O			
N_2	+	+	O	+	+
N_2^+	+	O	O	+	+
N	+	+	+	+	+
N^+	O	O	O	O	O
O_2				O	
O_2^+		O		O	O
O	+	+	+	+	+
O^+	O	O	O	O	O
C_3	O	O	O	O	O
C_2		+	+	+	+
C		+	+	+	+
C^+	O	O	O	O	O
C^{++}	O	O			
CN	O	+	+	+	+
CN^+		O	O	O	O
CO_2	O	O			
CO				O	
CO^+				O	O
NO	O	O		O	
NH	O	O			
NH^+	O	O			
OH	O	O			
OH^+	O	O			
Cu	+	O		+	
Cu^+	O	O		+	

* Graphite.
+ Found.
O Sought but not found.
Blank: Not looked for.   In general, these species were not looked for because of CN and C_2 interference.

Chemical Species

The chemical species present in the argon, air, and nitrogen plasmas were identified with the aid of the standard spectrographic reference compilations [8, 12, 17-19, 23, 25]. All the wavelengths used for identification are listed by Carrigan [4] for all of the species observed. The qualitative identification of the various species observed in the plasmas is summarized in Tables III and V.

Argon Plasma

The argon plasmas were run with 10- to 40-ft^3/hr flow rate, 12 to 30 volts, and 430 to 800 amperes. In all cases the neutral argon and singly ionized species, Ar and Ar$^+$, were predominant in the spectra. When the tungsten cathode and copper anode were used, N, N$_2$, N$_2$$^+$, and O also were observed. The nitrogen and oxygen species were presumed to be due to entrainment of air into the plasma, even when the environmental chamber was used. In all instances the spectra of these two contaminants were very weak. When graphite electrodes were used, the spectra for C, C$_2$, and CN were also observed. The species that were sought but were not detectable within the experimental limits are listed in Tables III and V.

Neutral Argon, Ar

Eighty-four lines were identified in the 3500 to 8700 A region with excitation potentials varying from 13.07 to 15.13 eV.

Ionized Argon, Ar$^+$

Twenty ionized argon lines were observed in the 4200 to 5200 A region with excitation potentials from 19.22 to 21.35 eV.

Atomic Nitrogen, N

A rich spectrum of atomic nitrogen lines was easily identified because of the numerous strong lines in regions where no interfering spectra were present. Thirty-two lines were recorded in the region from 4100 to 8800 A, ranging from 11.76 to 13.70 eV.

Atomic Oxygen, O

In addition to the oxygen triplet, 7771.93, 7774.14, and 7775.43 A, nineteen additional lines were observed in the 4300 to 8900 A region. The excitation potentials varied from 10.73 to 14.10 eV.

Excitation of the spectrum of atomic oxygen in arcs is known to be strongly influenced by the electrode material and surrounding atmosphere. Fassel et al. [6, 7] observed that the 7771-A oxygen triplet is only weakly emitted by DC arcs between metal or carbon electrodes in air, and often is not detectable at all. Relatively low-lying electronic excited states, ionization potentials, and dissociation energies of the atomic and molecular species present in such arcs depress the excitation of the oxygen triplet, which has an upper level of 10.73 eV. Despite the abundance of carbon species due to the graphite electrodes, sufficient energy was present to produce intense emission of the oxygen triplet.

Atomic Carbon, C

The 2478.57-A line was very prominent. Six other lines observed in the wavelength region studied were also quite intense. The excitation potential range for these lines is from 7.69 to 10.38 eV.

Diatomic Carbon, C_2

When graphite electrodes were used, the C_2 and CN spectra were dominant in all spectrograms and constituted a major source of interference with respect to detecting other species. Sixteen bandheads from the 3300 to 6200 A region were identified.

Cyanogen, CN

In some cases the CN bands were intense when graphite electrodes were used and interfered with identification of the presence of other species. In other instances they were weak, but distinctly present. Twenty-four bandheads were observed in the 3400 to 4600 A region.

The transition probabilities for these species, C_2 and CN, are relatively high; thus, they exhibit intense spectra even though they may be present in very low concentration [2, 9].

Molecular Nitrogen, N_2

Five bands with bandheads at 5653.5, 5660.8, 5733.6, 5748.1, and 5755.2 A were observed for the first positive system of molecular nitrogen. In general, these bands were very weak. The bandhead, 3894.6 A, of the second positive system was observed in two spectrograms.

Molecular Nitrogen Ion, N_2^+

Three bandheads were identified at 3308.0, 3532.6, and 3914.4 A. The last was well resolved, but useful only when the CN band did not interfere. This species was easily detected and identified when the CN bands were absent. Thus, it was detected only when the copper anode was used.

Air Plasma

The air plasmas were run with 7.6- to 20-ft^3/hr flow rate, 42 to 56 volts, and 160 to 210 amperes. When graphite electrodes were used, N, O, C, C_2, and CN were observed, but N_2, N^+, O^+, C_3, C^+, and CN^+ could not be detected. When the copper insert anode was used in place of the graphite anode, N_2, N_2^+, Cu, Cu^+, and Ar were observed in addition to those listed above. The species not detected are listed in Tables III and V. The source of the Cu and Cu^+ was the slight erosion of the anode. The N_2 and N_2^+ were probably present when the graphite electrodes were used, but their spectra would have been obscured by the intense C_2 bands.

Line Spectra

The line spectra for nitrogen, oxygen, and carbon were very similar in the air and argon plasmas. These are major species which confirm the prediction of the thermodynamic calculations by Hilsenrath et al. [13]. The hydrogen alpha line of the Balmer series was observed in several spectrograms. Atomic argon lines were also observed.

Molecular Nitrogen, N_2

This species was observed in the air plasma generated only when the copper anode was used. Bands in the first and sometimes in the second positive system were observed, as was the case with the argon plasma. Although N_2 should be the predominant species at the plasma temperatures used in this work, it was not observed when carbon was present. All the N_2 bands are relatively weak and are easily obscured by the C_2 bands which overlap many of the N_2 bands. This is especially true in the region where the first positive system of N_2 occurs.

Molecular Nitrogen Ion, N_2^+

Spectra for this species also were observed only when the copper anode was used. The same bands listed previously for the argon plasma were the ones observed. When graphite electrodes were used, the intense CN bands prevented the N_2^+ detection.

Thermodynamic equilibrium calculations by Hilsenrath et al. [13] indicate that the N_2^+ species was present in very low concentration (< 0.01 mole %) at air plasma temperature. Since its spectrum was readily observable, it is spectroscopically quite sensitive. The long radiative lifetime of the band with its head at 3914.4 A, $6.58 \cdot 10^{-8}$ sec [1], indicates a low transition probability.

Diatomic Carbon, C_2

Intense bands were observed when graphite electrodes were used. The same bands were observed that were found in the argon plasma.

Cyanogen, CN

The intense bands of the violet system were also observed in the air plasma and seriously interfered with identification of other species. Distinct red-degraded bands with heads at 7872.7 and 7894.7 A were noted in the air and nitrogen plasmas when graphite electrodes were used, which under similar conditions appeared very faintly in the argon plasma spectra. These bands have been identified with the CN molecule by Merrill [16]. Another relatively unknown series of weak CN bands [14, 18] with bandheads at 3127.9, 3142.9, 3160.2, 3180.4, 3203.7, 3296.3, 3322.3, 3360.1, 3380.4, 3404.9, 3433.0, and 3465.5 A was observed.

Nitrogen Plasma

The nitrogen plasmas were run with 6- to 10-ft^3/hr flow rate, 44 to 46 volts, and 160 to 190 amperes. Only graphite electrodes were used. The results were essentially the same as those found for the air plasma using the copper anode; i.e., N_2 and N_2^+ bands could be observed. In the nitrogen plasma, the higher concentration of nitrogen resulted in more intense N_2 and N_2^+ bands which could be observed in the CN band spectra.

Species Not Observed

A number of species were sought in each of the plasmas but not found (Tables III and V). The highest energy observed in the air plasmas was 13.7 eV (4109.98 A) from atomic nitrogen. Since ionized argon was not observed, the energy of the air plasmas was less than 19.22 eV. Thus, certain species such as ionized nitrogen, N^+, and oxygen, O^+, may have been present but were not observed because not enough energy was available to raise them to an excited state involving allowed transitions. They both have low-lying energy levels at 1.9 and 3.3 eV, respectively, but these are forbidden transitions. Their lowest allowed transitions which would have been observable in the wavelength regions we used are at 20.6 and 25.6 eV, respectively, above the ground state of the ion. On the other hand, ionized carbon, C^+, has a low-energy excited state at 9.3 eV. But this occurs in the vacuum ultraviolet region and would not have been observed by us. Energies of about 16 eV above the ground state of the ion would have been necessary for us to observe this species in the wavelength regions we used.

The molecular oxygen ion, O_2^+, was not observed, and no data were found with regard to its expected abundance. Molecular oxygen, O_2, was not detected, even when longer exposure and the envelope of cooler gases adjacent to the jet were examined. This species is predicted to be present in major quantities in the cooler part of the plasma and to be a minor constituent in the reasonably high temperature regions. In a pilot experiment in search of the Schumann-Runge bands with an ordinary carbon arc in air, the OH band at 3063.6 A was clearly observed. However, the O_2 band did not appear in the nonluminous or luminous region of the plasma or carbon arc. The nondetection of any O_2 bands is probably the result of weak emission and interference by CN bands rather than the result of the removal of molecular oxygen by chemical reactions.

Bands for nitric oxide, NO, were not observed under any conditions although high equilibrium concentrations are predicted thermodynamically. The gamma bands observed by Guillery [11] in air arcs between metallic electrodes required spectroscopic exposures of 4 hr. Thus, with the short exposures used in this work, it is not surprising that these bands were not seen. However, had longer exposures been made, the C_2 and CN bands would doubtlessly have obscured any of the NO bands that might have been present.

Bands for carbon monoxide, CO, which should exist in appreciable concentrations in the air plasma and has a widespread band system, were not detected.

Douglas and Routly [5] have reported spectra for the molecular cyanogen ion, CN^+. These were not observed on any of the plates prepared during this work.

Temperature Measurements

Argon Plasma

Spectroscopic temperatures of the argon plasmas were determined from the electronic levels of Ar(I) by examining the relative intensities of the lines at 8424.65, 8408.21, 8264.52, 8103.69, 8014.79, 7514.65, and 7503.87 A according to the method published by Pearse [22]. Since these lines exhibited considerable Stark and pressure broadening, the intensities were obtained from the areas under the microdensitometer tracings of the line profile.

The enthalpy and the temperature calculated from the enthalpy and from spectroscopic data are shown in Table IV. Although both types of temperature measurements are admittedly coarse, the large discrepancy between the two types of measurements as yet is unexplained. Two possible contributing factors are lack of thermodynamic equilibrium and increased heat capacity of argon at the high temperatures. The use of recently published revised values of argon transition probabilities [20] would increase the calculated temperatures by about 50% and make them comparable with the enthalpic temperatures. It is to be noted that thermodynamic equilibrium is usually attained in a high-pressure 1-atm plasma. Wachman et al. [27, 28] assumed that the heat capacity of argon is constant up to 10,000 K. The presence of Ar$^+$ leads one to believe that the true temperature is between the two extremes, but closer to the high values. Stokes et al. [26] calculated a temperature range from 9000 at 2 kW to 19,000 K at 14 kW, which they believed to be high.

Air Plasma

The spectroscopic temperatures of the air plasma were determined from the P and R branches of the 0-1 (4216 A) vibrational band of the $^2\Sigma^+ \rightarrow {}^2\Sigma^+$ electronic transition of the CN molecule according to the method reported by Greenshields [10]. The temperature is determined from the positions of the isointensity points of the P and R branches. Above 5700 K the intensity ratios of individual lines are used because the isointensity point is obscured by the 4197 A bandhead. The temperatures determined under various air-plasma conditions are listed in Table IV. The temperatures varied from 2700 to 6500 K. These temperatures, which are not particularly high for a plasma device, occurred because low power inputs and relatively high gas flow rates were used. The jet has been run at higher powers and lower flow rates, but the temperatures were not measured in these experiments. In most cases the spectra were derived from the integrated light of the whole plasma, and the temperature measured represented a weighted average. The stigmatic spectra allowed line-of-sight observations of temperature to be made at specific areas of the plasma. A representative temperature distribution is shown in Fig. 4.

The radial temperature gradient from the axis to the edge of the plasma at 7 and 15 mm above the orifice is about 550° and 350°, respectively. This comparatively small difference indicates that even though the observed temperature is an "average," because of contributions from the gradient along

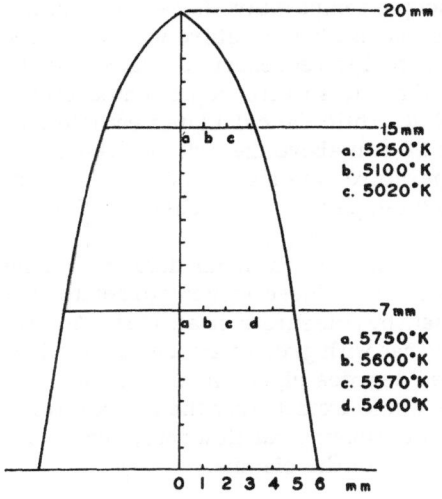

Fig. 4. Temperature distribution in a representative air plasma.

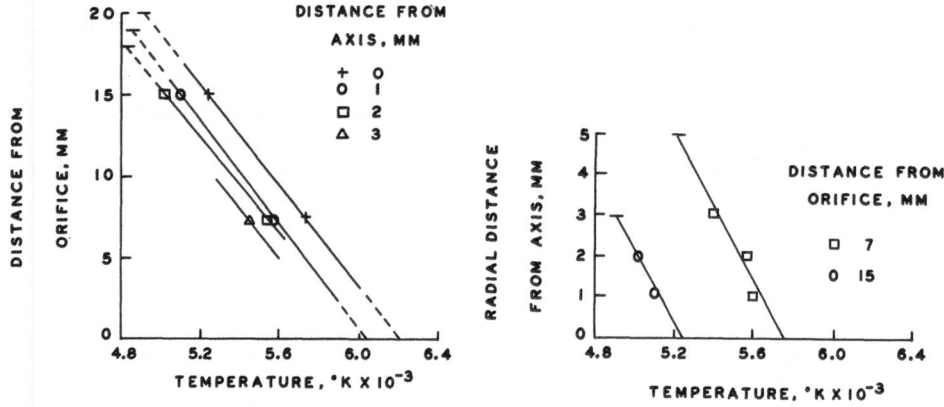

Fig. 5. Axial and radial temperature distribution.

the radial line of sight, it is probably as accurate as can be expected from this type of measurement. The temperature measurement at the edge is not an "average," because the gradient is negligible due to the short light path. The very steep temperature gradient just at the plasma boundaries allows one to extrapolate the temperature to the edge of the plasma radially and to the orifice and tip axially, as shown by the dashed portions in Fig. 5. The lowest temperature observed by extrapolation was 4800 K and occurred at the edge of the plasma 18 mm above the orifice and 2 mm from the axis. The highest temperature was 6200 K, on the axis at the orifice.

The temperature of the air plasma (Table IV) was also calculated from the enthalpy measurements, using data from the compilation of Hilsenrath et al. [13]. The difference between the temperatures calculated from the spectroscopic data and from the enthalpy data are much less for the air plasmas than for the argon plasmas. Considering the various conditions, the agreement for air plasmas is quite satisfactory. The enthalpy temperature was always lower than the spectroscopic temperature for several reasons. The primary reason is that the spectroscopic temperature was measured at a hot point in the plasma and the enthalpy represents an "average" temperature. Thus, Fig. 5 illustrates that the spectroscopic temperatures in this case varied from 4800 to 6200 K, while the calculated enthalpic temperature was 5200 K. However, at 12 mm above the orifice, which corresponds to the point where the temperature was measured on the other air plasmas (Table IV), the spectroscopic temperature was 5500 K. This is, of course, also higher than the enthalpic temperature.

The enthalpic temperatures were calculated from data by Wachman et al. [27, 28] based on uncontaminated air. These values correlated within 200 K with those calculated from the data by Hilsenrath et al. [13]. Additional calculations with their data indicated that with greater air contamination the temperature would be several hundred degrees higher in each case.

In general, it was found, as one would expect, that the temperature increased with increasing power input, decreasing gas flow rate, and decreasing distance from the orifice and the axis of the plasma.

TABLE VI
Species in Air at 1 atm Pressure

Species	% at 4000 K	% at 6500 K
N_2	66	40
N	–	29
O	25	30
NO	4	0.7
Ar	0.85	0.8
O_2	3	–

SUMMARY

The chemical species observed and not observed in the argon, air, and nitrogen plasmas are summarized in Table V. In the argon plasma, the species Ar, Ar^+, O, N, N_2, and N_2^+ were observed, the latter four being due to air contamination. When graphite electrodes were used, the carbon species C, C_2, and CN were also found.

Similar results were obtained with the air and nitrogen plasmas, except that argon spectra arising from the normal content of the air were noted in only a few instances, and the C_2 and CN bands dominated the spectra. The latter made identification of many species difficult because of their obscuring effect.

According to Table VI, at 4000 K the major species in air would be N_2 and O, and at 6500 K the major species would be N_2, N, and O. This table was prepared by interpolation of thermodynamic data by Hilsenrath et al. [13] for air at 1 atm. In this research the species NO and O_2 were not observed but N_2^+ was observed. However, it was found that NO and O_2 were difficult to excite, and that N_2^+ was readily excited but probably obscured by the intense C_2 and CN bands when graphite electrodes were used.

Ionized argon, Ar^+, was not observed in the air plasmas, but it was found in the argon plasmas where the initial argon concentration was greatly enriched, and where more intense excitation conditions existed. Other ionized species that were sought but not found include: Ar^{++}, N^+, O^+, O_2^+, C^+, C^{++}, CO^+, CN^+, NH^+, and OH^+.

The behavior of nitrogen and oxygen species in air plasmas in this laboratory is in agreement with results obtained in other laboratories. Pearse [22] implied that N could not be detected in the jet from a gaseous-air stabilized arc between graphite electrodes when a water-cooled plenum chamber was used. Temperatures in the range from 2240 to 4160 K were reported. Numerous N lines were present in the spectra of air plasmas used in this study, but neither N^+ nor O^+ was observed. On the other hand, Bonin et al. [3], working with an air plasma jet operating at 5000 amperes and 275 volts, observed both N^+ and O^+ at temperatures up to 12,000 K. Thus, the spectral results of all three laboratories are consistent for the reported temperatures.

REFERENCES

1. D. R. Bates, editor, Atomic and Molecular Processes, Pure and Applied Physics 13:73, 1960.
2. W. S. Benedict and E. K. Plyler, National Bureau of Standards, Circular 523, 1954.

3. J. H. Bonin, C. F. Prince, and D. E. Taylor, Wright Air Development Center, Tech. Rept. 59-87, Part I, p. 99, July, 1959.
4. R. A. Carrigan, E. Raisen, and K. Schmude, "Experimental Studies to Determine the Chemical Species Prevalent in Plasmas of an Air Arc and the Boundary Layers Adjacent to Ablating Materials," Wright Air Development Division, Tech. Rept. 60-359, 1960.
5. A. E. Douglas and P. M. Routly, Astrophys. J. 119:303 (1954).
6. V. A. Fassel, W. A. Gordon, and R. W. Tabeling, "Emission Spectrometric Determination of Oxygen in Metals," American Society Testing Materials, Special Technical Publication No. 221, p. 3, Philadelphia, 1958.
7. V. A. Fassel and R. W. Tabeling, Spectrochim. Acta 8:201 (1956).
8. A. Gatterer, Grating Spectrum of Iron, Specola Vaticana, 1951.
9. A. G. Gaydon and H. G. Wolfhard, Proc. Roy. Soc. 201:570 (1950).
10. D. H. Greenshields, "Spectrographic Temperature Measurements in Carbon-Arc-Powered Air Jet," NASA Tech. Note D—169 (1959).
11. M. Guillery, Z. Phys. 42:121 (1927).
12. G. R. Harrison, M.I.T. Wavelength Tables, (Wiley, New York, 1939).
13. J. Hilsenrath, M. Klein, and H. W. Woolley, "Tables of Thermodynamic Properties of Air Including Dissociation and Ionization from 1500 to 15,000 K," AEDC—TR-59-20, p. 32 ff. (1959).
14. R. C. Johnson and H. G. Jenkins, Phil. Mag. 2:621 (1926).
15. Sidney Katz, Edwin J. Latos, and Elliott Raisen, Ind. and Eng. Chem. 52:289 (1960).
16. P. W. Merrill, Lines of the Chemical Elements in Astronomical Spectra, Carnegie Institute of Washington, Publication No. 610, p. 23 (1958).
17. C. E. Moore, "A Multiplet Table of Astrophysical Interest," Revised Ed., Contributions from the Princeton University Observatory, No. 20, 1945.
18. C. E. Moore, "An Ultraviolet Multiplet Table," National Bureau of Standards, Circular 488.
19. C. E. Moore, "Atomic Energy Levels," National Bureau of Standards, Circular 467 (1949).
20. H. N. Olsen, J. Quant. Spect. Rad. Trans. 3:59 (1963).
21. R. W. Pearse and A. G. Gaydon, The Identification of Molecular Spectra, 2nd Ed. (Wiley, New York, 1950).
22. W. J. Pearse, "Plasma Jet Temperature Study," Optical Measurements of High Temperatures, compiled by P. J. Dickerman, (University of Chicago Press, Chicago, 1960) p. 125 ff.
23. B. Rosen, "Tables de constantes et données numeriques," Vol. 4, Donnees spectroscopiques concernant les molecules diatomiques, Hermann et Cie, Paris, 1951.
24. B. Rosen, "Tables de constantes et donnée numeriques," Vol. 5, Atlas de longueurs d'onde caracteristiques des bandes d'emission et d'absorption des molecules diatomiques, Hermann et Cie, Paris, 1952.
25. A. N. Saidel, V. K. Prokofiev, and S. M. Raiski, "Tables of Spectrum Lines," VEB Verlag Technik (Berlin, 1955). English translation.
26. C. S. Stokes, W. W. Knipe, and L. A. Streng, J. Electrochem. Soc. 107:35 (1960).
27. H. Y. Wachman and M. J. Linevsky, "Equilibrium Composition and Thermodynamic Properties of Air-Carbon Mixtures," General Electric Co., T.I.S. No. R595D344, May, 1959.
28. H. Y. Wachman, M. J. Linevsky, and J. H. McGinn, "The Effects of Electrode Contamination on the Properties of Air-Arc Plasmas," General Electric Co., T.I.S. No. R595D427, September, 1959.

Spectrographic Determination of Trace Impurities in High-Purity Gold

A. J. Lincoln and J. C. Kohler

Research and Development Division
Engelhard Industries, Inc.
Newark, N. J.

A DC arc optical emission spectrographic method for the quantitative determination of 26 impurities in the range of 0.1 to 100 ppm is described. In this method, the gold sample is dissolved in aqua regia and the gold solution is then reduced by the addition of a weighed portion of high-purity graphite with the aid of hydrazine hydrate. The solution is evaporated to dryness and a 200-mg portion of the gold on graphite is pressed into a pellet which is burnt in a 20-A DC arc using a Stallwood Jet with an atmosphere of 70% argon–30% oxygen. The preparation of synthetic standards is described. Analytical lines used for the various elements, together with concentration ranges covered, are tabulated. Typical working curves for selected elements are shown. The overall precision for a total mean concentration of impurities at 280 ppm was found to be 5.1% expressed as coefficient of variation.

INTRODUCTION

Gold is probably one of the earliest metals known to man and it has always been valued very highly. Despite its early discovery and its important role in history, it has had only limited industrial application until recently. Its industrial uses in the last 25 years have been expanding at a rapid rate and have more than doubled in the last 10 years. An increasing portion of this stepped-up industrial demand has been due to numerous space-age products which require a very high-purity gold for use in the various components. To meet this need, a quantitative spectrographic procedure has been developed for the determination of 26 impurites in gold for concentrations ranging from 0.1 to 100 ppm.

Qualitative spectrographic techniques have been reported by a number of investigators [2, 11, 18] for impurities in gold and gold ore. Raikhbaum [10] described a total-combustion and moving-plate technique for the quantitative determination of impurities in gold-ore flakes. Kheifitz and Cherezova [6] reported an addition technique for the quantitative determination of silver, copper, iron, and silicon in high-purity gold. Esterhuizen [4] has described a gold-bead method for the quantitative determination of copper and silver in gold. Strasheim and co-workers [16] have reported a globule-arc quantitative technique for the simultaneous determination of copper, silver, palladium, iron, lead, and zinc in gold.

The application of a direct-current arc technique for low-level determination of impurities in gold was considered the most practical approach in this study.   While sensitivity was a major objective, an acceptable level of precision was also an important consideration in the development of the technique.   The direct burning of metallic gold in both deep and shallow electrodes was rejected due to the extremely poor arcing characteristics which gold exhibits.   The stability of the arc was improved considerably when finely divided gold was mechanically blended with high-purity graphite powder and burnt in a deep crater electrode. Although this technique improved both sensitivity and precision, the precision obtained was still not suitable.

It was found that in mechanically blending finely divided gold with graphite, the gold tended to ball-up. This resulted in nonhomogeneous distribution of the gold, which had a marked effect on the precision. Homogeneous dispersion of gold on graphite was accomplished by the use of a solution technique.   Gold was dissolved in aqua regia and high-purity graphite was then added to the chloroauric acid solution. The gold in solution was reduced on the graphite by the addition of hydrazine hydrate, and the resulting product was evaporated to dryness.   The dry gold on graphite powder was found to give improved precision when burnt in a high-amperage direct-current arc.

The use of controlled atmospheres to suppress cyanogen bands, minimize background interference, and increase sensitivity in direct-current arc excitation has been described by Thiers [17] and Hammaker and co-workers [5].   The application of a Stallwood Jet [15] with various atmospheres has been reported by Rupp and co-workers [12] and Shaw and co-workers [13] to increase both precision and sensitivity in direct-current arc excitation procedure.   The use of pure argon as an atmosphere in the analysis of gold was found to improve the precision; however, the desired sensitivity was not obtained.   Studies using mixtures of argon and oxygen as an atmosphere were made .and it was found that satisfactory precision and sensitivity could be obtained with a mixture of 70% argon-30% oxygen.

It has been reported in a previous study [8] that when the gold on graphite preparation was packed into a crater of a high-purity graphite electrode and burnt in a Stallwood Jet with an atmosphere of 70% argon-30% oxygen, good sensitivity, and precision may be improved even more when the gold on graphite preparation is pressed into a pellet and burnt in a friction fitting crater of a platform-type electrode.

It was not possible to obtain chemically analyzed standards since available analytical techniques do not permit the accurate chemical determination of most elements in gold below 100 ppm.   For this reason, the preparation of synthetic standards was investigated. The addition of impurities as chlorides and nitrates to a solution of chloroauric acid was found to give the most satisfactory standards.

EXPERIMENTAL PROCEDURE

Apparatus and Reagents

The equipment used in this study consisted primarily of commercially available equipment and is tabulated in Table I. The pelleting press and die assembly was designed and constructed in our own laboratories using stand-

TABLE I
Apparatus and Reagents

Spectrograph . . . . Applied research laboratories two meter
with 24,000 lines/in. grating-dispersion
5.2 A/mm
Excitation . . . . . ., ... ARL multisource Model #5700
Densitometer . . . . ARL Model #5400
Mixer mill. . . . . Spex Industries, Model #5000
Pellet press and
die assembly . . . . Engelhard Industries, Inc.
Stallwood Jet. . . . . Spex Industries water-cooled Model #9027
Flow meter . . . . ., ... Brooks Rotar R-2-15C with steel ball
Nitric acid . . . . Mallinckrodt transistor grade
Hydrochloric acid. . . . . Mallinckrodt transistor grade
Hydrazine hydrate . . . . Fisher purified

ard components wherever possible. The reagents used were the highest
purity commercially available.

Preparation of Standards

The standards were prepared from gold having a purity of 99.999+%. The
gold was dissolved in aqua regia and the volume was adjusted so that 10 ml
contained 2 g of gold. The various impurities were added as chlorides, with
the exceptions of iron, thallium, silver, and indium, which were added as
nitrates, silicon as sodium silicate, and boron as boric acid.

Since 26 impurities were involved in the synthetic gold standards, the
impurities were divided into the two groups listed in Table II. The selection
of the impurities for each group was made on the basis of the possible line

TABLE II
Impurities Added to Gold Standards

Group A	Group B
Cadmium	Aluminum
Calcium	Bismuth
Copper	Boron
Indium	Chromium
Iridium	Cobalt
Iron	Manganese
Lead	Nickel
Magnesium	Platinum
Palladium	Rhodium
Ruthenium	Silver
Silicon	Thallium
Tellurium	Zinc
Tin	
Titanium	

interference at the higher concentration. A series of standards was prepared in concentration ranging from 0.1 to 100 ppm. The combined impurities shown in Groups A and B were used in each single standard for concentration levels up to and including 10 ppm. Concentration levels above 10 ppm were prepared by adding only those impurities in Group A or B in a single standard.

The impurities as solutions were added to the previously prepared chloroauric acid solution. The volume was then adjusted, using distilled water for the low standards and hydrochloric acid for the two highest standards, to give a concentration equivalent to 100 g of gold per liter. The solution standards were then bottled and stored until ready for use.

Preparation of Sample

A 1-g portion of gold was dissolved in 6 ml of hydrochloric acid, 2 ml of nitric acid, and 2 ml of distilled water in a quartz evaporating dish on a hot plate. After the gold was completely dissolved, the solution was cooled to room temperature and 500 mg of high-purity graphite powder was added. The solution was gently agitated to ensure that all of the graphite was completely covered.

The gold was precipitated on the graphite by the addition of 20 ml of a 34% solution hydrazine hydrate. The solution was then evaporated for 1 hr under a battery of infrared lamps. The evaporation was continued on a hot plate with the aid of infrared lamps until completely dry. The sample was

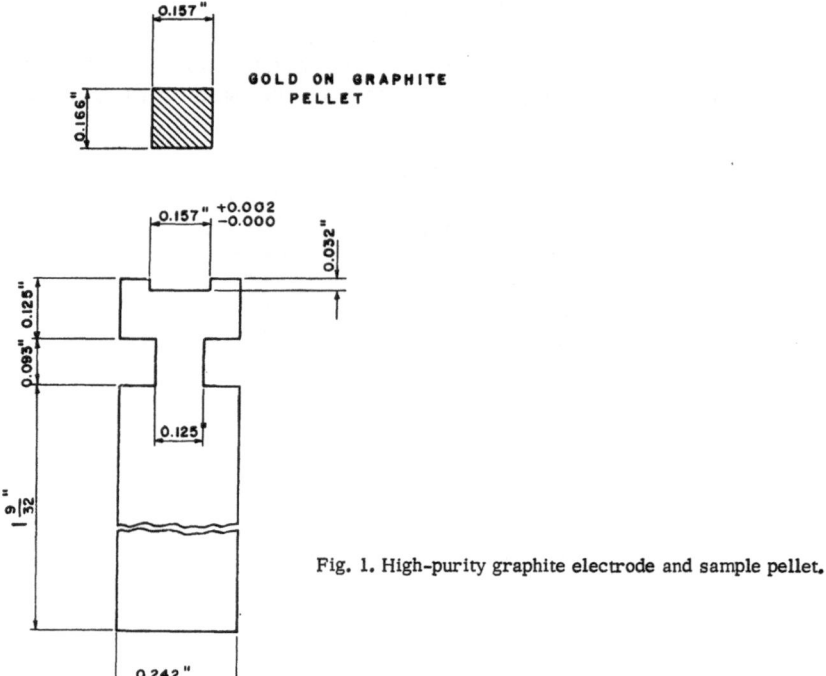

Fig. 1. High-purity graphite electrode and sample pellet.

cooled and transferred to a polyethylene vial and then blended in a mixer mill for a period of 1 min. The blended sample was then stored in a polyethylene vial in a desiccator.

A 200-mg portion of gold-impregnated graphite was pressed into a pellet with a load of 80 tons using the pellet press and die assembly which has been previously described [7].

Electrode System

The pellet was pressed into the precision-machined crater of a high-purity undercut-type electrode similar to that shown in Fig. 1. It is important that the pellet have a friction fit in the electrode since looseness between the pellet and electrode can have a marked effect on the precision.

The pellet and electrode assembly was placed in a drying oven at 200 C for $\frac{1}{2}$ hr prior to use to ensure that the pellet is completely dry. Failure to follow this procedure may result in the pellet breaking up during the arcing procedure.

The pellet and graphite electrode assembly was placed in a water-cooled Stallwood Jet which forms the lower electrode (anode) in the system. The quartz igloo dome [14] was placed over the Stallwood Jet containing the lower electrode, and a $\frac{1}{8}$-in. flat-face high-purity graphite upper electrode was passed through the center of the dome. The system was adjusted for correct

Fig. 2. Spex Stallwood Jet with quartz igloo.

analytical spacing. Figure 2 shows the photographic arrangement of this system.

The Stallwood Jet assembly was flushed with a mixture of 70% argon–30% oxygen for a period of 1 min prior to excitation. The argon–oxygen was maintained at a flow rate of 4 liters/min during the excitation. The conditions for excitation and photography are tabulated in Table III.

DATA AND DISCUSSION

A single gold standard containing all of the impurities covered in this study was prepared to study the effect of argon and mixtures of argon and oxygen in the Stallwood Jet. All samples were burnt in a 20-A DC arc for 30 sec under identical conditions. The gas was maintained at a flow rate of 4 liters/min during excitation. The signal-to-noise ratios obtained on selected analytical lines in the various atmospheres are tabulated in Table IV. An examination of these data shows that while no one condition was ideal, the most satisfactory data with respect to the majority of the impurities were obtained using the 70% argon–30% oxygen mixture.

Moving-plate studies were made to determine the optimum exposure time. The samples were run using the Stallwood Jet with 70% argon–30% oxygen flowing at 4 liters/min in a 20-A DC arc. Table V lists the analytical lines used together with the intensity ratios obtained for the various time intervals. The intensity for the first 10 sec of the exposure was assumed to be one for the purpose of determining ratios at the other time intervals. It may be seen that the ratios of the majority of the impurities do not significantly change after the first 30 sec of total exposure.

Table VI lists the various analytical and internal standard line pairs together with the concentrations covered. Filter factors for each analytical and internal standard line are included.

TABLE III
Excitation and Photography

Voltage . . . . 265 V DC
Current . . . . 20 A DC
Spectral region . . . . 2100–4600 A
Slit width . . . . 25 μ
Analytical gap . . . . 3 mm
Filter system . . . . Primary filtering at slit
                             split field 50T/100T at
                             camera with selected
                             lines independently fil-
                             tered
Exposure . . . . 30 sec
Atmosphere . . . . 70% Argon–30% Oxygen
                             at 4 liters/min
Emulsion . . . . Eastman SA-1
Development . . . . D-19 5 min at 65 F
Fixer . . . . Eastman X-ray

TABLE IV
The Effect of Argon and Mixtures of Argon-Oxygen on the Signal-to-Noise Ratio of the Analytical Lines of Various Impurities in Fine Gold

Analytical line, A	Pure argon	Argon—10% oxygen	Argon—20% oxygen	Argon—30% oxygen	Argon—40% oxygen
Ag 3382.9	12	30	42	50	50
Al 2373.1	ND	ND	1.4	1.9	2.0
B 2497.7	ND	1.9	1.5	1.6	1.8
Bi 3067.7	ND	ND	ND	ND	ND
Ca 4226.7	3.5	6.4	8.6	10.7	10.2
Cd 2288.0	1.5	1.8	2.6	1.6	3.0
Co 3405.1	1.1	3.6	3.8	5.3	4.8
Cr 2835.6	1.3	3.8	5.0	6.2	4.7
Cu 3274.0	ND	8	15	20	23
Fe 2599.4	1.3	5.4	5.8	6.1	7.8
In 3039.4	0.8	2.1	2.7	3.1	2.4
Ir 3220.8	ND	1.8	3.0	3.6	4.1
Mg 2779.8	1.2	1.6	3.4	4.3	4.0
Mn 2801.1	ND	2.8	4.8	6.6	6.7
Ni 3437.3	ND	1.6	1.9	2.4	2.2
Pb 3683.5	1.5	6.7	8.8	11.2	7.9
Pd 3421.2	1.2	3.0	8.8	10.7	6.5
Pt 2659.4	ND	3.4	5.6	6.3	6.6
Rh 3434.9	ND	1.4	2.6	3.0	2.7
Ru 3436.7	ND	2.1	5.2	5.4	4.6
Si 2987.6	ND	1.8	1.4	2.6	2.2
Sn 2840.0	ND	1.7	1.6	1.7	1.8
Te 2385.8	ND	1.7	1.7	1.3	1.8
Ti 3349.0	1.7	2.8	5.4	7.1	7.0
Tl 3519.2	2.0	5.6	5.6	7.0	5.2
Zn 3345.0	2.8	4.6	4.4	4.4	3.4

The relative transmittances of the line pairs obtained on a nonrecording densitometer were converted to log relative intensity ratios using an emulsion calibration curve [1]. Background corrections were made whenever necessary. The working curves were constructed by plotting the average log relative intensity ratio obtained from nine runs made over a period of three days on each standard against the corresponding log concentration.

The elements Al, Ca, Cu, Fe, Mg, Ni, and Si were known to be introduced into both the standards and samples in low concentrations from both the reagents and the graphite powder used, despite the fact that they were all of the highest available purity. Corrections on the various working curves were made by standard spectrographic procedures [3, 9] and were made only for residual impurities known to be present in the gold used to prepare the various standards and are listed in Table VII. It was assumed that impurities

TABLE V
Impurity Volatilization Rate

Element	Wavelength, A	Intensity ratio					
		10 sec	20 sec	30 sec	40 sec	50 sec	60 sec
Au	3105.3	1	0.9	1.0	1.0	1.0	1.0
Ag	3382.9	*	*	*	*	*	*
Al	3082.2	1	0.41	0.26	0.24	0.23	0.21
B	2497.7	1	0	0	0	0	0
Bi	3067.7	1	0	0	0	0	0
Ca	3158.9	1	0.35	0.29	0.29	0.30	0.27
Cd	2288.0	1	0.37	0	0	0	0
Co	3453.5	1	0.35	0.25	0.23	0.23	0.20
Cr	4254.3	1	0.24	0.14	0.14	0.14	0.12
Cu	3274.0	1	0.58	0.42	0.39	0.35	0.33
Fe	3220.6	1	0.27	0.21	0.18	0.17	0.15
In	3256.1	1	0.21	0.15	0.13	0.13	0.12
Ir	3220.8	1	0.64	0.59	0.59	0.58	0.54
Mg	2779.8	1	0.28	0.21	0.16	0.20	0.18
Mn	2801.1	1	0.28	0.20	0.19	0.17	0.16
Ni	3414.8	1	0.57	0.44	0.39	0.38	0.33
Pb	3683.5	1	0.22	0.15	0.12	0.13	0.11
Pd	3404.6	1	0.52	0.39	0.38	0.35	0.31
Pt	2659.4	1	0.72	0.72	0.74	0.77	0.72
Rh	3434.9	1	0.47	0.36	0.36	0.36	0.30
Ru	3436.7	1	0.44	0.34	0.33	0.32	0.28
Si	2519.2	1	0.53	0.44	0.44	0.44	0.42
Sn	2840.0	1	0	0	0	0	0
Te	2385.8	1	0	0	0	0	0
Ti	3361.2	1	0.45	0.37	0.40	0.42	0.40
Tl	2567.9	1	0	0	0	0	0
Zn	3345.0	1	0.32	0.25	0.22	0.20	0.17

*Line intensity too great to read.

introduced by the reagents and the graphite powder would be reasonably constant in both standards and samples.

Typical working curves for representative elements are shown in Figs. 3, 4, 5, and 6. Examination of these working curves show the slopes are satisfactory and are, for the most part, linear for the concentration range covered.

The lower limit of detection for B, Bi, Cd, Co, Cr, In, Ir, Mn, Pd, Pt, Rh, Ru, Sn, Te, Ti, Tl, and Zn is the lowest value reported for each element in Table VI in the column on concentration range since the analytical line beyond this point goes to extinction. For the balance of the impurities, the limit of detection is dependent upon the residual impurities in the gold used to prepare the standards, the reagents used, and graphite powder. With improved purity in all of these areas, it should be possible to extend the limit of detection to lower levels.

TABLE VI
Line Pairs and Analytical Ranges

Element	Analytical line, A	Filter, %T	Internal standard line, A	Filter, %T	Concentration range, ppm
Ag	3382.9	0.6	Au 3147.6	3	5 - 50
	3382.9	0.3	Au 3147.6	3	15 - 100
Al	3082.2	16	Au 3105.3	16	0.1- 10
	2373.1	32	Au 3105.3	16	10 - 100
B	2497.7	32	Au 3105.3	16	1 - 25
	2497.7	16	Au 3105.3	16	25 - 100
Bi	3067.7	16	Au 3105.3	16	10 - 100
Ca	3158.9	16	Au 3105.3	16	1 - 50
	3158.9	6	Au 3147.6	3	25 - 100
Cd	2288.0	32	Au 3105.3	16	0.5- 10
	2288.0	16	Au 3105.3	16	10 - 100
Co	3405.1	32	Au 3105.3	16	0.1- 1
	3453.5	16	Au 3105.3	16	1 - 10
	3405.1	1.6	Au 3105.3	16	10 - 100
Cr	4254.3	32	Au 3105.3	16	0.1- 1
	2835.6	16	Au 3105.3	16	1 - 25
	2835.6	3	Au 3147.6	3	25 - 100
Cu	3274.0	1.6	Au 3105.3	16	0.5- 10
	3274.0	0.3	Au 3147.6	3	10 - 100
Fe	3020.6	16	Au 3105.3	16	5 - 50
	2788.1	16	Au 3105.3	16	25 - 100
In	3256.1	16	Au 3105.3	16	1 - 25
	3039.4	16	Au 3105.3	16	10 - 50
Ir	3220.8	16	Au 3105.3	16	5 - 50
	2924.8	16	Au 3105.3	16	10 - 100
Mg	2779.8	32	Au 3105.3	16	0.5- 10
	2776.7	16	Au 3105.3	16	10 - 100
Mn	2794.8	16	Au 3105.3	16	0.1- 5
	2801.1	16	Au 3105.3	16	1 - 25
	2799.8	16	Au 3105.3	16	25 - 100
Ni	3414.8	16	Au 3105.3	16	0.5- 10
	3437.3	3.2	Au 3105.3	16	10 - 50
	3437.3	1.6	Au 3105.3	16	25 - 100
Pb	3683.5	32	Au 3105.3	16	1 - 10
	3683.5	16	Au 3105.3	16	5 - 25
	2614.2	16	Au 3105.3	16	10 - 100
Pd	3404.6	16	Au 3105.3	16	0.1- 5
	3404.6	3.2	Au 3105.3	16	1 - 10
	2447.9	32	Au 3105.3	16	10 - 100
Pt	2659.4	16	Au 3105.3	16	5 - 50
	2930.8	16	Au 3105.3	16	10 - 100
	2659.4	3	Au 3147.6	3	100 - 500
Rh	3434.9	16	Au 3105.3	16	0.5- 10
	3396.8	3.2	Au 3105.3	16	5 - 50
	3396.8	1.6	Au 3105.3	16	25 - 100

TABLE VI (continued)

Element	Analytical line, A	Filter, % T	Internal standard line, A	Filter, % T	Concentration range, ppm
Ru	3426.7	16	Au 3105.3	16	1 - 25
	3436.7	8	Au 3105.3	16	10 - 50
Si	2519.2	16	Au 3105.3	16	1 - 50
	2987.6	16	Au 3105.3	16	25 - 100
Sn	2840.0	32	Au 3105.3	16	5 - 25
	2840.0	16	Au 3105.3	16	25 - 100
Te	2385.8	32	Au 3105.3	16	10 - 100
Ti	3361.2	16	Au 3105.3	16	0.5- 5
	3349.0	16	Au 3105.3	16	1 - 10
	3349.0	3	Au 3147.6	3	10 - 100
Tl	3519.2	16	Au 3105.3	16	5 - 25
	2767.9	16	Au 3105.3	16	10 - 50
Zn	3345.0	32	Au 3105.3	16	5 - 50
	3345.0	16	Au 3105.3	16	10 - 100

PRECISION AND ACCURACY

The precision of the method was determined on a specially prepared sample to which the various impurities were added. Eight runs were made on the sample over a period of four days. Table VIII lists the mean concentration of each element together with the precision of the determination. The overall precision of the method calculated on the basis of a total mean concentration of impurities of 280 ppm was found to be 5.1% expressed as coefficient of variation.

No accuracy data are available since conventional chemical techniques are not suitable for the determination of the various impurities in gold at the low level required. It is assumed that the accuracy is comparable to the precision based on the method of preparing standards.

TABLE VII
Residual Impurities Con-
tained in Gold Blank

Element	Concentration, ppm
Ag	5.0
Ca	1.0
Cu	0.5
Fe	5.0
Mg	0.5
Pb	1.0

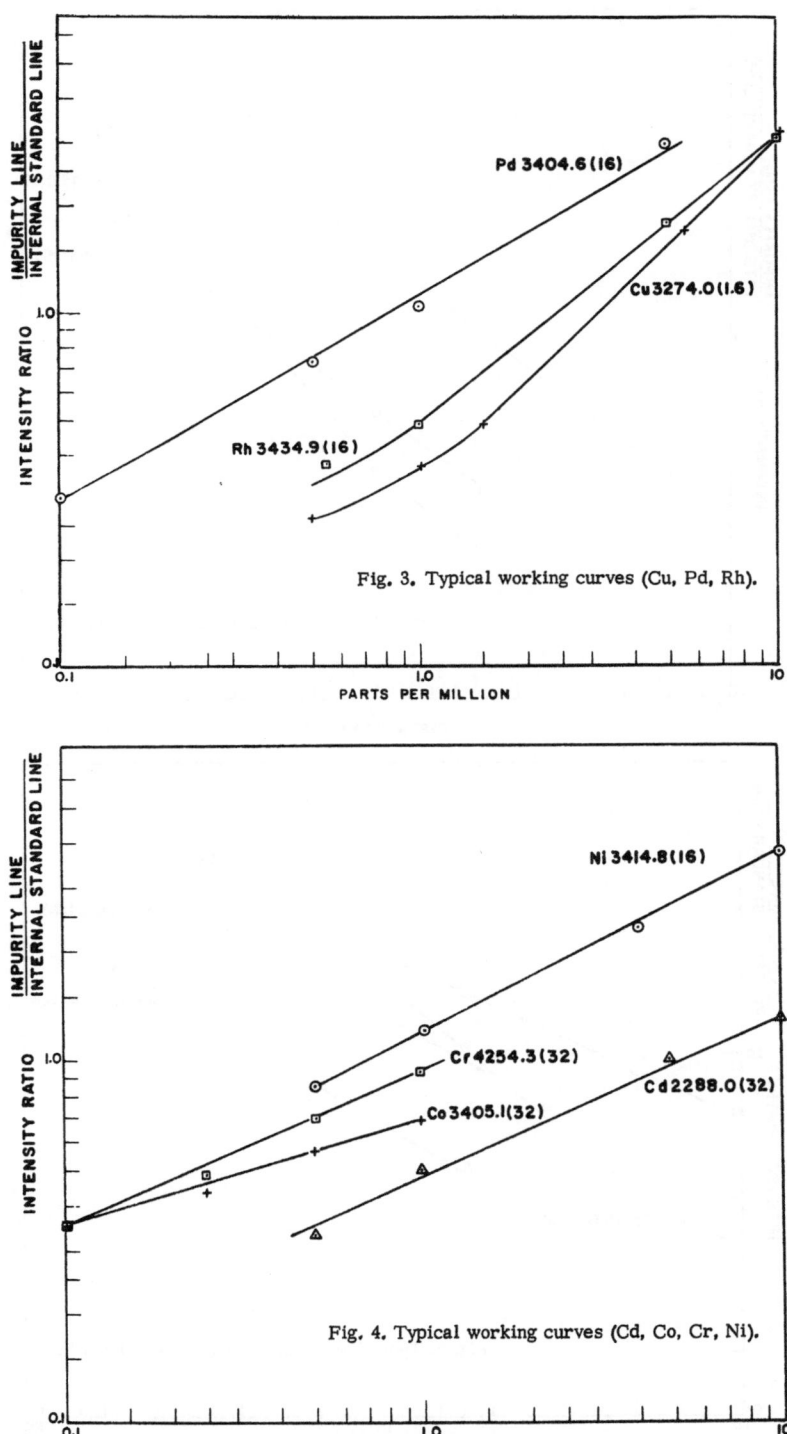

Fig. 3. Typical working curves (Cu, Pd, Rh).

Fig. 4. Typical working curves (Cd, Co, Cr, Ni).

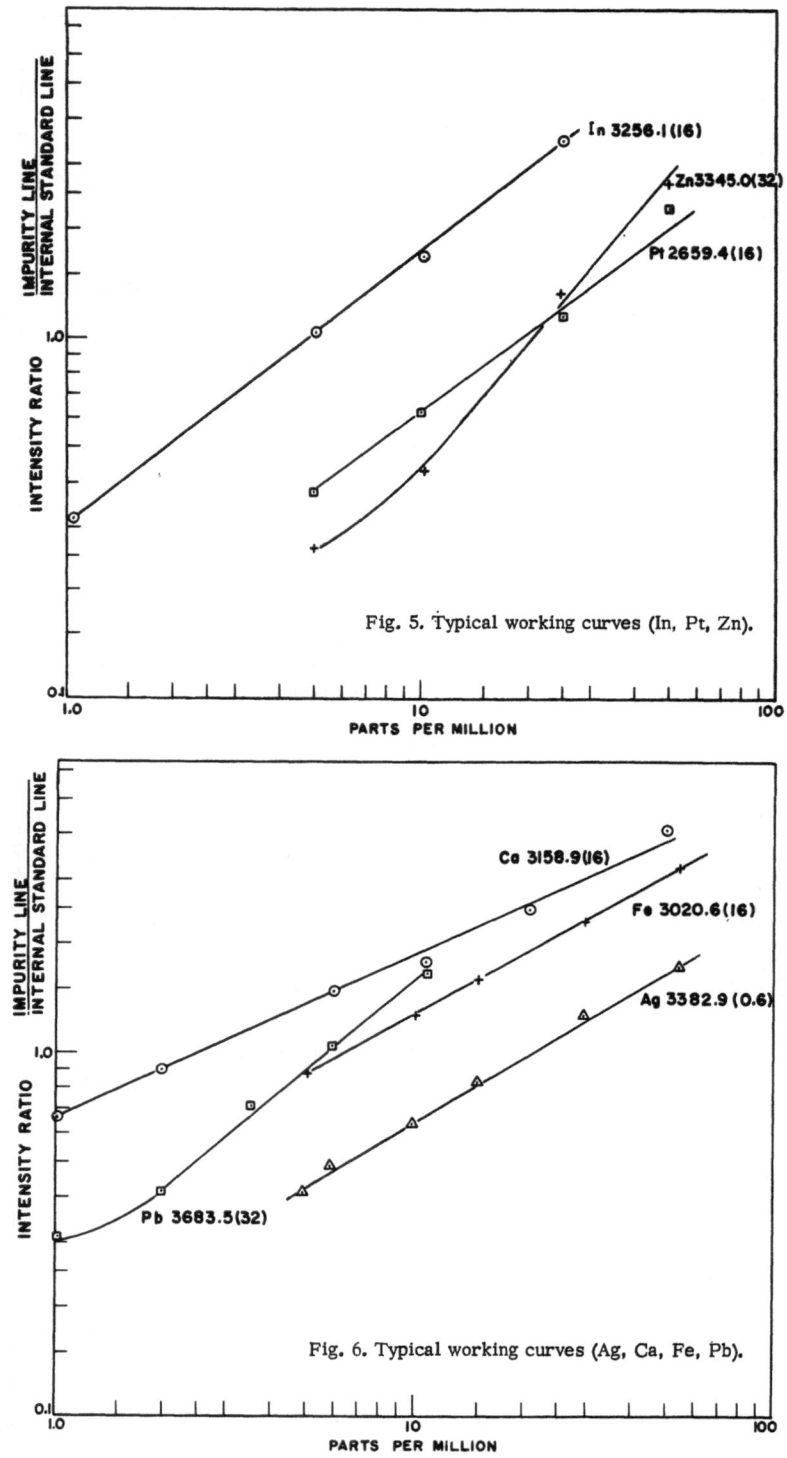

Fig. 5. Typical working curves (In, Pt, Zn).

Fig. 6. Typical working curves (Ag, Ca, Fe, Pb).

TABLE VIII
Data on Precision of Method

Element	Mean concentration, ppm*	Coefficient of variation, %
Ag	35.0	7.3
Al	12.0	14.2
B	1.8	52.8
Ca	3.8	26.3
Cd	1.2	25.8
Co	1.9	9.5
Cr	4.0	11.8
Cu	2.7	11.1
Fe	33.0	16.7
In	23.0	13.5
Ir	20.0	12.5
Mg	4.1	12.9
Mn	2.0	18.5
Ni	16.0	8.8
Pb	7.4	13.5
Pd	0.4	13.6
Pt	25.0	17.6
Rh	3.9	15.1
Ru	6.2	3.9
Si	8.0	41.2
Sn	6.5	14.0
Te	12.0	16.7
Ti	4.2	12.8
Tl	13.0	13.8
Zn	38.0	9.7
Overall precision for total impurities		5.1

*Eight determinations.

ACKNOWLEDGMENT

The authors are indebted to H. Robinson and R. Walyus for their preparation of the many standards required in this study.

REFERENCES

1. J. R. Churchill, Ind. Eng. Chem., Anal. Ed. 16:653 (1944).
2. W. J. Crook, Trans. Am. Inst. Min. Met. Eng. 152:255 (1943).
3. O. S. Duffendack and R. A. Wolfe, Ind. Eng. Chem., Anal. Ed. 10:161 (1938).
4. A. L. Esterhuizen, J. S. A. Inst. Min. Met. 62:739 (1962).
5. E. M. Hammaker, G. W. Pope, G. Ishidaan, and F. Wagner, Appl. Spectroscopy 12:161 (1958).
6. A. L. Kheifitz and L. N. Cherezova, Izvest. Akad. Nauk S.S.S.R., Ser. Fiz. 12:462 (1948).
7. A. J. Lincoln and J. C. Kohler, Anal. Chem. 34:1247 (1962).
8. A. J. Lincoln and J. C. Kohler, Paper presented at Pittsburgh Conference on Analytical Chemistry (March 1962).
9. W. C. Pierce and N. H. Nachtrieb, Ind. Eng. Chem., Anal. Ed. 13:774 (1941).

10. Y. D. Raikhbaum, Bull. Acad. Sci. U.R.S.S., Ser. Phys. pp. 246-9 (1941). (English Summary.)
11. Y. D. Raikhbaum, Zavòdskaya Lab. 10:168 (1941).
12. R. L. Rupp, G. L. Klecak, and G. H. Morrison, Anal. Chem. 32:931 (1960).
13. D. M. Shaw, O. Wickremashinghe, and C. Yip, Spectrochim. Acta 13:197 (1958).
14. Spex Industries, Inc., based on independent designs developed by D. L. Mash (Bell Telephone Laboratories), and R. Susman (Republic Aviation Corporation).
15. B. J. Stallwood, J. Opt. Soc. Am. 44:171 (1954).
16. A. Strasheim, D. B. de Villiers, and D. Brink, J.S.A. Inst. Min. Met. 62:728 (1962).
17. R. E. Thiers, Appl. Spectroscopy 7:157-63 (1953).
18. H. B. Warren and R. M. Thompson, Econ. Geol. 39:457 (1944).

Basic Techniques of Making Measurements on the Continuum and Determination of the Relative Intensities of Stellar Spectral Lines

R. I. Johnson

Adler Planetarium and Astronomical Museum
Chicago, Illinois

The accurate measurement of intensities in stellar spectra, both relative and absolute, must be made in all observable spectral regions in order to obtain the basic data for the computation of model stellar atmospheres. The effects of interstellar and atmospheric extinction, absorption, and scattering in the instrument must be especially considered in the work of absolute spectrophotometry. Differential atmospheric extinction and instrumental absorption may be neglected if the various points on a line are measured relative to the continuum. The determination of the continuum, however, involves many problems. Sensitive photographic and photoelectric techniques record a stellar spectrum in order to allow the determination of the distribution of energy with wavelength. The low intensity level of most stellar sources has encouraged the development of photoelectric spectrophotometry, which in certain applications obtains better results than photographic methods. However, the present slow rate of scan, wavelength by wavelength, even with several photocells simultaneously in use, indicates that photographic procedures will undoubtedly be the most used for several years to come. The methods of calibration of the spectrograms, the use of the "densitometer" (generally called "microphotometer" by astronomers) and the correction of its reproduction of intensity data are the basic techniques applied to gain the fundamental data for different examples of stars and stellar systems. The determination of line profiles, both theoretically and by measurement, their importance to the theory of model atmospheres, the importance of high-dispersion studies of selected line profiles, and related procedures are further problems. The intensity distribution of the continuous spectrum for stars of different spectral types, the classification of stars by spectra, the theoretical interpretation of the measurements, the problems resulting from uncertainties in the determination of the continuum and stellar equivalent widths to approximately 10%—a permissible accuracy—completes the acquisition of data for the description of the atmosphere's physical structure.

This paper deals with astrophysical reasearch, an area of study which is probably quite unfamiliar to most spectroscopists. It is intended to provide a general survey of current procedures used by astronomers in attempting to determine quantitatively the chemical composition and physical structure of the outer layers of stars from studies of stellar spectra. The paper is necessarily wide in scope because of the comprehensiveness of the problem of interpretation, and the fact that many astrophysicists seeking the answers to problems of stellar evolution have recently made diverse contributions toward gaining its general solution.

The discussion will be restricted to some theory, the basic techniques of gaining spectrophotometric data by making measurements on the continuum, and the determination of spectral line intensities relative to the continuum. A few considerations of the problems of absolute spectrophotometry and of the astrophysical studies of stellar atmospheres will also be stated briefly.

A further restriction shall be that only normal stars (all of which, however, are in some way unique) which produce absorption spectra are used as examples. These stars are in general similar to the sun in composition and structure, but the sun will be omitted from this discussion because of differences between the techniques applied in solar and stellar research. The sun is unique inasmuch as the solar spectrum can be studied for different points on the sun's disc, but the spectra of the other stars can be formed only from their integrated energy. With the exception of the sun, all stars are point sources. Their radiation is therefore received from all parts of a hemisphere, and many different levels of the gaseous stellar atmosphere, all of which have different temperatures. Thus, the final estimates of the temperature variation in depth have to be derived from the integrated radiation of different strata of the star's atmosphere and the entire apparent surface.

This treatment also concerns only stars with luminosites that do not vary with time over astronomically short periods, stars whose atmospheric or photospheric layers are small in comparison with their radii, stars that are not characterized by peculiarities in their spectra due to unusual chemical composition or rapid mass motions in their atmospheres, and stars that do not have extremely high or low temperatures since the greater part of their energy flux lies outside of the optically observable range. The basic techniques described in the following are, however, also applied to unusual objects.

The optically opaque, most dense layer of a star's atmosphere is its photosphere. The photospheric layers emit almost all the radiant energy in the optical region, and are basically where the continuous spectrum of the star is formed. The absorption lines of the stellar spectrum are formed in approximately these same layers. The photosphere is the level which determines stellar luminosities, and it may be assumed to consist of stratified plane-parallel layers. Curvature and some other factors may be neglected in an elementary study.

In addition, in this brief consideration it must be assumed that all effects of the earth's atmosphere and interstellar matter on the stellar spectra have been removed. The problems of absolute spectrophotometry in astronomy are difficult since one must account for not only instrumental influences, but also the effects of differential atmospheric extinction in the earth's atmosphere and interstellar extinction.

A casual comparison of stellar and laboratory spectra shows that the closest parallel between them is that red stars show only strong lines of the low-temperature classes, stars of intermediate temperature (such as the sun) show fully developed arc spectra, and enhanced lines alone are found in the spectra of blue stars. This, however, indicates only that the spectra of elements in stars and in the laboratory are consistent, but does not contribute to our understanding of the formation of either. The temperature range in stellar atmospheres has also been found to be relatively the same as that which can be created for very short periods of time in terrestrial laboratories.

The determination of line profiles may be quite accurately accomplished when the chosen points on a line are measured relative to the continuum, thus approximately removing the effects of variable atmospheric extinction and instrumental absorption. But still, the determination of the position of the continuum where there are many lines and the exclusion of the instrumental profile (which determines to a great extent the shapes of many lines in stellar spectra, except where high dispersion is obtained) create many problems. For a few stellar spectrographs of highest dispersion, the instrumental profile has been determined photographically from the shapes of weak comparison emission lines.

The photographic plate is most often used where whole regions of a spectrum are to be studied. This is because of its ability to record an extended range of a stellar spectrum faster than can be presently done with photoelectric equipment. Though the time required for the reduction of the photographic plate is great, and measures from several spectrograms have to be averaged in order to gain accuracy comparable to photoelectric measures, they will continue to be used where spectral ranges have to be obtained. They are usable over a range of wavelengths from approximately 2900 A (our atmospheric cutoff) to 12,500 A.

The photographic plate is also a desirable astronomical tool for several other reasons. It integrates the light received over a given exposure time; the emulsion is stable over long periods after it has been developed; and though not efficient in converting energy into darkening of the emulsion, it is reasonably rapid.

The use of objective prism spectrographs, in which a prism or grating is placed over the objective of a telescope or camera, allows photographic surveys and discovery of previously unknown objects to be accomplished in a short period of time. Unfortunately, no one has yet devised an entirely suitable way of impressing a comparison spectrum on an objective prism plate (Fig. 1).

Since a photographic plate does not respond linearly to the amount of radiant energy it is exposed to, an appropriate calibration spectrum must be imposed on each plate along with the stellar spectrum. Its other disadvantage is that reciprocity failure and the intermittency effect make it necessary that the stellar and calibration exposure times be as close to equal as possible. Because almost all stars are weak sources and must be trailed on the spectrograph slit in order to broaden the spectrum, this equality is difficult to achieve. The photographic plate's variation of sensitivity with wavelength necessitates that a family of characteristic curves be obtained from the calibration spectra or spots. Each of the curves must correspond to a certain wavelength range recorded in the stellar spectrum. The wavelength used for a particular calibration in astronomical spectrophotometry is usually only required to be within 100 A of the wavelength being measured, except where the plate sensitivity is known to change rapidly when the difference must be less.

When fast emulsions of the various types (sensitive to different spectral regions), which are usually required for astronomical studies, are used, resolving power is reduced because of the large grain size. These emulsions generally have lower contrast than slow, fine-grained emulsions. They therefore decrease the slope of the characteristic curve allowing a wide range of intensities to be measured—though with decreased accuracy. No strong advantage can be gained in astronomy by using "fine-grain" developers because of the resulting decreased density.

The Eberhard effect, the greatest source of error incurred in development, must be prevented as much as possible. This is especially true where high- and low-density latent images are near each other, causing the de-

Fig. 1. A portion of a photograph of stellar spectra obtained by use of a Schmidt camera with an objective prism in place.

velopment to proceed at an unequal rate resulting in too dense an image in some areas and too weak an image in adjacent regions.

Other effects must also be considered, particularly if absolute measurements are to be attempted. The latent-image effect requires that the calibration and stellar spectra be imposed simultaneously, or that development be delayed for a few hours after the second exposure if simultaneous exposures are not possible. It is also desirable that the two spectra be positioned close together on the plate in order to avoid variations in the emulsion or glass. Generally, fogging effects can be easily avoided. Then it is possible to achieve the ultimate accuracy of measurement.

For relative rather than absolute measurements, it can be assumed that two regions of equal density on the same photographic plate were created by energy of the same intensity, if they were given the same exposure and development. This is especially true if they correspond in wavelength and polarization and if the same or equivalent light and optical paths were used. But in no case should a characteristic curve be extrapolated beyond its well-determined length.

One solution to the problem of placing both calibration and stellar spectra on the plate simultaneously has been to use two identical spectrographs with comparable dispersions. All scattered light should be eliminated by stops and diaphragms within the auxiliary spectrograph if one is used.

The source for the calibration spectrum may be an arc, a tube of pure gas, or an incandescent lamp behind a suitable glass screen and filter producing a wavelength distribution similar to that of the stellar spectrum to be examined. The slit must be evenly illuminated by the source.

There are many methods which are in use for impressing intensity calibrations, as required for suitable photographic calibrations. Two of the commonly used procedures of calibration and some of the requirements for proper photographic calibration might now be considered.

Step-filters have been made by sputtering different thicknesses of substances such as platinum on optical quartz plates. These are very suitable since they deviate only slightly from neutrality over lower wavelengths, and are elsewhere quite stable. Photographic plates have also been used because of their stability and neutrality, but some scattering in the emulsion occurs and photoelectric methods should be used to calibrate and test the plates while they are in the spectrographs.

The rotating step-sector is often used since its steps can be cut mechanically with high accuracy for the desired intensity ratios. The greatest step of intensity is produced by radiation on the plate during a full turn of the sector, and cuts can be made over successively smaller angles to a minimum limit of approximately 1°. A step ratio of 1.58 (0.2 in the logarithm) has proved satisfactory. Sectors can be constructed to cover a range of more than 100 in intensity. (This exceeds the accuracy obtainable from most photographic plates.) The intermittent exposures of the light source cause the darkening of the emulsion to approach asymptotically that produced by continuous exposures when the speed of the sector is great. The frequency of exposures required to gain approximate equality also is dependent on the size of the photographic grains and the intensity of the continuous source. Frequencies of approximately 1000 rpm are satisfactory for most astronomical photographic plates.

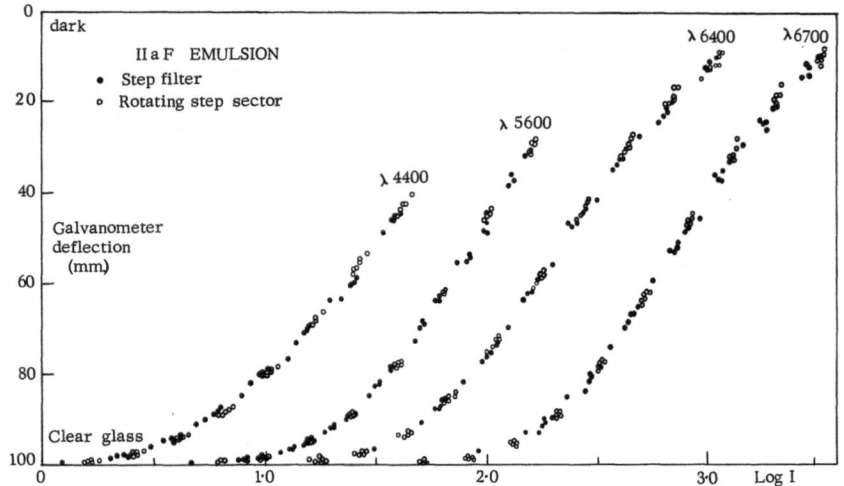

Fig. 2. Characteristic curves obtained at Victoria for the IIa-F emulsion, using rotating step-sector and step-filter.

Figure 2 shows a comparison of characteristic curves gained by using a rotating step-sector and a step-filter of sputtered platinum on quartz at the Dominion Astrophysical Observatory in Victoria, British Columbia. The results based on six plates, of IIa-F emulsion, are shown as galvanometer deflections plotted against the log of the intensity. The tests indicate that for this emulsion the rotating step-sector produces a calibration the same (within the probable error) as that of the step-filter. If rotating sectors are used, however, they should be tested against other methods for each type of emulsion in order to ensure that the calibrations agree.

The stability and accuracy of a calibration system should be checked by impressing two or more sets of calibration indices on a given plate with all conditions the same except for the intensity of the light source. Identical calibration curves at each wavelength, though the intensities of each mark differ, indicate that the system is adequate. If the characteristic curves are not in agreement, the source of error must be established and removed.

The calibrating devices described above were chosen on the assumption that discrete densities are to be obtained which, when determined with a microphotometer, are to be plotted against known relative intensities to gain the characteristic curve needed for the spectrum being examined.

When the spectrogram is first examined the true positions of the lines and their identification must be established. This is accomplished with the aid of the comparison spectrum that was impressed upon the plate while it was in the spectrograph attached to the telescope. Both the narrow and widened spectra of the star Arcturus with comparison spectra from a titanium spark above and below, are shown in Fig. 3.

From a comparison of the positions of the stellar and comparison spectrum lines the relative radial velocity of the star is determined from the Doppler shift. Then the true wavelengths of the stellar lines are calculated. The identification of these lines can then be made with the aid of a list of atomic spectral lines such as the "Revised Multiplet Table" by Charlotte

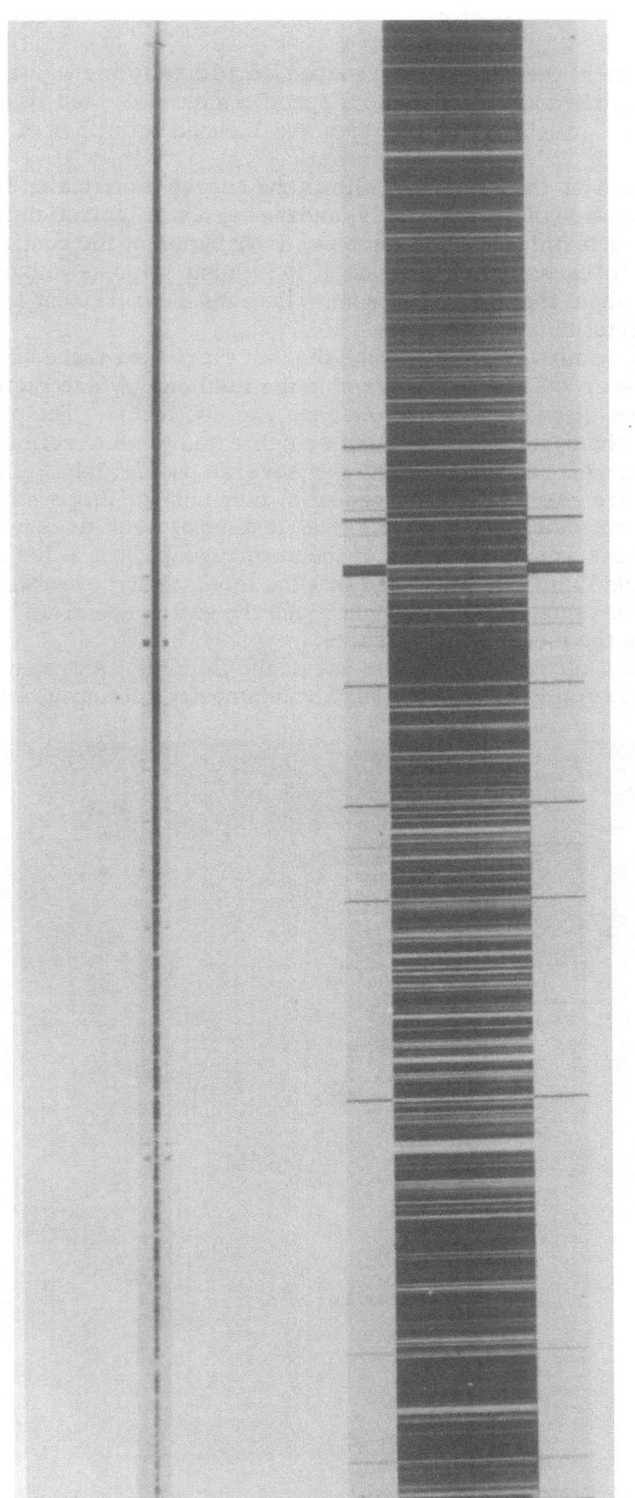

Fig. 3. The narrow and widened spectra of Arcturus with titanium spark comparison spectra above and below each.

M. Sitterly, in which the lines are grouped according to multiplet. If a certain line is identified as belonging to a specific multiplet, then stronger lines must also be present (though they are blended with lines of other elements).

The spectrogram of the star then allows the energy distribution in the spectrum, the population of energy levels, and the degree of ionization in the atmosphere to be determined. The intensity distribution of the continuous spectrum must also be established since it is in many ways as important, if not more so, than the line spectrum. Both its measurement and its theoretical interpretation are difficult.

The equivalent widths and profiles of the lines are next traced by use of a microphotometer. This is to determine the total energy subtracted in terms of equivalent angstroms of the neighboring continuum. The profile derived from the photographic plate must be corrected for diffraction, resolving power, perhaps scattered light, and several photographic effects. Equivalent widths are essentially independent of instrumental dispersion and resolving power, or can be made so. The centers of weak lines may be filled in by the finite resolving power of the spectrograph, but at the same time the width of the line is increased so that the total amount of subtracted energy remains the same. Scattered light along the entire spectrum has to be corrected for in the measured intensities.

A rather typical microphotometer in use at the Dominion Astrophysical Observatory is shown in Fig. 4. The constant luminosity continuous source

Fig. 4. The microphotometer at the Dominion Astrophysical Observatory, Victoria, B. C.

A, is focused on slit B, which is set perpendicular to the direction of the dispersion of the spectrogram D, on which the light from A is focused by lens C.   The spectrum is then focused by lens E on the analyzing slit F, behind which there is a photocell.   An amplifying system then increases the photoelectric current and causes a galvanometer deflection which is recorded by a beam from light source H on the galvanometer mirror, which is focused by a cylindrical lens K on a roll of photographic paper enclosed in a light-tight box.   When the spectrum is moved in the direction of the dispersion the paper passes the cylindrical lens at a rate which is magnified 20 or 100 times by means of a gear train.   An auxiliary system allows viewing of the spectrum so that correct alignment may be achieved and the desired spectral region selected.   When calibrations, as well as clear glass and zero deflections, are repeated after as much as a run of an hour no differences in observed deflections can be detected on the tracing—when conditions are favorable.

In working with astronomical spectra the accuracy that can be attained in physical laboratories cannot be matched.   Reproduction of intensity data with an error approaching less than 2% is not possible.   When this degree of accuracy is gained it is because the characteristic curve was obtained under ideal conditions, and from spectra calibrated by the same source as the spectra being studied.

The principle sources of error in the microphotometer are because the slits B and F are possibly too wide.   If B is too wide there may be scattered light from the emulsion and an error results from the light reflected by the lenses (which varies with the density of the emulsion).   If slit F approximates the width of the spectrum line there may be integration effects.   A long slit should also be used so that the grain effect on the microphotometer tracing is minimal.   The adjustment of the focus and too rapid motion of the plate are also possible sources of error.

The Victoria microphotometer with its amplifying system records high densities well, and is quite suitable for studying intense absorption line spectra.

The reduction of microphotometer tracings to intensity tracings, by the instrument shown in Fig. 5, is accomplished by converting the continuous microphotometer trace into an intensity trace by following the characteristic curve.   The Victoria intensitometer can be used to obtain direct-intensity traces, or rectified intensities (by first using a logarithmic characteristic curve and then repeating the process with an antilogarithm curve).   This feature is useful when it is desired to separate two stellar spectra, or in order to change the scale of a tracing.   (A direct-intensity recording microphotometer incorporating an X-Y recorder is now in use at Victoria, the California Institute of Technology, and other places.)

The characteristic curve is determined by means of the calibration marks on the microdensitometer tracing, and reproduced on an illuminated ground-glass screen, G.   The image of the curve is then projected onto the tracing. As the logarithmic tracing in box B is driven along track T the operator keeps the projected characteristic curve at the intersection of the microphotometer tracing and a standard index line I by means of the wheel W.   The moving paper roll at the right then has intensities recorded on it by a pen which moves with the plate containing the characteristic curve.

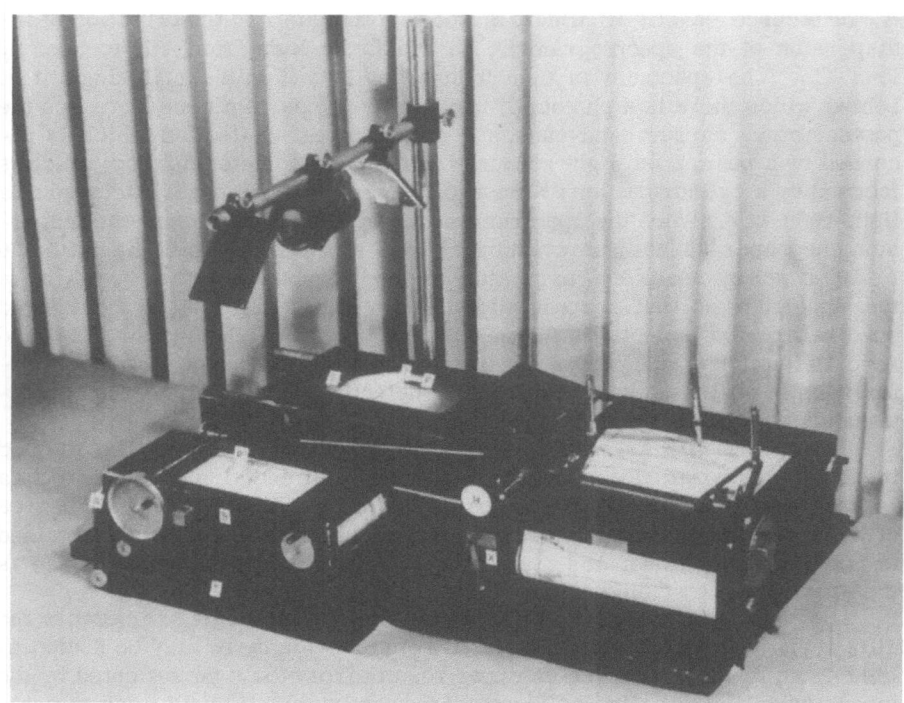

Fig. 5. The intensitometer at the Dominion Astrophysical Observatory, Victoria, B. C.

The scale of the tracing is a function of the scale of the characteristic curve and the magnification of the intensity tracing which can be varied by a gear train X. Other transformations can be gained by moving the box B containing the microphotometer tracing or by substituting straight lines on the ground-glass screen for the characteristic curve.

Examination of the intensity tracings obtained for representative stellar spectra indicates the great difficulty in locating the continuum for stars cooler than approximately 6500°K. The continuum may be derived from determination of the high points on the intensity profiles, but many of these cannot be detected because the dispersion is usually low.

A true stellar absorption line profile is a function of the absorption coefficient at that wavelength; it is therefore essential for the determination of the distribution with depth in the stellar atmosphere of the number of active atoms producing a line. It is even difficult to determine the true profile in stellar spectra with sharp lines according to one distribution of temperatures and pressures for given abundance element ratios and surface gravity. Only the theory of model atmospheres makes possible better than approximate calculations.

The empirical determination of line profiles is influenced by the shape of the spectrographically observed line. Instrumental effects are almost absent from broad lines. However, the spectra of some stars contain absorption lines so sharp that they are defined by the instrumental profile. If the instrumental profile of a spectrograph has been determined, $h(x)$, the true profile

$f'(x)$ and the observed profile $f(x)$ are related by the integral equation

$$f(x) = \int_{-\infty}^{\infty} h(x-y) f'(x) dy \qquad (1)$$

However, if the half-width of $h(x)$ is comparable to $f'(x)$ the true profile is difficult to ascertain.

The profiles of the measured absorption lines have to finally be interpreted theoretically. The equivalent widths of the stronger lines may be measured and placed on a relative scale. The relationship between the equivalent width and the number of active atoms producing a line has been developed, and is known as the "curve of growth." Its application entails the use of a large quantity of observational data to deduce parameters such as mean temperatures, pressures, and chemical compositions. The simple assumption which is fundamental to the "curve of growth" principle is that the characteristics of a spectral line may be calculated as if they were formed at a fixed height in a stellar atmosphere. This approximation does not result in great percentage errors when only lines of atoms or ions in the same stage of excitation and ionization are compared. The curve of growth can be used to establish an average value of the temperature and density in a stellar atmosphere upon which a model of the stellar atmosphere may be calculated.

The solution to the problem of absorption line formation must be gained from the spectrum of a stellar atmosphere composed of gases which absorb varying degrees of radiation at all wavelengths. For example, the photospheric layers are appreciably opaque. Hydrogen, primarily because of its abundance, is the chief source of the opacity of the stellar atmospheres of solar-type stars. Negative hydrogen ions absorb quanta in the region less than 8000 A. Free electrons may pass near neutral hydrogen atoms, which then absorb energy in the infrared ("free–free" transitions). These are the principal sources of absorption in the far infrared.

The coefficient of continuous absorption for the hotter stars may be calculated if the temperature and electron pressure are known.

The general temperature distribution of a normal star varies over a great range from its outer tenuous envelope to its dense center. For example, the sun's photosphere has a temperature of about 6000°K; then beyond 40,000 km above the photosphere, in the solar corona, temperatures range from 1,000,000 to 2,000,000°K. (The corona is transparent in the optical region and does not affect radiation from the photosphere.) The chromosphere is a turbulent region in which the rise of temperature evidently averages 50 deg/km. Temperatures range in the chromosphere, separating the photosphere from the corona, from about 4500 to about 30,000°K. The corona and chromosphere are important only in the far ultraviolet beyond the limit of the Lyman series, and in the microwavelength region.

Inside the star, the temperature monotonically increases toward the center. But an average temperature, that of approximately the midpoint on the radius, may be determined and found adequate for some approximations yet inadequate for many others. The interior temperatures of normal stars are usually measured in the low tens of millions of degrees. The general range of temperatures in the deepest interior of the sun, for example, is from

1,000,000 to 20,000,000°K. The nuclear processes which release energy in the stellar interior do not take place to any significant degree in normal stellar photospheres or atmospheres. The exchange of thermal energy in the photosphere is essentially by radiative transfer.

Excitation temperatures are found from comparison of the relative populations of different levels of excitation. The numbers of atoms capable of absorbing the various lines are determined from spectral line intensities. This value may not necessarily agree with the local temperature defined by the gas-kinetic motions, but if local thermodynamic equilibrium exists, it closely agrees with the ionization temperature as well as with the kinetic temperature. The following general equation may then be used:

$$\log N_j - \log N_i = \log (\lambda' k f' g' / \lambda k' f g) - (5040/T) (\chi' - \chi) \tag{2}$$

In equation (2) the primed and unprimed symbols refer to quantities of levels j and i, respectively; N_j and N_i are obtained from knowledge of the equivalent widths of two lines and the curve of growth for lines of wavelength λ' and λ, respectively, whose relative f values (oscillator strengths) are known, and whose excitation potentials are χ' and χ, respectively. The absorption coefficients k and k', the statistical weights g and g', f', and the coefficient for $1/T$ are all known. (Unfortunately, f values are known only for a few transitions.) Several lines may be grouped according to χ, and a least-squares solution for T obtained.

By use of equation (3), a Planckian curve for some color temperature can be matched to a limited spectral region of a stellar spectrum. Color temperatures derived by this type of comparison usually differ from one spectral region to another, and do not necessarily have a close relationship to any other type of temperature.

$$\log(F_{\lambda_1}/F_{\lambda_2}) = \frac{\lambda_2^5 \, [\exp (hc/\lambda_2 k T_{c_2}) - 1]}{\lambda_1^5 \, [\exp (hc/\lambda_1 k T_{c_1}) - 1]} \tag{3}$$

where F_{λ_1} and F_{λ_2} are the measured fluxes at λ_1 and λ_2.

The color temperature may be expressed by a "spectrophotometric gradient." The relative flux distribution of the star is compared to that of a standard source or star of known energy distribution in order to gain this quantity.

$$G \equiv \frac{d}{d(1/\lambda)} \ln (F_\lambda / F^0_\lambda) \tag{4}$$

The relative gradient, as given by equation (4), can be obtained without great difficulty by making a long series of observations and comparing them with standard stars which have been studied extensively and in detail. The color temperature of stars determined from data on both sides of the Balmer discontinuity is probably the most accurate.

If the absolute flux at different wavelengths can be determined, the brightness temperature can be established. In equation (5) F_λ is the emergent flux density at the stellar photosphere, not the same as the F_λ's in equations (3) and (4), which are measured.

$$F_\lambda \Delta \lambda = (2\pi hc^2/\lambda^5) \frac{\Delta \lambda}{\exp(hc/\lambda k T) - 1} \tag{5}$$

TABLE I
Suggested Effective-Temperature Scale
(Main-Sequence Stars, Luminosity Class V)*

Spectral class	T_e, °K	Spectral class	T_e, °K	Spectral class	T_e, °K
O5	45,000	A0	9450	K0	5170
O7	40,000	A2	8900	K1	5000
O9	37,000	A5	8220	K3	4660
B0	33,500	A7	7800	K5	4350
B1	27,500	F0	7200	K7	4000
B2	23,500	F2	6850	M0	3520
B3	20,500	F5	6500	M1	3400
B4	19,000	F6	6400	M4	3150
B5	17,500	F8	6200	M5	3050
B6	16,300	G0	5970	M6	2950
B7	15,700	G2	5780	M8	2700
B8	13,500	G5	5570		
B9	11,600	G8	5350		

*As given by L. H. Aller in "Optical Spectrometric Measurements of High Temperatures" (The University of Chicago Press, 1961).

For this quantity to be found, the angular size of the star and amount of flux arriving at the earth in absolute units must have been determined.

The effective temperature derived from the flux establishes the luminosity, which is a necessary datum required in order to interpret the physical structure of a star. The effective temperature is that of an absolutely black spherical surface of the star's diameter radiating an exactly equal amount of energy per each unit area. The luminosity is given by

$$L = 4\pi R^2 \sigma T_e^4 = 4\pi R^2 \int_0^\infty F_\lambda dx \qquad (6)$$

Thus, effective temperatures cannot be measured directly. Only calculation of a model stellar atmosphere, in which the variation of temperature and density with depth is specified, allows the effective temperature to be determined accurately. The mechanism of energy transport is usually assumed to be that of radiative equilibrium in the outer layers, and at times, convective equilibrium in the deeper layers. From these models the fluxes are computed and compared with the observations. (This has verified that the layers from which radiation escapes directly are to a good approximation in radiative equilibrium.)

The theoretical effective temperatures, T_e, can then be given to the spectral classifications for the normal stars.

At any given level in a stellar atmosphere the temperature varies, and in general the atmosphere is not in local thermodynamic equilibrium. Therefore inaccuracies in the estimation of temperatures based on formulas involving local thermodynamic or hydrostatic equilibrium conditions are incurred. The deviation from thermodynamic equilibrium is greatest near the top of the atmosphere, but decreases in intermediate layers. It depends essentially on the energy transfer mechanisms in action. The temperatures in stellar atmospheres have been defined in terms of the processes occurring within them. Ionization, excitation, and dissociation are the commonly used bases for temperature scales. The deviation of the atmosphere from thermodynamic equilibrium prevents one temperature from describing all physical characteristics of the spectrum. None of the several definitions of stellar temperatures allows unique determination of the total range of stellar spectral features. The "effective temperature," which is defined in terms of the total energy emitted by the star, must be determined from observations and through the computation of model atmospheres. It is not directly observable because of our inability to observe the low- and high-temperature wavelength ranges of stellar spectra.

Evidence indicates that throughout the most strongly emitting region of the stellar atmosphere the velocity distributions of electrons and ions closely approximate a Maxwellian distribution at some gas-kinetic temperature. Therefore the temperature defined in terms of radiation density can be assumed equal to the gas-kinetic temperature only for thermodynamic equilibrium.

Ionization temperatures determined from the successive stages of ionization of the same element showing different intensity maxima of its lines also allow an approximate value of the electron pressure to be determined along the temperature scale.

With knowledge of the energy levels and the "transition rules," the relative numbers of atoms that produce the lines of a multiplet can be calculated. Plotting the measured equivalent widths of lines against relative numbers of atoms gives a portion of a curve of growth. By combining the results from a large number of multiplets, a fairly accurate curve of growth can be constructed. It then provides an estimate of the total number of atoms that are effective in producing each line, provided their efficiency of absorption is known. (This must be gained by laboratory experiment.) In practice, because of terrestrial atmospheric absorption, the sum of the number of atoms

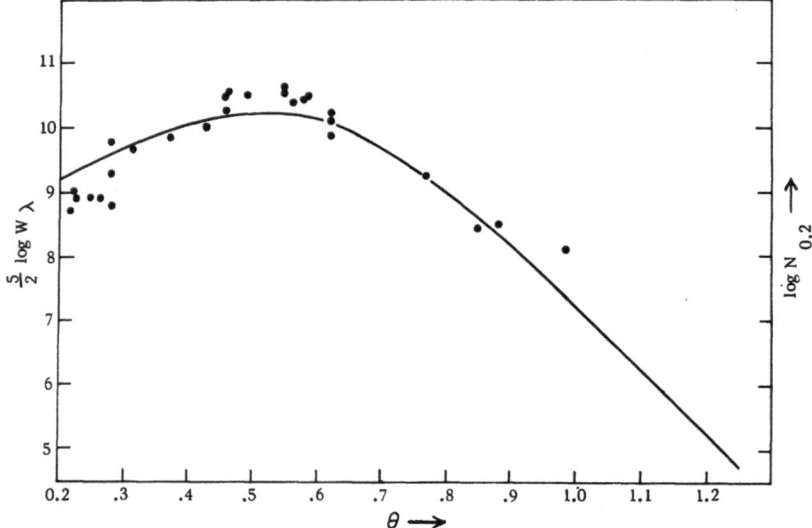

Fig. 6. The theoretical curve showing the variation with θ (an inverse function of the temperature) of the $n = 2$ level of hydrogen and the observed equivalent widths of H_γ.

that produce all observable lines of an element is most often not the total number, but only a percentage. The theory of ionization and excitation permits the percentages to be calculated, and therefore the total number of atoms of each element in a square centimeter column of a stellar atmosphere to be estimated.

Figure 6 shows the predicted variation with T ($\theta = 5040/T$) of the population of the $n = 2$ quantum level of $H\gamma$ compared to the observed intensities of $H_{i\gamma}$ (equivalent widths in mA) indicated by dots.

This approximate quantitative chemical analysis of the spectrum allows the abundances of the various elements that have observable lines to be estimated. The elements show no distinct variation in amount from hot to cool normal stars. The average chemical composition of these stars is approximately 75% hydrogen by weight and about 24% helium. Sixty-seven other elements have been found with oxygen and neon outweighing any other. In spite of the great numbers of strong metallic lines in stellar spectra, the metals represent only a fraction of 1% of the total composition.

The general study of stellar structure and evolution requires that stars of different types be classified in an orderly sequence according to age. These classifications are based in part on the estimates of stellar surface temperatures and gravities gained from spectrograms.

Usually particular elements in more than one stage of ionization can be recognized in bright stars. In comparing hot (45,000°K) to cool (2700°K) stars the differences in the spectra are found to arise essentially because of differences in temperature, and to an extent, in atmospheric density. Figure 7 shows the variation in intensity of absorption lines for several elements throughout the spectral classifications of normal stars.

The appearance of hydrogen lines throughout the spectral sequence serves as an example of the excitation and ionization effects of temperature. In the hotter stars, since more Lyman a is present and collision energies are greater, more atoms are excited by both processes. The Balmer series is then absorbed by the excited hydrogen atoms.

In the cooler stars few hydrogen atoms are in the second level and the Balmer series is weak, although there has been no decrease in the amount of hydrogen present. The hydrogen absorption lines steadily strengthen from

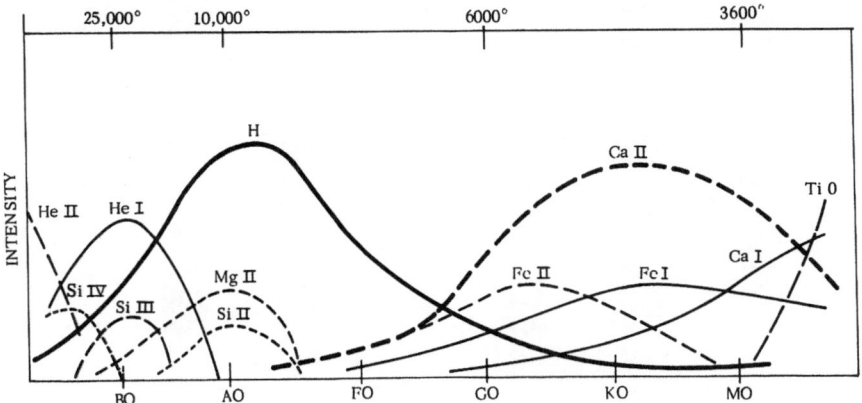

Fig. 7. The intensities of various absorption lines through the spectral sequence (after L. H. Aller).

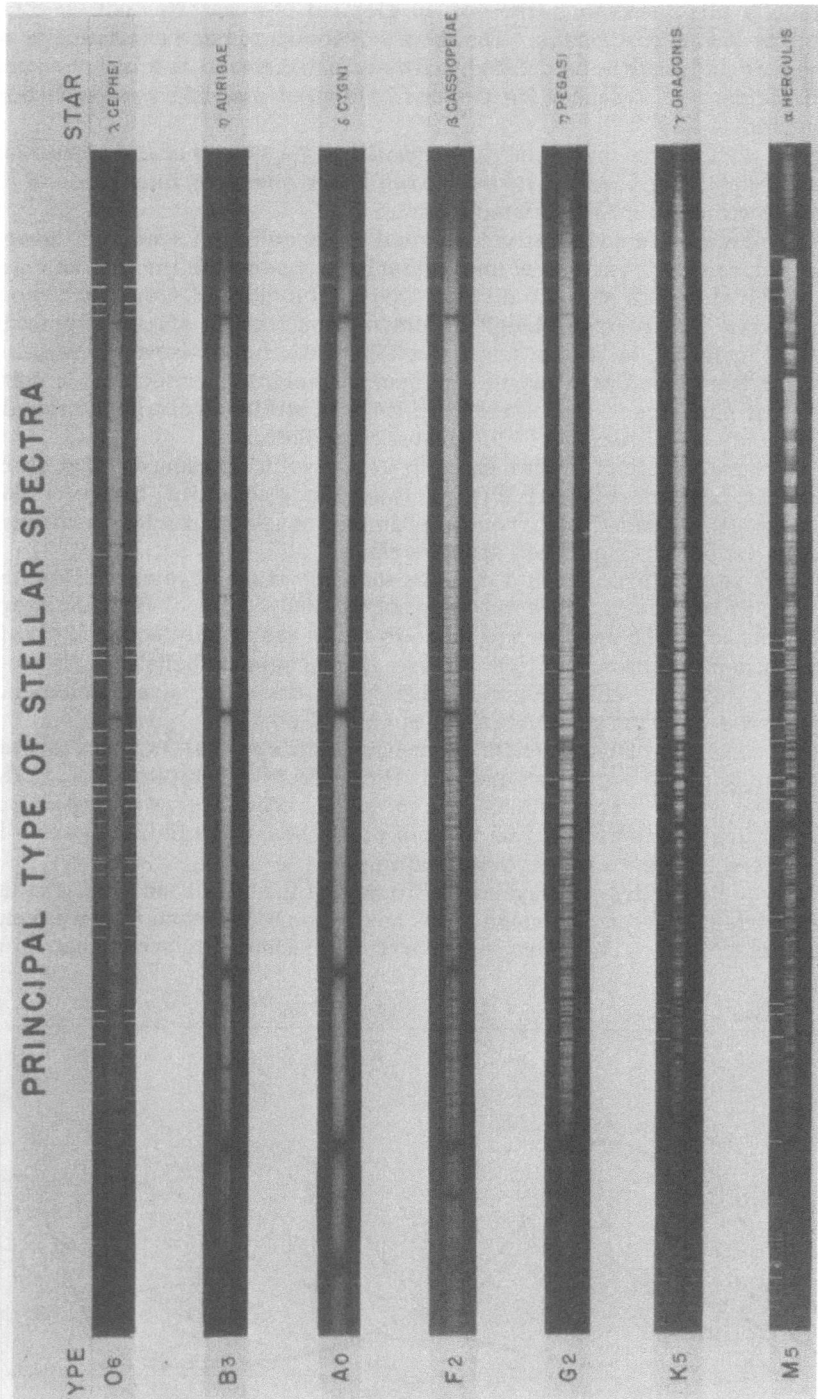

Fig. 8. Stellar spectra of classes O, B, A, F, G, K, and M showing continuous change in many absorption line intensities through the sequence. (Mount Wilson and Palomar Observatories photograph.)

the cooler classes to class A0 (9450°K). Then through the high-temperature end of the sequence rapidly increasing ionization decreases the number of excited hydrogen atoms, and the Balmer series becomes weaker through classes B and O.

The lines of helium, beginning with cool M stars and progressing toward class O, are not found until near class A0. When increasing temperature matches the ionization potential of helium many more helium atoms are excited and these lines intensify until class B3 is reached. Ionization now begins to dominate and the lines weaken in hotter stars. However, faint lines of ionized helium begin to appear in class B0 and strengthen until class O5, in which no trace of neutral helium appears. The hottest known type of star which produces a normal absorption spectrum is an O5 star.

Metallic lines apparently behave differently from those of hydrogen and helium because the metallic lines of low excitation lie mainly in the observable part of the spectrum. In classes of cool stars metallic lines are strong; in stars of higher temperature, lines of high excitation increase in strength. And, as ionization increases, all the lines of neutral metals weaken through classes G (approximately 5000 to 6000°K) to F (approximately 6000 to 7000°K). Ionized metallic lines (enhanced lines) appear in class K and strengthen to maxima in classes G and F. These lines then weaken in hotter stars and disappear near class B8 (13,500°K) because second ionization becomes complete at this point. Doubly ionized atoms have lines in the observable region of the spectrum, but these are few in number. A few lines of iron III occur in spectra of class B. The lines of most doubly or triply ionized metals are in the far ultraviolet.

Silicon is exceptional in showing four different stages of ionization in stellar spectra. Neutral silicon has strong lines in intermediate temperature stars, silicon II has its strongest lines in class A0, silicon III reaches maximum in class B2, and silicon IV in class O9.

Strong molecular bands are found in the cool stars in the classes lower than G, where the temperatures are less than 6000°K.

Now a brief look might be taken at the physical sources of line distortion with which the astrophysicist must contend. Turbulent motion of the gases in a stellar atmosphere produces Doppler broadening of the lines. When absorption over the width of the Doppler core is complete a large increase in the depth of gas occurs without a proportional rapid gain in line intensity. Finally wings begin to develop and the line strength again increases. While the Doppler core is increasing in intensity, the line's strength is proportional to the number of atoms. Then the intensity increase almost ceases with the addition of more atoms. When the wings increase in strength the line intensity becomes proportional to the square root of the number of atoms. In normal stars a slight amount of microturbulence is present, but it does not exceed the thermal motion of the atoms.

In the spectra of hot stars the intensities of the hydrogen lines are influenced by the Stark effect, causing the absorption to be spread over a considerable range of the spectrum producing "wings." These wings constitute a great portion of the intensity of the hydrogen absorption lines of the stellar atmospheres containing a large number of ions. The rotation of stars can be detected and must be considered for hot stars and for certain classes of peculiar stars which rotate relatively rapidly. In some cases the Zeeman effect must also be considered and, in other extraordinary spectra, the reasons for the presence of emission lines must be found.

These general considerations, presented here unfortunately with lack of details, have provided the basis for our understanding of stellar atmospheres.

The mass of a normal stellar atmosphere is of the order of 10^{-10} the total mass of the star. Within the rather large probable errors in the determination, the chemical composition of the star's interior can be considered to be approximately the same as that of the atmosphere, if evolutionary changes can be ignored.

When reliable and precise observations of a normal star and accurate measurements of its spectra are available, the extensive calculations required for the production of a model atmosphere, or a family of models, is possible. The determination of the structure of the model then allows calculation of the observable parameters of a spectrum. These are then compared to the measurements of the spectrum in order to gain improved estimates of the star's surface gravity, effective temperature, and chemical composition.

In order to investigate the structure of a stellar atmosphere, the magnitude (measure of brightness), color, and distance of the star must be known. The observable parameters that are also gained are (1) the energy distribution in continuum, (2) possibly the limb darkening as a function of wavelength, (3) the absorption line spectrum of the star (from which the profiles and equivalent widths are obtained), and (4) the mass and surface gravity (therefore the radius must be known). These parameters have been obtained for many of the brighter stars.

The major problems which the theory of stellar atmospheres attempts to solve are (1) to find the variation of temperature, density and other physical characteristics such as electron and gas pressure (and in hotter stars, radiation pressure) with depth in the atmosphere, (2) to explain the properties of the continuous spectrum, and (3) to examine the law of variation of brightness over the stellar disc. (The solution to the first of these problems determines those of the other two.)

Figure 9 shows a comparison of an observed monochromatic energy curve (the solid line), the results of a model atmosphere (the dashed line), and the observed flux corrected for line blanketing.

The absorption line spectrum allows the profiles and equivalent widths, chemical composition, and perhaps the effective gravity of the star to be determined. If the gravity and temperature are known, two or more models of the star may be calculated. The star under examination may fall between these models, but then they may be interpolated.

Estimates of the surface gravities, hydrogen to helium ratios, and ratios of other elements in the star all must be considered in gaining an accurate model atmosphere to predict the features of the continuous spectrum and the properties of the line spectrum. The theoretical profiles of the hydrogen lines, the relative intensities of lines of certain elements in successive stages of ionization (for example, silicon) allows computation of these quantities for a network of model atmospheres. The surface gravity and effective temperature of the actual star can then be determined by interpolation. Lines that must be used in these calculations are those formed in the intermediate layers, where deviations from thermodynamic equilibrium are not excessive.

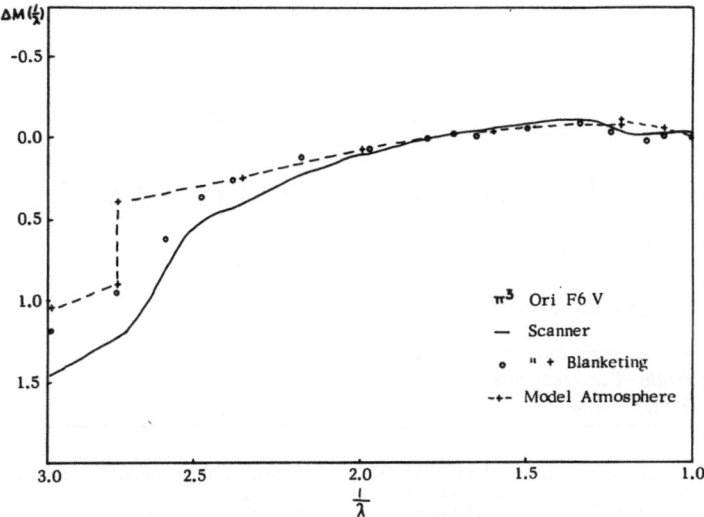

Fig. 9. Comparison of the observed monochromatic flux, with and without correction for line blanketing, with a model atmosphere constructed by Swihart in 1956.

In summary, accurate models of stellar atmospheres can be calculated after having gained values for a surface gravity, effective temperature, and chemical composition.   The emergent flux and comparison to the star's spectrum are the end products of this analysis since the differences are thus established, and finally the physical structure of the stellar atmosphere determined by the interpolation.

ACKNOWLEDGMENT

This material was prepared to be an educational paper presenting aspects of astrophysical spectrophotometry in which relatively few spectroscopists are engaged.   The author gratefully acknowledges the permission of K. O. Wright of the Dominion Astrophysical Observatory and L. H. Aller of the Department of Astronomy, University of California to use much of their published work.

REFERENCES

1. Lawrence H. Aller, The Atmospheres of the Sun and Stars (Ronald Press, New York, 1953).
2. Philip J. Dickerman, ed., Optical Spectrometric Measurements of High Temperatures (University of Chicago Press, Chicago, 1961).
3. Jesse L. Greenstein, ed., Stellar Atmospheres (University of Chicago Press, Chicago, 1960).
4. W. A. Hiltner, ed., Astronomical Techniques (University of Chicago Press, Chicago, 1962).

The Spectrochemical Determination of Boron in Molybdenum

James H. Muntz

Aeronautical Systems Division
Wright–Patterson Air Force Base
Dayton, Ohio

A spectrochemical method for the determination of boron in molybdenum was developed to aid in a metallurgical study on the effect of this element on the fabricability of molybdenum. Boron is determined in the 50 to 500 ppm range with a precision of $\pm 10\%$ using the 2496.78-A line. The range can be extended to 5 to 50 ppm and perhaps lower by using the 2497.73-A boron line and eliminating the filters that were used. The molybdenum is dissolved with hydrogen peroxide and the boron complexed with mannitol. Excitation is by low-amperage DC arc. The boron spectrum is enhanced by the use of sodium fluoride, and graphite is used to reduce the molybdenum spectrum. It is possible to use the technique for the determination of other impurities in molybdenum.

INTRODUCTION

Currently, an internal research program is being conducted by the Physical Metallurgy Branch, Metals and Ceramics Laboratory, Directorate of Materials and Processes, to study the effects of selected grain refining additions on the structure of molybdenum and other refractory metals and alloys. One major difficulty in processing refractory metals has been the coarse columnar structure present in the as-cast ingots. Nucleation methods have produced fine-grained equiaxed cast structures in other alloy systems. It was predicted that certain borides would be the most effective nucleating compounds for the systems under investigation, and the addition of boron and boron compounds were studied. The cast ingots were to be evaluated by macrosection examination, correlation of measured average grain diameters with boron content, and correlation of fabricability. Since there have been no methods published for the spectrochemical determination of boron in molybdenum, it was necessary to develop a method in order to analyze these materials for boron content.

The spectrochemical determination of boron in molybdenum presents two major problems. The first problem involves the complex spectrum and intense continuum produced when molybdenum is arced. The second involves the possible loss of boron if chemical treatment of the sample is necessary. Also to be considered in any spectrochemical method is the problem of obtaining standards that correspond to the sample to be analyzed. The first problem is common in the analysis of any material that has a complex spectrum. Line interference and intense background make the determination of impurities in these materials difficult. To overcome this problem, either the impurities must be separated from the matrix, or a technique which will depress the matrix spectrum without depressing the spectra of the impurity elements must be used. An example of the first approach is the technique of Anderson [1], which removes the base metals of iron and nickel alloys by use of a mercury cathode. Examples of the second approach which might be applicable to the problem are (a) the technique of Dyck and Veleker [2], in which the refractory molybdenum carbide is formed in the arc—this depresses the molybdenum spectrum but enhances the more volatile impurities, and (b) the technique of Paterson and Grimes [5], where the impurity elements of interest are converted to a very volatile form, in this case fluorides of boron and silicon. The method described follows somewhat the technique of Paterson and Grimes. This approach was favored due to the simplicity and convenience of the sample and standard preparation. Sodium fluoride was used since the low excitation potential of sodium tends to keep the temperature of the arc down, helping to reduce the excitation of the molybdenum.

In the chemical treatments necessary in preparing the molybdenum for this technique, the second problem, the possible loss of boron, must be considered. Feldman [3] studied the loss of boron when an acid solution was evaporated and its control by the use of mannitol. However, acids that are usually used to dissolve molybdenum will destroy mannitol during evaporation. Murau [4], in his study on the dissolution of tungsten by hydrogen peroxide, made comparison tests on molybdenum which indicated its rapid dissolution. From these considerations, the following sample treatment was developed: (1) dissolve the molybdenum in hydrogen peroxide; (2) maintain the volume while expelling the excess hydrogen peroxide; and (3) finally complex the boron with mannitol to prevent its loss. In this treatment, no strong acid is used, and the hydrogen peroxide can be eliminated so that the mannitol will not be destroyed. Also, according to Feldman, boron is not lost until the volume is reduced to 2 or 3 ml, so by maintaining the volume at 10 ml, no loss should occur while the hydrogen peroxide is expelled.

The method involves two phases, the preparation phase, where the volatility of the boron is carefully kept at a minimum to prevent its loss,

TABLE I
Excitation Conditions

Direct-current arc	230 V, 7 A (short circuit)
Grating	30,000 lines/in.
Emulsion	SA-1 spectrographic plate
Filter	25% neutral filter
Analytical gap	3 mm
Spectral region	2400 to 3400 A first order
Slit width	0.01 mm
Slit width	2.5 mm
Exposure time	20 sec

and the second or excitation phase, where the boron volatility is purposely increased to enhance its spectrum.

APPARATUS

The apparatus used in these experiments was as follows: Bausch and Lomb dual grating spectrograph with illuminator external optics. National Spectrographic Laboratories arc-spark stand; Jarrell-Ash Co. Varisource DC arc portion; National Spectrographic Laboratories comparator-densitometer; Applied Research Laboratories photo processing equipment; and Cresent Dental Mfg. Co. "Wig-L-Bug" amalgamator.

PROCEDURE

Sample Preparation

Place 100 mg of molybdenum chips in a 30-ml polyethylene beaker. Add approximately 10 ml of distilled water and 2 ml of 30% H_2O_2, and warm to effect dissolution. Maintain the volume at approximately 10 ml by periodically adding water while heating to expel the excess hydrogen peroxide. After the hydrogen peroxide has been driven off, add 50 mg of mannitol and take to dryness. Do all heating on a water bath. Remove the residue from the beaker and place in vial with $3/8$ in.-diameter plastic ball. Grind with the Wig-L-Bug for 30 sec. Mix 100 mg of the residue with 50 mg of graphite (100 mesh) and 7.5 mg of NaF for 30 sec in a vial but without the plastic ball. Weigh duplicate charges of 10 mg, place in the electrodes, and pack with a $1/8$-in.-diameter graphite rod.

Preparation of Standards

Place 100 mg of boron-free molybdenum in each of five 30-ml polyethylene beakers. Add aliquots of 0, 1, 2, 4, and 10 ml of a solution containing 5 μg of boron per ml (as boric acid), and enough distilled water to make the volume approximately 10 ml. Add 2 ml of 30% hydrogen peroxide and continue as for sample preparation. This provides four standards with 50, 100, 200, and 500 ppm boron in molybdenum and a "blank."

Electrode System

Use a $3/16$-in.-diameter "necked" electrode with a $1/8$-in. diameter by $3/16$-in.-deep cup for the sample (Ultra Carbon Corp. 105-D is adequate) and a $3/16$-in. diameter rod $1\frac{1}{2}$ in. long pointed on one end with a pencil sharpener for the upper counter electrode. Make the sample the anode.

Excitation

Arc the sample and the standards using the conditions listed in Table I.

Photographic Processing

Process the plate using the following conditions: develop—3 min in D-19 at 68 F, shortstop—20 sec, fix—Kodak Rapid Fixer 2 min, wash—10

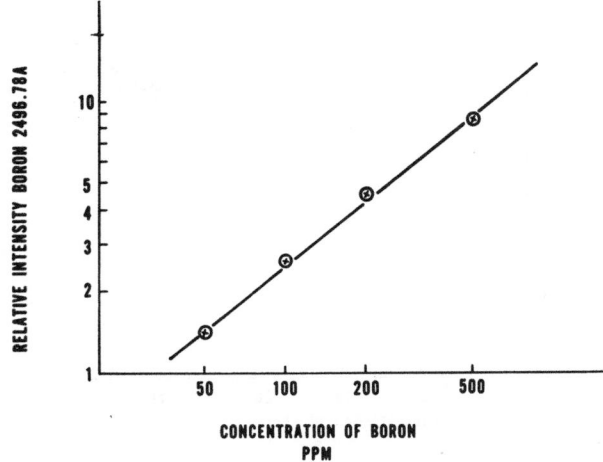

Fig. 1. Analytical curve—boron in molybdenum, 50 to 500 ppm range.

min with running water, and dry—Remove excess water with chamois and dry under hot air for 2 min.

Photometry and Calibration

Photometer the 2496.78-A boron line and convert the transmittance values to intensities using an emulsion calibration curve derived from the two-step method. Prepare the analytical curve by plotting intensities vs. concentrations for the standards. Obtain the concentration of boron in the sample by relating the intensity to the analytical curve. Figure 1 shows a typical analytical curve.

DISCUSSION

Moving-plate studies indicated that, with the matrix used and in the concentration range of 50 to 500 ppm, boron was completely volatilized in 20 sec. For much higher concentrations, longer arcing is necessary, but not practical since the intensity of the molybdenum spectrum begins to increase markedly after 25 sec. To extend the range to higher concentrations it is necessary to dilute the sample with the "blank" material. A lower range can be obtained by using the stronger boron line 2497.73 A and removing the neutral filter (see Fig. 2).

Trials with molybdenum and tin as possible internal standards gave results that were more variable than when only the intensity of the boron line was used. This was probably due to the difference in the volatility of these elements and boron in this matrix. Further attempts to find an element of similar volatility were not made since the reproducibility of the method was considered adequate without the use of an internal standard. The

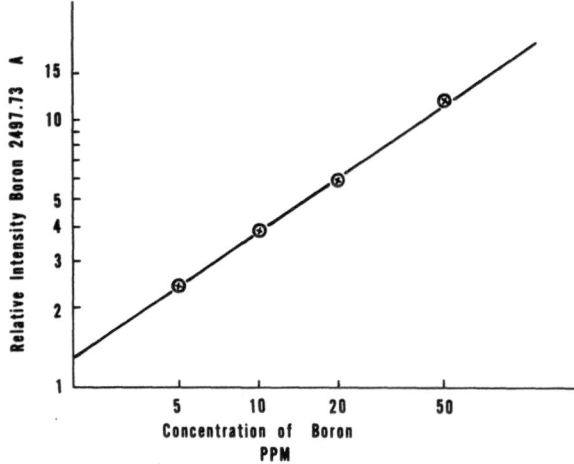

Fig. 2. Analytical curve—boron in molybdenum, 5 to 50 ppm range.

precision of the method is indicated by the data presented in Table II. Listed under "Determinations" are 10 values obtained for the same prepared sample, showing the reproducibility of the arcing procedure. Values listed under "Runs" represent the results obtained from duplicate determinations of a sample prepared four times. This shows the reproducibility of the sample preparation. Table III gives the comparison between wet chemical and spectrochemical values. Although there is as much difference as 100% between wet chemical and spectrochemical results on one sample, the values are of the same order of magnitude and, since higher, indicate that there was probably no loss of boron by this technique of sample preparation.

TABLE II
Reproducibility of Boron Determination

Determinations		Runs	
Determination No.	ppm Boron	Run No.	ppm Boron
1	180	1a	245
2	165	b	255
3	165	2a	285
4	160	b	255
5	165	3a	255
6	160	b	265
7	155	4a	295
8	150	b	265
9	165		
10	160		
Average	163		265
Relative Standard deviation	4.9%		6.5%

TABLE III
Comparison of Spectrochemical and Wet Chemical Analysis

Sample No.	ppm Boron	
	Wet chemical	Spectrochemical
1	163	169
2	282	265
3	73	115
4	54	110
5	1728	1850*
6	330	430
7	451	460
Bottom	17	<10
Top	285	330

*This sample was diluted to bring it within the concentration range of the method.

To determine if added boron could be recovered, 50, 100, and 150 ppm boron were added to a metal sample. The sample was dissolved and processed in the usual manner. Another study was made to determine if all the boron in the sample was being dissolved and recovered in the residue. To accomplish this, the appropriate amounts of metal chips from a sample and the boron-free metal were mixed to prepare 0, 25, 50, 75, and 100% with respect to the sample. The chips were dissolved and processed. The results obtained from these two studies are given in Table IV and indicate that the recovery is within the precision of the method.

A preliminary test with tungsten indicated the possible application of this technique for the determination of boron in this metal also. Evaluation

TABLE IV
Recovery Studies

	ppm Boron		
	Found	Theoretical	%
Sample	161		
Sample +50	200	211	95
Sample +100	269	261	103
Sample +150	322	311	104

% Sample	% Blank	Found	Theoretical	%
0	100	0	0	
25	75	120	119	101
50	50	260	238	110
75	25	390	355	110
100	0	475		

of this technique for the determination of other impurities in molybdenum is in progress.

REFERENCES

1. R. F. Anderson, Appl. Spectroscopy 14:123 (1960).
2. R. Dyck and T. J. Veleker, Anal. Chem. 31:1640 (1959).
3. C. Feldman, Anal. Chem. 33:1916 (1961).
4. P. C. Murau, Anal. Chem. 33:1125 (1961).
5. J. E. Paterson and W. F. Grimes, Anal. Chem. 30:1900 (1958).

The RF Discharge at Atmospheric Pressure and Its Use as an Excitation Source in Analytical Spectroscopy

R. Mavrodineanu and R. C. Hughes

Philips Laboratories
A Division of North American Philips Co., Inc.
Irvington-on-Hudson, New York

Radio-frequency fields can be used to produce excitation of the spectral emission of various chemical species at low pressure, or at atmospheric pressure. The use of RF excitation under low pressure has a long history dating back to the works of J. J. Thompson in 1891. Successful results have been obtained since then in the excitation and measurements of samples in gaseous and solid form contained in electrodeless, evacuated glass vessels. The use in analytical spectroscopy of RF discharges of flamelike appearance occurring at atmospheric pressure is of more recent origin. This work is concerned with this type of discharge and presents a general review of the field, together with the results obtained with two RF discharges, one of 30 Mc and 250 W and the other of 2450 Mc and 2 kW. This last unit has been specially designed and constructed to be used on an optical bench in a manner similar to the usual spark and arc sources. From the several carrier gases studied, air, N_2, O_2, CO_2, He, and H_2, the last two were chosen for their relatively reduced background and for their ability to permit the excitation of numerous chemical species. Practically all compounds present in a dissociated state can be excited by the 2450-Mc discharge in helium as can be seen from the emission spectra of some 70 elements recorded thus far, producing atomic and molecular radiations. The RF discharge at atmospheric pressure possesses a high electron energy and a comparatively low thermal energy. Various procedures developed for the introduction of solid, liquid, and gaseous samples are discussed and the use of these procedures for analytical purposes are presented. The following chemical species have been investigated more closely with regard to quantitative measurement possibilities: Sn, Pb, Hg, B, Au, Cd, As, Sb.

INTRODUCTION

High-Frequency Discharges at Low Pressure

Excitation of the spectral emission of gases under low pressure by the action of radio-frequency fields (hereafter called RF) has a long history, dating approximately from the invention of the Tesla coil. One of the earliest published studies concerning the RF excitation of gases was made by J. J. Thompson [46]. The use of this discharge to produce spectral excitation of various solid or gaseous chemical species followed later through the works of Dunoyer [14-16], Bloch and Bloch [4], Vaudet [49], and Winans [52]. Excitation by the Tesla coil was adapted to practical spectrochemical analysis by Gerlach and Schweitzer in 1931 [25]. Further extensions of the technique were made in 1933 by Potapenko [40] and Goroncy and Urban [26], employing tube-driven, undamped oscillators. In 1941 Fenner [17] used an electrodeless discharge tube for the quantitative determination of traces of Cd in Se and of Hg in various gases. The sample in the evacuated tube was excited by a high-frequency generator producing a wavelength of 6 m.

The application of RF excitation to gases at low pressure, and to relatively volatile solids, was given further impetus by the work of Meggers [38] on monochromatic emissions as an ultimate standard of length. Further contributions have been made by Gatterer [20-23], who succeeded in exciting and determining the halogens, S, and Se in various samples containing as little as 0.001% of these elements, with an error of ±10%. He used for this purpose a generator producing a discharge of 40 to 100 Mc with an output at the sample tube of 600 W. A similar discharge has been used by Tomkins and Fred [47] for the determination of D_2 in heavy water.

The method continues to be refined, adapted, and applied, as evidenced by the work of Zelikoff et al. [54] on the excitation of the vapors of Pb, In, Ga, Sb, Bi, Hg, Cd, Tl, Zn, Ca, and Ba. Keller and Smith [33] used a 40-Mc discharge to excite and determine Cl, Br, and I at the level of 0.001%. The analytical samples were enclosed in an evacuated, electrodeless tube. A similar discharge was employed by Broida and Morgan [5] to analyze hydrogen-deuterium mixtures in the presence of air by exciting the gaseous sample contained in an electrodeless tube with a 150-Mc oscillator.

Corliss et al. [11] have succeeded in exciting the volatile halides of relatively involatile metals such as Be, Ti, Fe, Ni, Cu, Mo, U; and Tomkins and Fred [48] produced the spectra of rare earths and some heavy metals present as vapors of the halides. A magnetron-operated circuit was applied by Servigne et al. [44] to excite and measure impurities such as N_2 and CO_2 in rare gases. The need in optical pumping experiments for intense spectral lines free of self-reversal occasions continued interest in electrodeless RF discharge tubes with alkali metal fillings [3, 24]. Electrodeless discharges for the analysis of gases have been extensively explored and exploited by Gatterer and Frodl [20], White [50], White and Lovelace [51], Ishida [30], Frisch and Schreider [19], and Stolov [45].

More recently, Chakrabarti et al. [7, 8] have studied the emission spectra of N_2, H_2, NH_3 and that of mixtures of A-CO_2 excited in RF fields; their results are published in two issues of "Talanta" of this year. An excellent reviewing paper on RF discharges in gases has been published by Morgan [39], and covers this field up to 1953 with 70 references.

Radio-Frequency Discharge at Atmospheric Pressure

The foregoing discussion has been limited to RF discharges produced in electrodeless, evacuated vessels. Radio-frequency discharges of flame-like appearance occurring at atmospheric pressure can also be produced by using a suitable portion of an RF circuit [28]; such discharges can be considered as excitation sources in analytical spectroscopy. Asami and Hori have described the spectrum produced by the discharge in air [1]. One of us described the excitation of samples in solution by an RF discharge in 1954 [35]. Further developments of the method have been made by Bădărău et al. [2], who introduced the sample solution, as an aspirated aerosol, into the discharge in air and excited lines of the elements: Al, Ba, Be, B, Ce, Cs, Fe, Pb, K, Sn, V, and Zn. Korolev and Zheenbaev [34] have investigated the sensitivity of excitation of the elements Na, K, and Li. A similar work has been discussed by van Calker at the IXth Colloquium Spectroscopicum Internationale [6]. Using excitation sources similar to the one described by Scholz [43], Yamamoto and Murayama have examined the excitation of Ca and Mg [53], and Dunnken et al. [13] that of Ca, Sr, Ba, Mn, and Cr.   In both cases nitrogen and air were used as carrier gases. Cristescu and Grigorovici [12,27] and Zheenbaev [55] have investigated the discharge with particular regard to its temperature.   Zheenbaev [56] has demonstrated an improved sensitivity, obtained by hydrodynamic compression of the discharge, in the detection of a number of elements.

Apparatus designed specifically for the production of the RF brush discharge, for which the designation "Electronic Torch" has been generally applied, was developed by Cobine and Wilbur [9,10]; this source is magnetron-powered. A similar magnetron-driven apparatus has been described by Schmidt [42].   Scholz [43] has described a "hochfrequenz-Plasma-Flamme," powered by a triode oscillator tube, and employing a tubular-shaped "burner" through which a gas stream is passed for localizing and stabilizing the discharge.

APPARATUS

RF Generators

The work described here refers to RF discharges at atmospheric pressure [29,36,37]. We have employed both a triode-driven oscillator at 30 Mc

Fig. 1. 30-Mc source supplied with the analytical sample through pneumatic atomization. An all-glass indirect internal atomizer operated with helium is used. (Reproduced by permission from Spectrochimica Acta.)

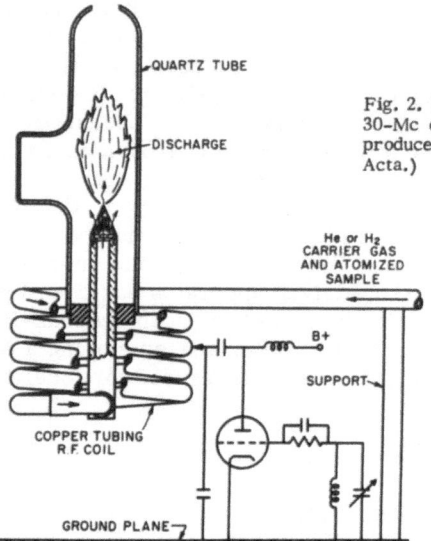

Fig. 2. Diagrammatic illustration showing the 30-Mc oscillator with the inductance coil. (Reproduced by permission from Spectrochimica Acta.)

similar in design to that of Scholz, and a magnetron-driven source, 2450 Mc, with torch of coaxial design, following the Cobine-Wilbur and Schmidt designs. The particular apparatus employed was designed and constructed by Amperex Electronic Corporation, and has been described by Roddy and Green [41].

The 30-Mc Oscillator

The 30-Mc oscillator, of approximately 250 W output power, is shown in Fig. 1. The oscillator circuit is shown in diagrammatic form in Fig. 2, in which the torch is also indicated. Figure 3* shows the "flame" of this torch, burning in a stream of helium. In these figures it will be seen that the discharge originates at a conical tip at the end of a conductor forming a part of the main inductor of the tank circuit. This tip is surrounded by a quartz cylinder. A gas stream is made to flow outward past the tip; this stream shapes, stabilizes, and determines the character of the discharge.

The 2450-Mc Oscillator

The 2450-Mc, 2-kW oscillator is powered by a magnetron. The power is conducted by a coaxial system terminated by the coaxial torch. Figure 4 shows this apparatus. The conductor and torch configuration is shown in Fig. 5, while the "flame" in helium is seen in Fig. 6a and b; Fig. 6b shows a flame containing sodium. The gas stream is made to flow outward through the coaxial space, establishing the discharge at the conical tip terminating the center conductor. A quartz cylinder encloses the flame to exclude air.

Magnetron-Powered Spectroscopic Source

A more compact unit, suitable for mounting on an optical bench, has been designed and constructed in collaboration with Amperex Electronic

*Figures 3, 6, 27, 28, 29, and 30 will be found following p. 326.

Fig. 4. 2450-Mc magnetron-operated source. (Reproduced by permission from Spectrochimica Acta.)

Corporation; this unit is described in Fig. 7a. The 2450-Mc, 2-kW magnetron is seen at the right with its water and air cooling arrangement. The horizontal coaxial element conducts the RF oscillations to the tip of the vertical torch provided with a quartz tube. Figure 7b is a diagrammatic description of this unit. The coaxial circuit can be tuned with an adjustable plug operated by the dial at the left. The helium gas is supplied to the tip of the torch through the central electrode lined with a Teflon tube. This unit, shown here with the protective cover removed, can be placed on an

Fig. 5. Diagrammatic illustration of the coaxial conductor and torch of the 2450-Mc oscillator. (Reproduced by permission from Spectrochimica Acta.)

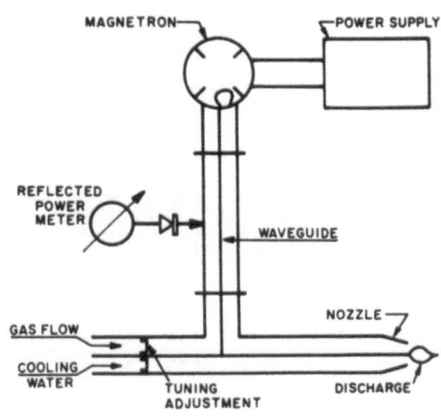

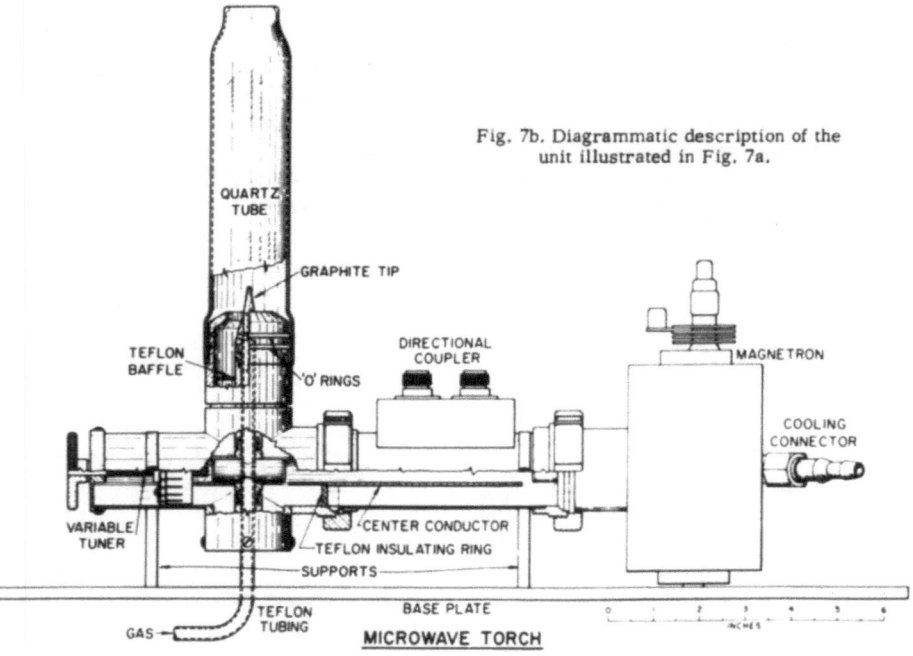

Fig. 7a. 2450-Mc magnetron-powered spectroscopic source placed on the optical bench.

Fig. 7b. Diagrammatic description of the unit illustrated in Fig. 7a.

optical bench in front of the spectrograph or monochromator and manipulated like an arc-and-spark source.

EXPERIMENTAL METHODS

Provisions were made for alignment of the RF sources with a Bausch and Lomb Medium Quartz Spectrograph or, in some instances, with a Baird 3-m grating (15,000 lines/in.) spectrograph. Eastman 1N plates were generally employed for recording the spectra. All spectrographic techniques employed were conventional, except as otherwise noted [29, 36, 37].

Various gases were employed to supply the torch atmosphere, and the spectra produced were recorded and examined. A Geissler discharge tube of quartz, continuously pumped and fed with a small stream of gas controlled by a variable leak, was employed to provide comparison spectra of a familiar nature.

Temperatures of the flames were investigated in a crude fashion by introducing wires of various elements and determining the highest melting-point element which could be melted. The temperatures attained at the discharge tip were determined by employing a graphite tip, somewhat isolated thermally, which could be observed by an optical pyrometer.

The introduction of samples into the discharge presents a problem not adequately solved by earlier workers. We have therefore given major attention to this problem. Of the various modes of introduction of liquids into the 30-Mc discharge, most were found to cause partial shorting of the discharge. In order to avoid partial grounding of the discharge through liquid films, we introduce the gas stream carrying the aspirated sample into the grounded end of the $3/8$-in. tubular copper conductor of the tank circuit inductor, allowing it to traverse the coil and exit from four small holes in the tip, as shown in Fig. 8. The 2450-Mc torch has been adapted for similar introduction of liquid samples, since presence of the dispersed liquid in the coaxial space causes a bad mismatch, and extinguishes the torch. Thus, when solutions are to be excited, a pneumatic atomizer operating with helium is used to produce a fine suspension of droplets which are sent through the Teflon-lined central conductor. Figure 9 is a close view of the tip of the torch showing the molybdenum nozzle provided with four openings for the flow of the helium gas. Another close view of the tip provided with a graphite electrode is shown in Fig. 10. In these cases the outer conductor has been dismantled.

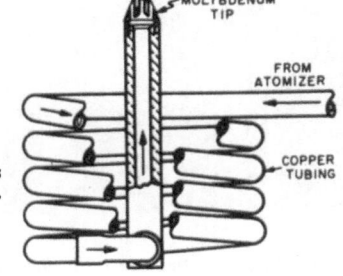

Fig. 8. Procedure for the introduction of a gaseous sample or atomized solution into the 30-Mc discharge.

Fig. 9. Introduction of a gaseous sample or atomized solution into the 2450-Mc discharge. A molybdenum tip is used here.

Fig. 10. Same as Fig. 9 except for the tip which is made here from an interchangeable graphite electrode.

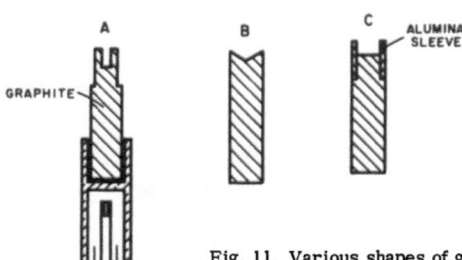

Fig. 11. Various shapes of graphite electrodes used to supply the 2450-Mc discharge with a solid sample. (Reproduced by permission from Spectrochimica Acta.)

By substituting various cratered tips, machined from $\frac{1}{4}$-in.-diameter graphite (spectroscopic grade) for the discharge tip, and loading solid samples into the crater, we find that the sample may be slowly volatilized into the flame. In some instances a nonconducting sleeve of silica or alumina surrounding the tip is convenient for confining the discharge site. A few configurations of electrodes are shown in Fig. 11. Figure 12 illustrates a method of supplying a gaseous sample or an atomized solution into the 2450-Mc discharge through the Teflon inner tube and a specially shaped

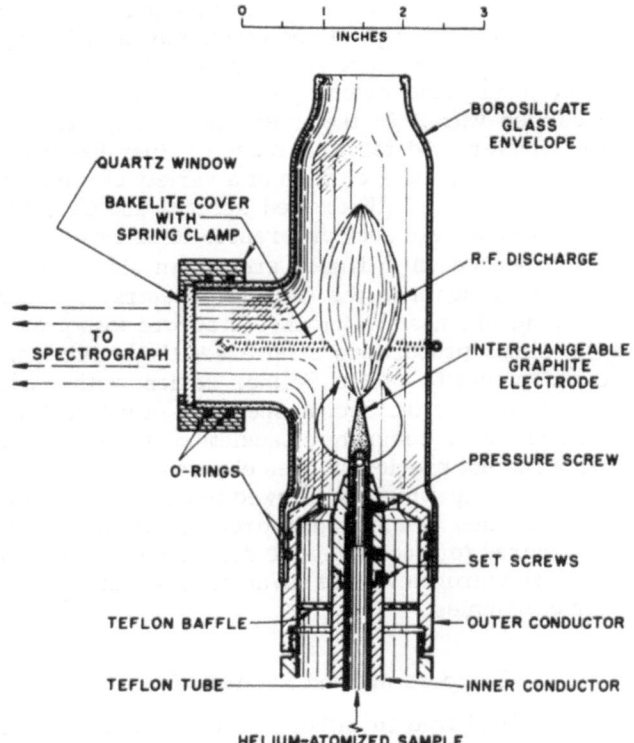

Fig. 12. Method used to supply the 2450-Mc discharge with a gaseous sample or atomized solution. In this arrangement the analytical sample is in contact only with Teflon and graphite.

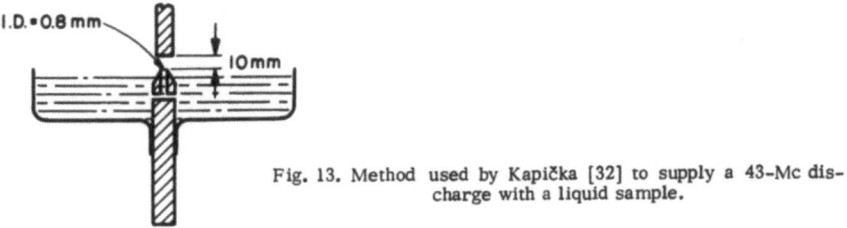

Fig. 13. Method used by Kapička [32] to supply a 43-Mc discharge with a liquid sample.

graphite electrode. In this case the sample is in contact only with the Teflon and graphite materials. A different procedure to supply a 43-Mc torch with a solution was used by Kapička [32] and is illustrated in Fig. 13. In this case a counter electrode is employed, the solution being introduced into the arclike discharge by capillarity through the lower electrode, which functions in a manner comparable to that of the well-known "Fulgurator."

EXPERIMENTAL RESULTS AND DISCUSSION

Sustaining Atmospheres

The spectra produced by the microwave discharge in several gases are shown in Fig. 14 [29, 36, 37]. The intense spectra produced by nitrogen and oxygen, extending almost throughout the entire wavelength range examined (2000 to 10,000 A) should interfere seriously with the spectra of any other elements introduced into the discharge. This effect has been a serious impediment in earlier work. On the other hand, the spectrum produced by helium is much simpler, and for this reason helium has been chosen and used as a carrier gas for the excitation of a variety of elements. No profound differences between the RF-excited spectra and Geissler-discharge spectra were detected, nor are such differences to be expected [18].

The spectra produced in hydrogen are shown also in Fig. 19. These spectra, corresponding again to the Geissler spectra, are comparatively simple, though individual lines are intense. In accord with previous findings [10], hydrogen in small proportions suppresses the helium emission. The hydrogen spectrum consists of the Balmer series, with a comparatively weak ultraviolet continuum; the second spectrum of molecular hydrogen is absent, indicating virtually complete dissociation of the hydrogen. The RF flame in hydrogen has not previously been observed. We are able to obtain it only by employing a tip which is allowed to come to incandescence in a helium discharge, and then switching to hydrogen. Apparently, an electron-emitting tip is required for sustaining the discharge in hydrogen. We conclude that helium and hydrogen will be advantageous atmospheres to employ for the excitation of samples.

Temperature and Excitation

The microwave discharge in either helium or hydrogen is capable of melting molybdenum, and fails to melt tantalum and tungsten. The neutral-gas temperature therefore lies between 2900 and 3300 K, agreeing well with the estimate of 3000 K made by Cobine and Wilbur for other gases. Our observed temperature in helium is unaccountably higher than that estimated

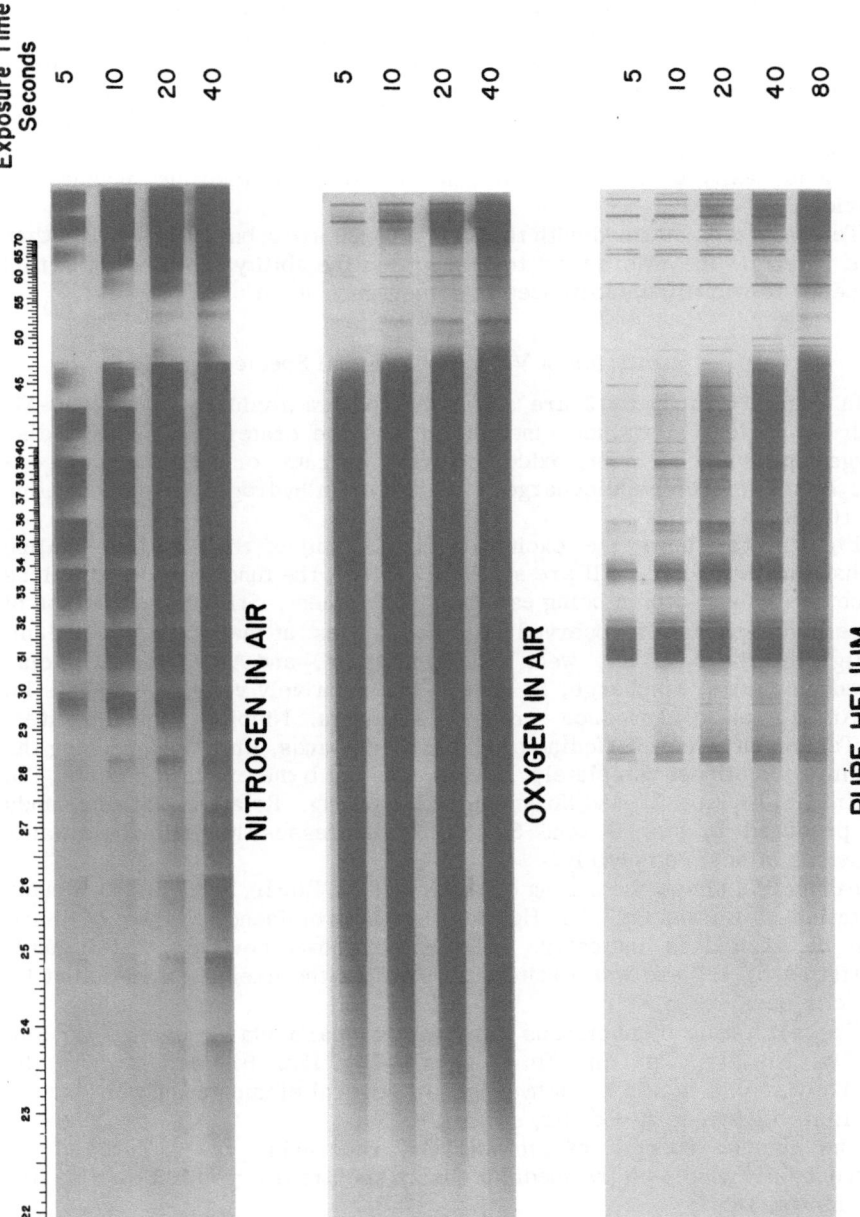

Fig. 14. 2450-Mc discharge in various gases and for various exposure times.

by Cobine and Wilbur, who observe in that gas a cool flame barely capable of scorching paper. In sharp contrast to the low neutral-gas temperature, the electron temperature of the discharge is extremely high, as evidenced by the ability to excite energetic levels such as those of helium.

The graphite tips have been observed to reach a temperature of about 1400 K in helium and about 2300 K in hydrogen. Hydrogen is, therefore, markedly more effective in volatilizing the more refractory substances. Another evidence as to the comparatively low temperature of the neutral-gas atmosphere of the discharge is seen in the strong excitation of band spectra of easily-dissociated molecules (sulfur, phosphorous, magnesium chloride).

Temperatures attained with the 30-Mc torch are substantially lower than those given by the microwave torch, so that the ability of the torch at this frequency to volatilize substances is somewhat limited.

Excitation of Various Chemical Species

In Figs. 15 through 19 are shown the spectra produced in the 2450-Mc torch by various substances introduced into the crater of a graphite discharge tip as the element, oxide, chloride, nitrate, or other salt, and excited by the microwave discharge in helium or in hydrogen. Exposure time was 10 sec.

Figure 15a shows the excitation (in helium) of the alkalies, alkaline earths, and beryllium. All are strongly excited, the fundamental radiations of the un-ionized atoms being especially prominent. Strong suppression of the helium spectrum is observed with the alkalies, at maximum for cesium. In Fig. 15b the relatively weak excitation of B_2O_3 and Al_2O_3, not well volatilized in the He discharge, is seen. The relatively volatile elements In, Tl, Au, Ag, and Cu produce strong line spectra. Niobium is well excited, but Ta and Hf are not. Iodine produces weak bands, while suppressing the He emission almost completely. In Fig. 16a and b the excitation of Mo, Pb, Sn, Zn, Fe, Ni, Co, Cd, and Sb is especially strong. Relatively strong bands are produced by As, P, and S. Sulfur suppresses the helium and water emissions almost completely.

Figure 17a shows the strong excitation of Pt, Ru, Ir, and Pd, and weaker excitation of Rh and Os. The Hg spectrum is prominent. Failure of Si and U to be excited is indicated. Figure 17b shows, however, that Si and U volatilized by HF are well excited. In Fig. 17c the intense excitation of Re and Sc is seen.

The excitation of numerous rare earths, particularly intense for Sm, Eu, Dy, Ho, Er, Tm, and Yb, is seen in Fig. 18a. Figure 18b shows the relatively strong excitation in hydrogen of several elements not well excited in helium, including Hf, Gd, Pr, and Nd.

The intense spectra of Sm and Dy, recordable at exposures of $\frac{1}{250}$ second by 1N plates on the medium quartz spectrograph with 30-μ slit, are seen in Fig. 18c.

Band spectra of some inorganic molecules are shown in Fig. 19a, in which the influence of the atmosphere is further evident. Magnesium chloride in He produces strong molecular bands, which are not exhibited in hydrogen.

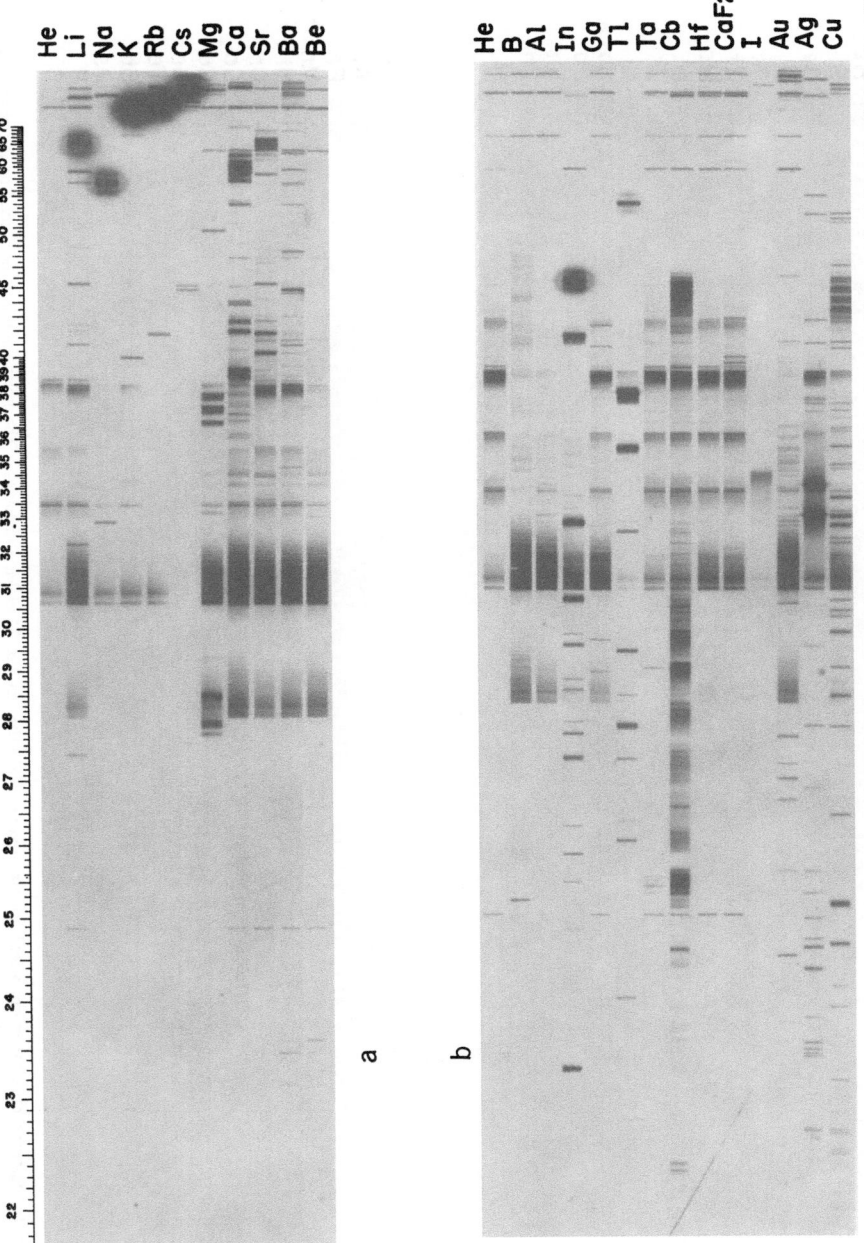

Fig. 15. Spectra of various chemical species excited in helium by the 2450-Mc discharge. The solid sample was placed on the tip of a graphite electrode.

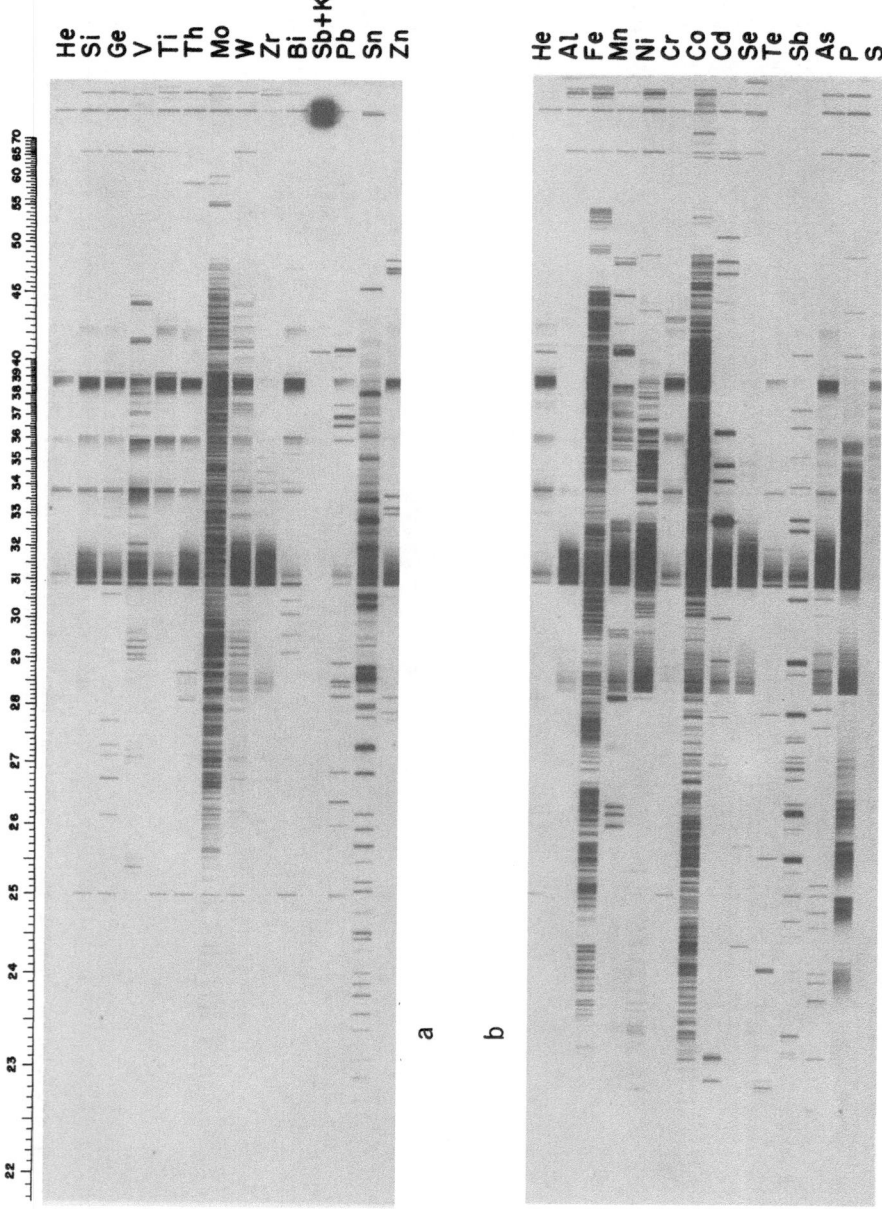

Fig. 16. Spectra of various chemical species excited in helium by the 2450-Mc discharge. The solid sample was placed on the tip of a graphite electrode.

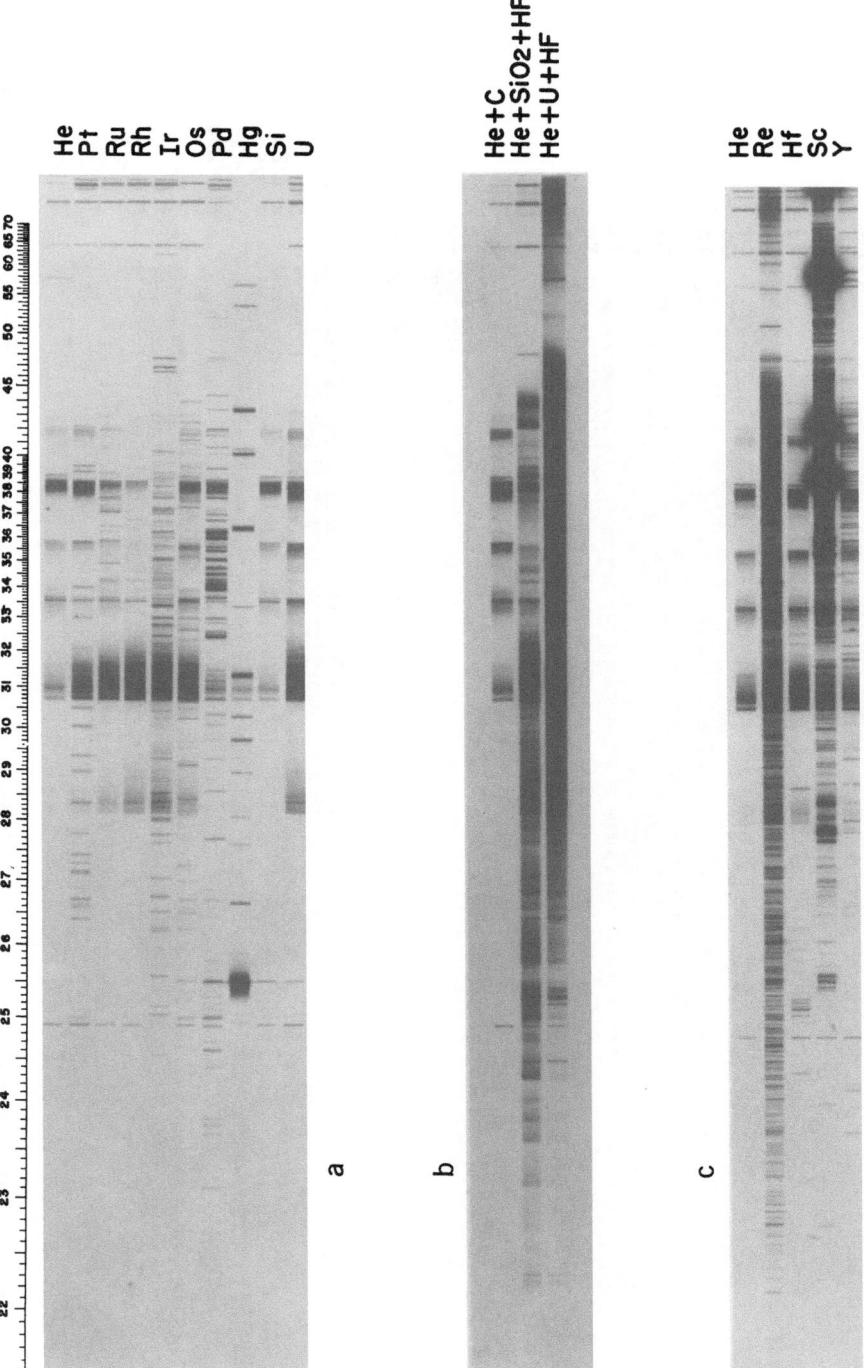

Fig. 17. Spectra of various chemical species excited in helium by the 2450-Mc discharge. The solid sample was placed on the tip of a graphite electrode.

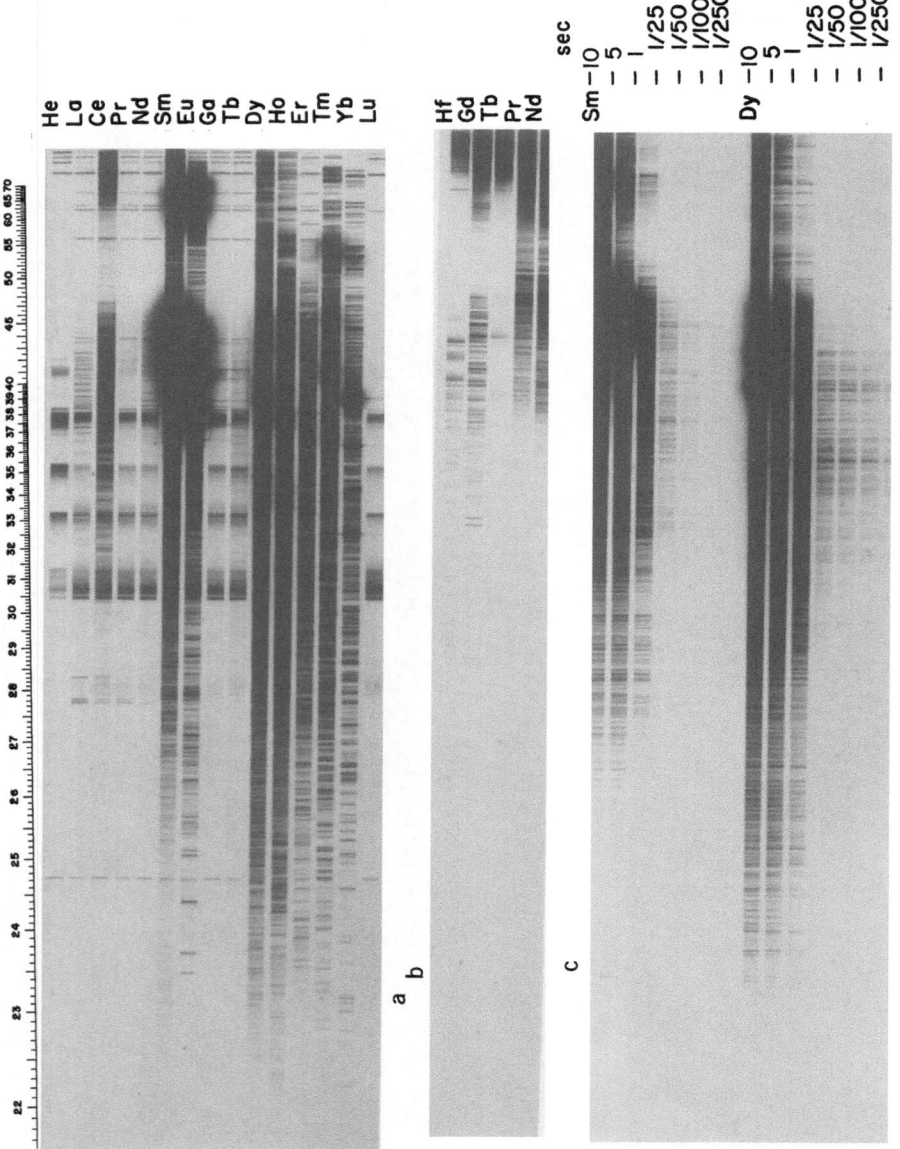

Fig. 18. Spectra of various chemical species excited in helium by the 2450-Mc discharge. The solid sample was placed on the tip of a graphite electrode.

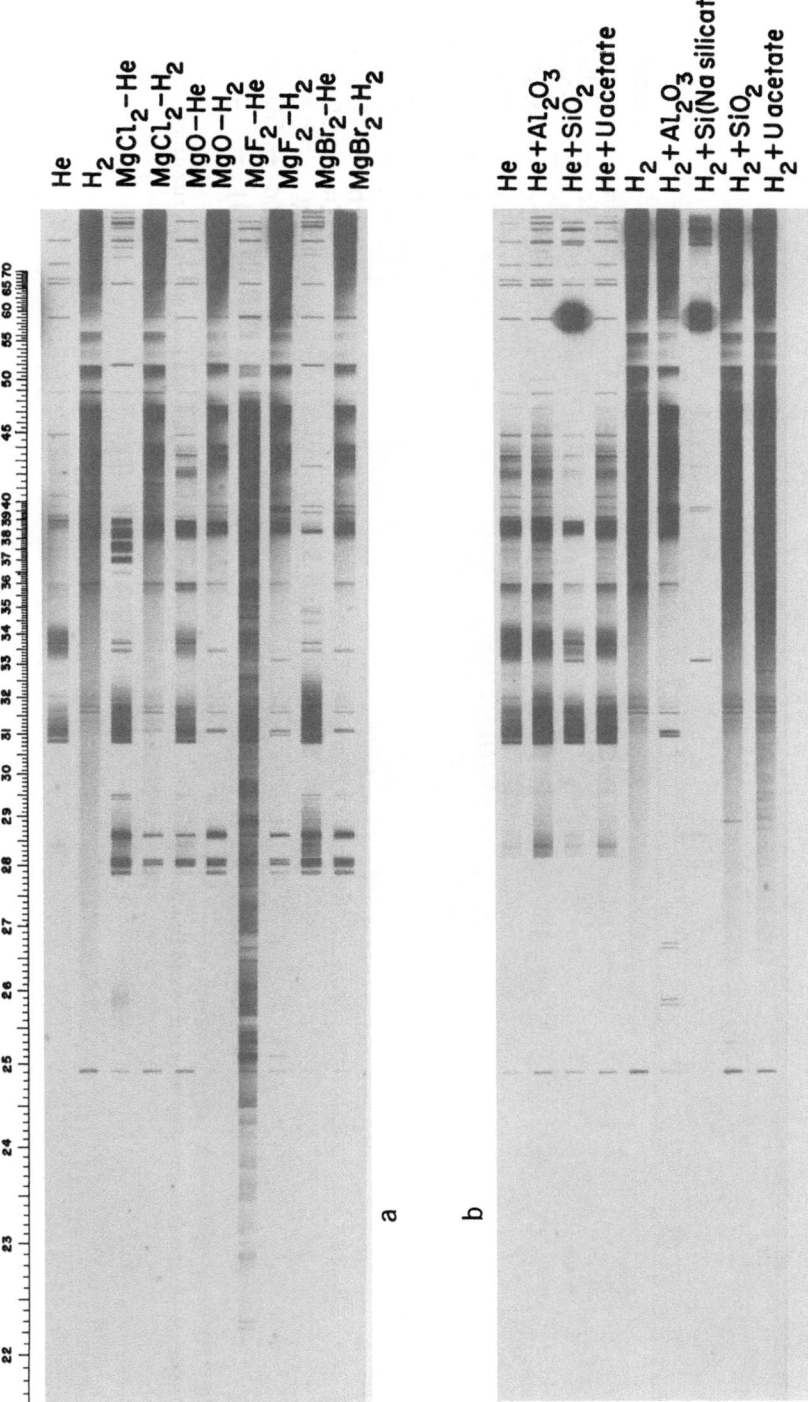

Fig. 19a. Spectra produced by various magnesium salts introduced into the 2450-Mc discharge produced in helium and hydrogen. The solid sample was placed on the tip of graphite electrodes.

Fig. 19b. Spectra of SiO₂, Al₂O₃, and U acetate excited in the 2450-Mc discharge produced in helium and hydrogen.

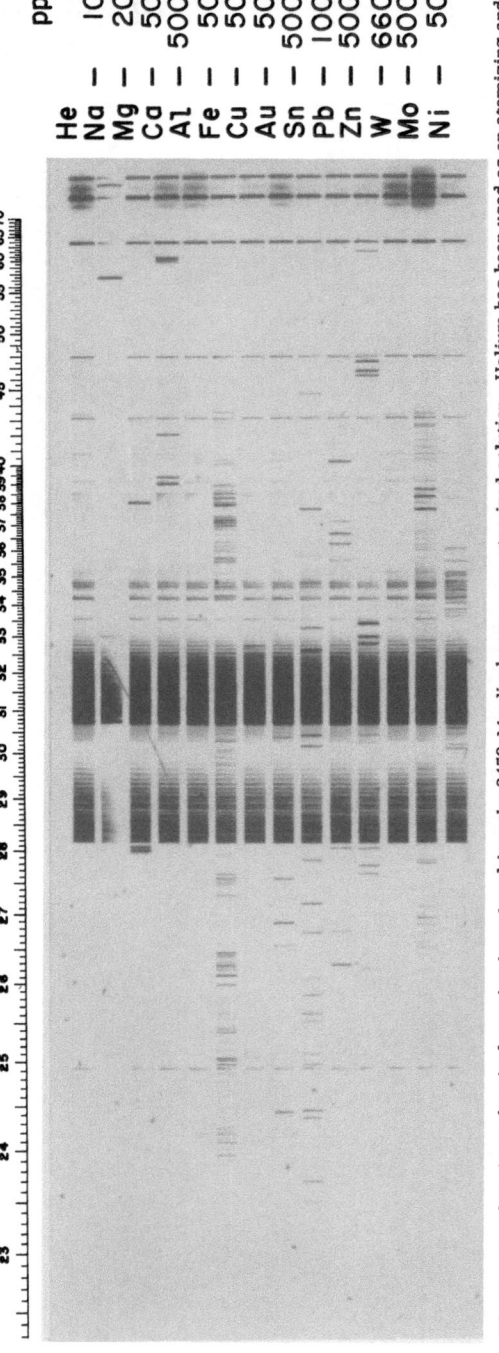

Fig. 20. Spectra of various chemical species introduced into the 2450-Mc discharge as an atomizing and carrier gas. Helium has been used as an atomized solution. Helium has been used as an atomizing and carrier gas.

Figure 19b illustrates the relative efficacy of the hydrogen discharge in volatilizing (and exciting) Al and U, present as the oxides. The preferential excitation of Na in sodium silicate is also seen.

The spectra discussed above were produced by placing a solid salt of the various elements on the crater tip of pure graphite electrodes.

Figure 20 illustrates the spectra obtained from various chemical species in aqueous solution introduced into the flamelike 2450-Mc discharge as a suspension of fine droplets produced through pneumatic atomization with helium. An indirect internal atomizer with atomization flask has been used for this purpose. The spectra of Mg, Fe, Au, Sn, Pb, Zn, Mo, and Ni are particularly intense.

The lower limits of detection have also been determined for a certain number of chemical species excited in the 2450-Mc discharge. The aqueous solutions were atomized with helium, and the spectra were recorded on Kodak 1N plates by a medium quartz Bausch and Lomb spectrograph; Table I presents these preliminary results.

Figure 21a and b, presents the spectra of a number of relatively volatile elements excited by the 30-Mc source, operating with helium. Compounds of the elements, in aqueous solution, were introduced by aspiration in the manner described above. Exposure time was 60 sec.

Figure 22 is presented to illustrate the pronounced interference which may be exhibited with this source. In helium, solutions containing 1000 ppm of Fe or Ni, present individually, are well excited. A solution of the same concentration of the two elements, present together, fails to be excited. This behavior is interpreted as being due to the formation of nickel ferrite,

TABLE I

Detection Sensitivity for Several Chemical Species Atomized in the 2450-Mc Discharge in Helium

Chemical species	Wavelength, A	Sensitivity, ppm	
		Measured	Estimated
As	2349.8	100	50
Au	2428.0	10	3
B	2497.7	10	1
Hg	2536.5	0.01	0.005
P	2554.9	100	
Sb	2598.1	100	10
Sn	2706.5	100	80
Pt	2733.9	100	50
Mg	2802.7	10	5
Cd	2980.6	10	5
Tl	3775.7	100	80
Mo	3798.3	100	80
Pb	4057.8	100	10
Zn	4722.2	100	10
Na	5895.9	1	0.8

*The spectra were recorded with a Bausch and Lomb medium quartz spectrograph on Kodak 1N plates with an exposure time of 60 sec.

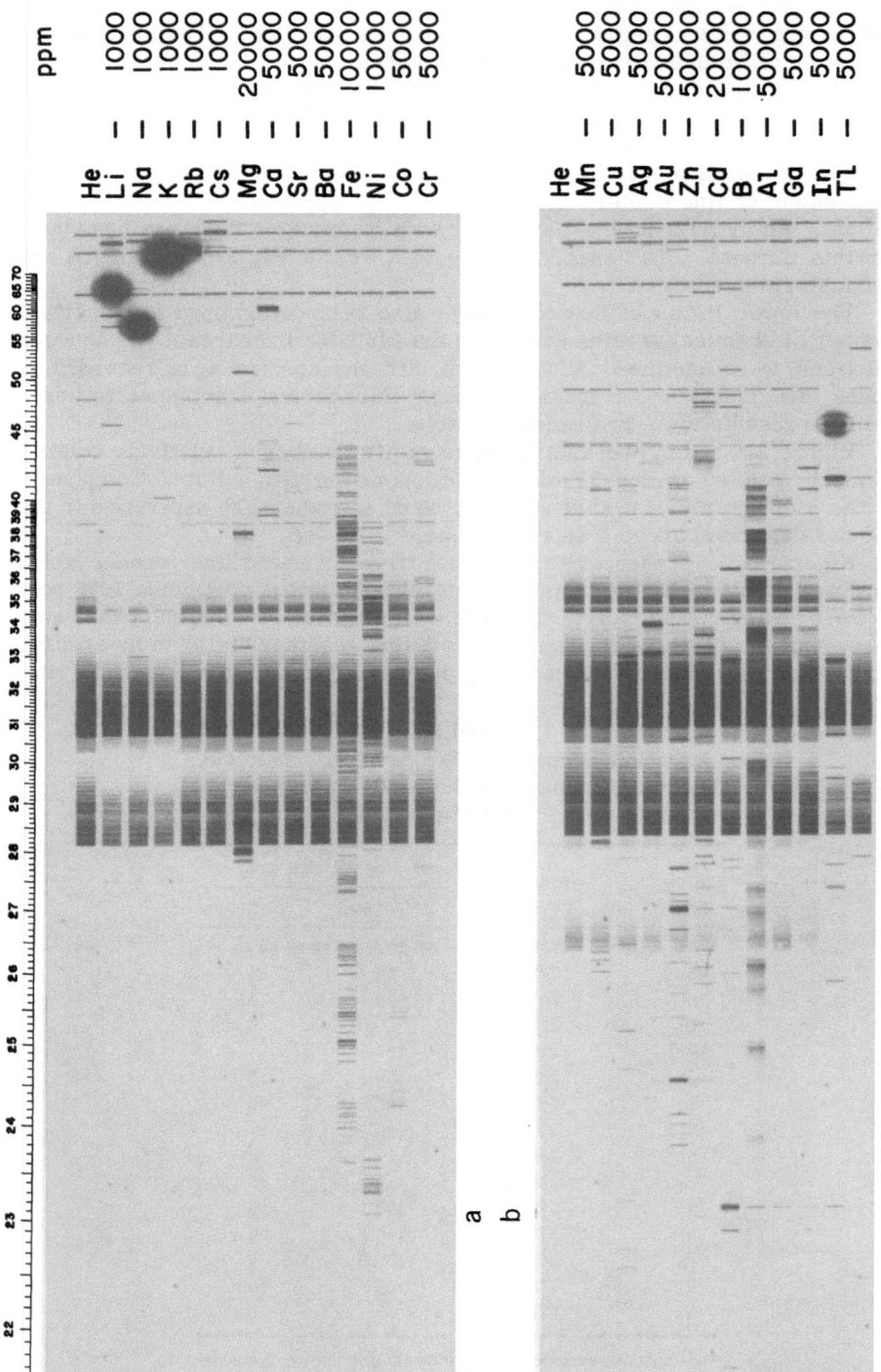

Fig. 21. Spectra of various chemical species supplied to the 30-Mc discharge in atomized form. Helium has been used as the atomizing and carrier gas.

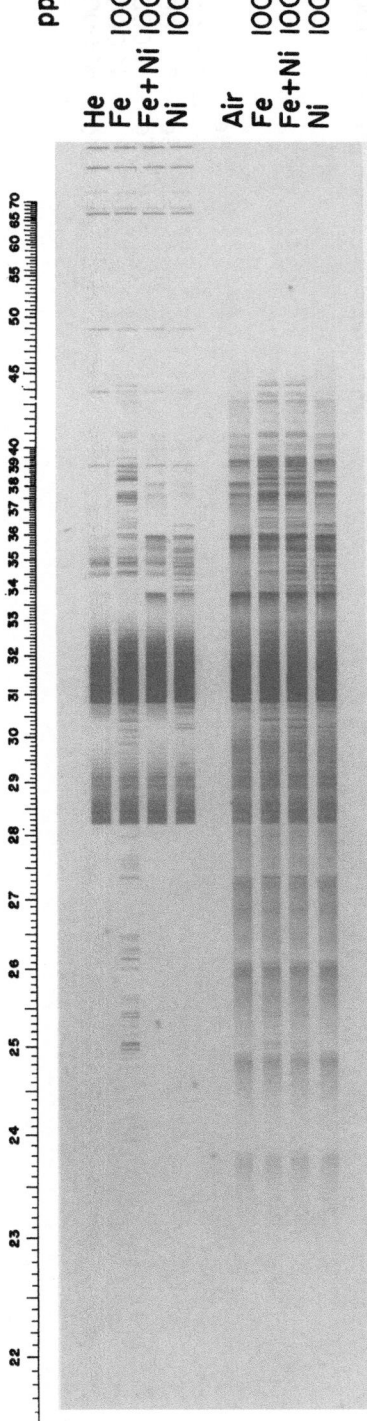

Fig. 22. Pronounced interference phenomenon occurring in the 30-Mc discharge. Iron and nickel salts have been atomized with helium and air.

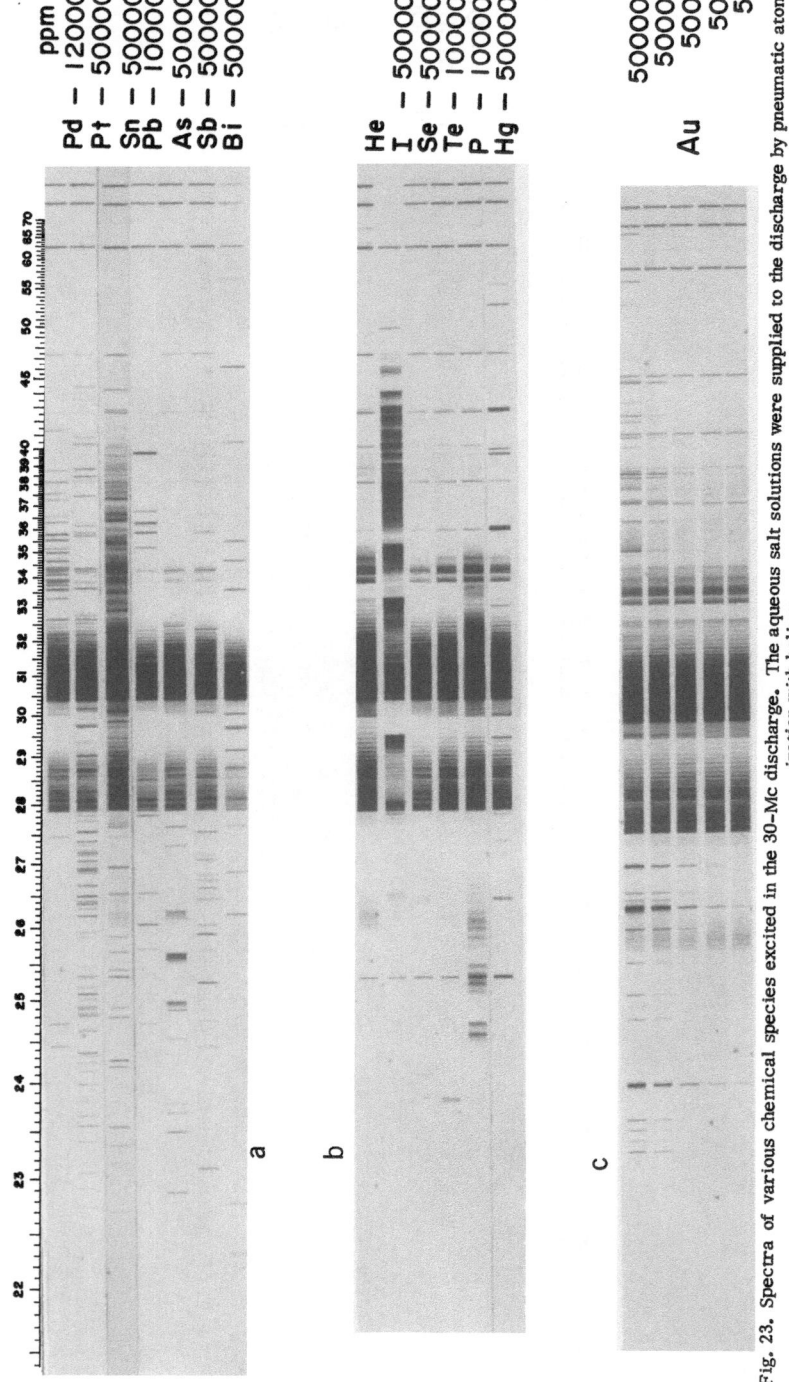

Fig. 23. Spectra of various chemical species excited in the 30-Mc discharge. The aqueous salt solutions were supplied to the discharge by pneumatic atomization with helium.

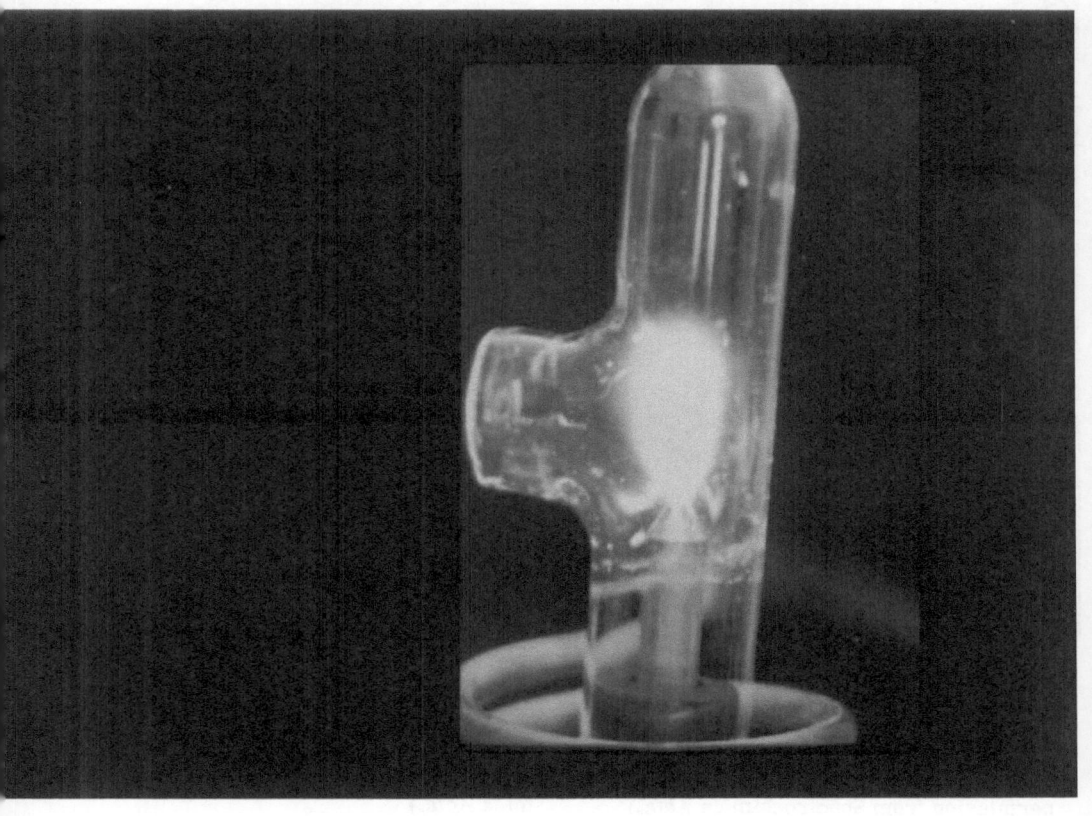

Fig. 3. Flamelike discharge produced in helium at the molybdenum tip of the 30-Mc generator.

Fig. 6a. Flamelike discharge produced at the graphite tip of the 2450-Mc torch. The sustaining gas is helium. (Reproduced by permission from Spectrochimica Acta.)

Fig. 6b. The discharge illustrated in Fig. 6a is supplied with sodium chloride. (Reproduced by permission from Spectrochimica Acta.)

Fig. 27. Premixed city-gas–air flame burning at the molybdenum tip of the 30-Mc inductance coil.

Fig. 28. Premixed city-gas–air flame burning at the molybdenum tip of the 30-Mc inductance coil on which the RF discharge has been superimposed.

Fig. 29. Same flame as in Fig. 27 in which a lithium salt has been atomized with air.

Fig. 30. Same flame plus the 30-Mc discharge as in Fig. 28, but in which a lithium salt has been atomized with air.

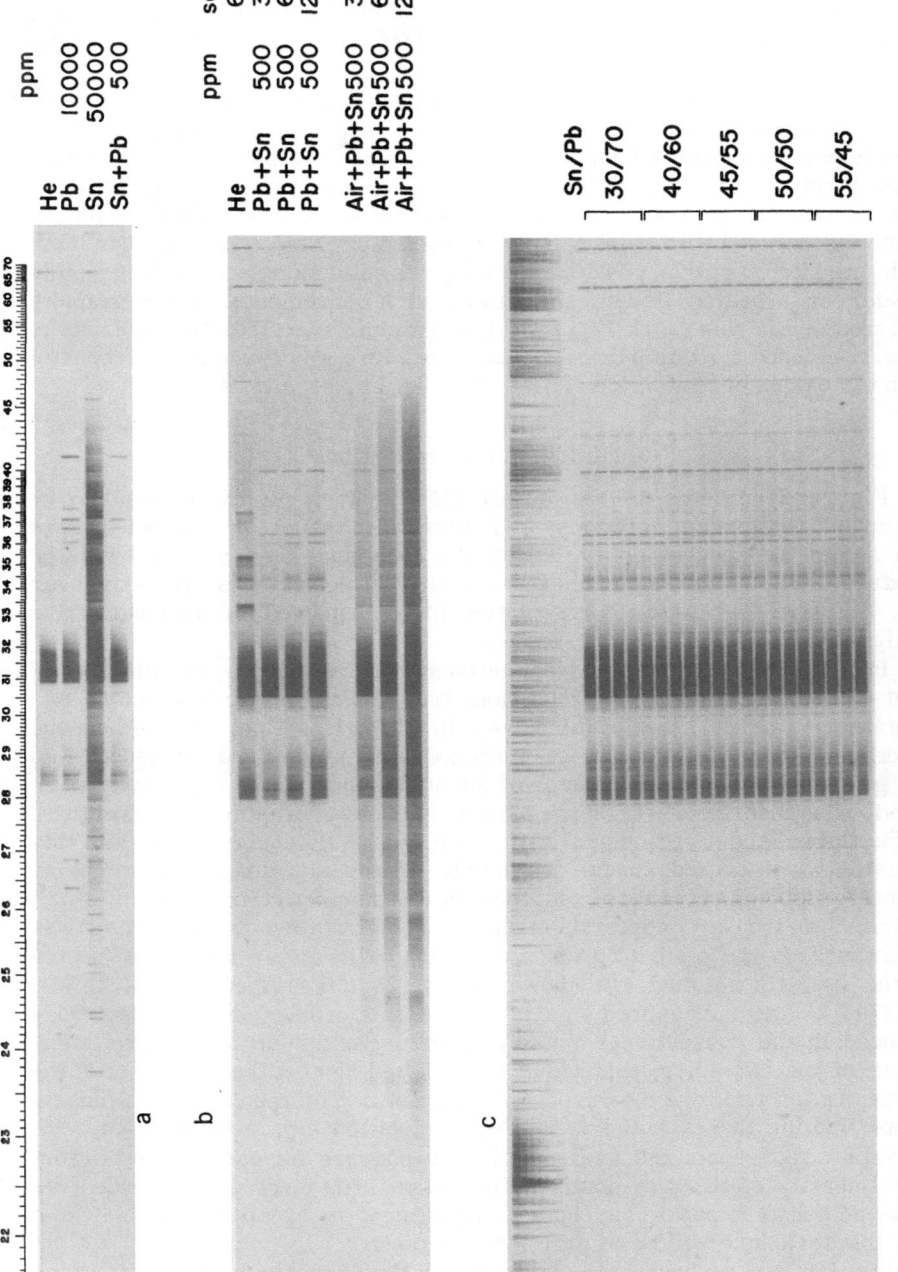

Fig. 24a. Spectra of tin and lead excited in the 30-Mc discharge supplied with the aqueous sample through pneumatic atomization with helium.

Fig. 24b. Helium and air have been used to atomize tin and lead aqueous solutions into the 30-Mc discharge.

Fig. 24c. Illustration of a spectrogram employed for the quantitative analysis of a tin–lead solder excited in the 30-Mc discharge.

relatively less volatile than the individual constituents. However, in the presumably hotter discharge in air, each constituent of the mixture is excited.

The following spectra show the emissions from certain elements atomized with helium in the 30-Mc discharge. Thus Figs. 23a, b, and c, illustrate the strong emission obtained from Pd, Pt, Sn, Pb, Sb, Bi, I, and Hg. On this same figure one can see a series of spectra produced by Au in solutions containing 50,000 to 5 ppm Au.

The examination of the spectra presented above demonstrates that the 2450-Mc discharge in helium or hydrogen is capable of exciting practically any element or compound which can be obtained as a vapor in the discharge. Certain refractory compounds can be dissociated and volatilized by the addition of convenient reagents, such as hydrofluoric acid. Another means which can be used with solid samples of refractory oxides is to reduce them with graphite or carbon powder by mixing the solid sample with the graphite powder and loading it into the crater of a conveniently shaped graphite electrode. In this work Ultra Carbon graphite electrode No. 105D cut to 1 in. was used for this purpose. This procedure was especially successful with the oxides of Si, Ge, Zr, Ti, Ta, Gd, Tb, Lu, Al, and Hf.

Quantitative Determinations

Figure 24 relates to the quantitative analysis of tin—lead solder by excitation of aqueous solutions, in helium, with the 30-Mc source. Figure 24a shows the spectra obtained with the individual elements, and with the mixture. Figure 24b indicates that a helium atmosphere is preferred over air. Figure 24c is a representative plate employed for the quantitative analysis.

For these quantitative investigations, we have prepared solutions of lead—tin solder, ranging in proportions from 30Sn—70Pb to 55Sn—45Pb, and containing 1 g of sample (dissolved in $HCl + H_2O_2$) per liter of aqueous solution. These solutions were atomized with helium, and introduced into the grounded end of the inductor of the 30-Mc unit, in the same manner as previously described. Light was taken over the approximate central third of the flame height, and central half of the width. By double projection, this radiation was caused to fall uniformly on the slit of the medium quartz spectrograph. Spectra were recorded on Eastman Spectrum Analysis No. 1 plates. An iron-arc spectrum taken through a rotating stepped sector was impressed on the plate to provide emulsion calibration. Replicate spectra of the lead—tin solution are shown in Fig. 24c. The lines Pb 2614.2A and Sn 2706.5A were measured on a Jarrell—Ash microphotometer. Data were reduced in the conventional way to give the logarithm of the relative intensities of the two lines; this value was plotted against the logarithm of the concentration ratios of the respective elements. The resulting plot, closely approximating the expected straight-line relationship, is shown in Fig. 25. This plot represents the average of 15 exposures for each concentration. The linearity of the plot obtained from exposures involving all concentrations at constant power and optical adjustment is excellent, and indicates that this technique will be of practical interest.

Similar results have been obtained with the 2450-Mc discharge in which the following chemical species in aqueous solutions were atomized with helium: Hg (3-100 ppm), B (3-100 ppm), Au (10-1000 ppm), Cd (10-1000 ppm),

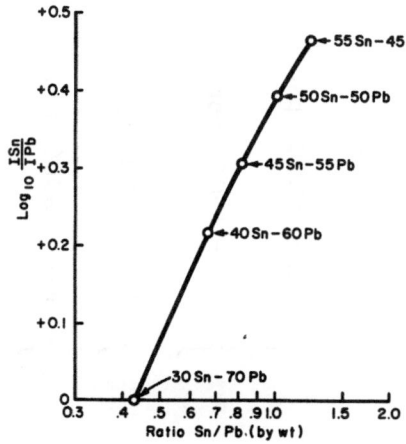

Fig. 25. Calibration curve used in the analysis of tin-lead solder. (Reproduced by permission from Spectrochimica Acta.)

As (300-10,000 ppm), and Sb (100-10,000 ppm) using as internal standard B, As, Sb, Sb, Sb, and Au, respectively.

A particularly interesting analytical measurement possibility is offered by Hg because of the extreme sensitivity of detection of this element. For this determination 10 ml of solutions containing 0.01 to 1 μg Hg have been electrolyzed in Teflon vessels to deposit the Hg onto small Cu wire cathodes. The Cu cathodes were then placed in the crater of a graphite electrode and submitted to the helium discharge (2450 Mc). The spectra recorded indicated that 0.01 μg Hg produced a measurable density. The problems of the production of a mercury-free blank and of introduction of an internal standard remain unsolved. Alternatively the sample containing mercury can be mixed with graphite powder, to which a convenient internal standard, such as antimony, has been added, and this mixture loaded into the graphite electrodes.

The production of reliable quantitative results depends on the constancy and reproducibility of the excitation conditions. Some of these factors have been examined in conjunction with the 2450-Mc magnetron-powered spectrographic unit, and the results are summarized in Fig. 26a and b. The data presented in Fig. 26a were obtained by exciting 0.1 μg of Hg supplied to the discharge as a solid sample mixed with graphite. A graphite electrode provided with a convenient crater was used to receive the sample (50 mg). The influence of the change in helium flow supplied to the discharge is clearly visible. Figure 26b illustrates the influence of power variations. In this case 100 ppm calcium was atomized with helium into the discharge.

Fig. 26a. Effect of helium variation on the transmittance of mercury excited in the 2450-Mc discharge. The mercury was supplied to the discharge as a solid sample (HgCl) mixed with graphite power and loaded into the crater of a graphite electrode.

Fig. 26b. Effect of power variation on the transmittance of calcium excited in the 2450-Mc discharge. The calcium was supplied to the discharge as an aqueous solution atomized with helium.

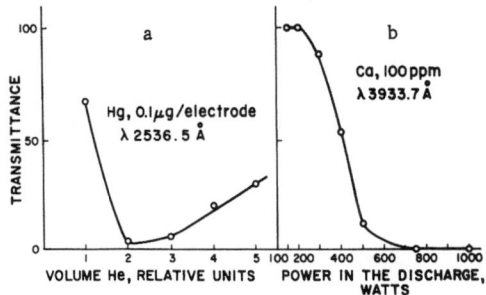

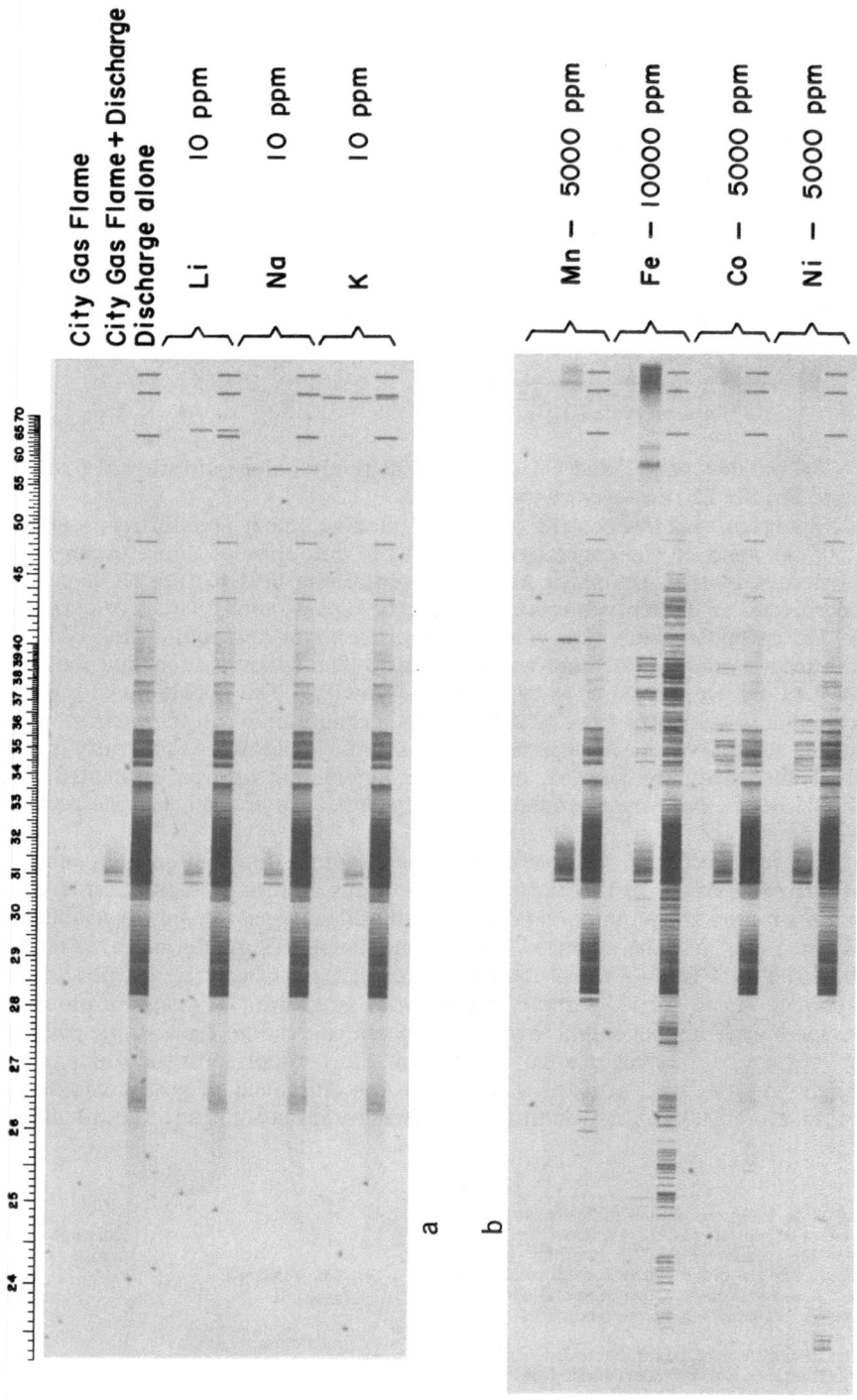

Fig. 31. Excitation of various chemical species atomized with air from aqueous solutions in the city–gas–air flame, city–gas–air flame and superimposed 30–Mc discharge, and in the 30–Mc discharge alone.

To estimate the precision obtainable in the present working conditions with the 2450-Mc discharge, 16 calibration plates were taken using a series of Ca solutions over the concentration range of 10-300 ppm and examined, employing Sr at a concentration of 1000 ppm as the internal standard. The solutions were atomized with helium. The plot of log intensity ratios vs. log concentration of calcium was linear, and has approximately the theoretical slope. Within the examined concentration range, the standard deviation of a single determination was approximately 5%. No effort was made to adjust all working conditions to the optimum values, which would be expected to lead to a further improvement in precision.

Similar results were obtained for mercury in the range of 0.01 to 1 μg at the electrode. In this case the mercury salt (HgCl) was diluted with graphite powder, and 50 mg of this mixture was added into the crater of a graphite electrode as mentioned above. Antimony was used as an internal standard (50 μg/electrode).

RADIO-FREQUENCY DISCHARGE ASSOCIATED WITH A COMBUSTION FLAME

An attempt was made to associate the high electronic-excitation energy of the RF discharge with the relatively high thermal-excitation properties of combustion flames. Figure 27 illustrates the aspect of a city-gas—air flame burning at the molybdenum tip of the 30-Mc torch, and Fig. 28 shows the aspect of the combustion flame plus the discharge. Figure 29 shows the radiations of Li excited in the combustion flame alone, while Fig. 30 illustrates the aspect of the city-gas—air flame supplied with the Li salt plus the 30-Mc discharge. These figures were taken with the same exposure time and it can be seen that an increase in radiation is obtained from the association of the combustion flame and the RF discharge.

The spectra produced by various chemical species in this combined excitation source were recorded with a medium quartz Bausch and Lomb spectrograph on Kodak 1N plates and are shown in Fig. 31a and b. The first spectrum of the city-gas—air flame alone is practically invisible and is followed by that of the combined combustion flame plus the 30-Mc discharge; the third spectrum is that of the 30-Mc alone. The same order was maintained for the groups of spectra which follow and which illustrate the excitation obtained for Li, Na, K, Mn, Fe, Co, and Ni. It can be seen that an increase in intensity is obtained from this association for Li, Na, K, Mn, and Co. At the same time, it is interesting to note the aspect of the spectrum produced by the combustion flame plus the discharge in air and that of the discharge in helium. More work is needed in this direction to assess the practical value of such an association, which might produce more interesting results when hotter flames, such as the acetylene—air mixture, are used.

SUMMARY AND CONCLUSION

The radio-frequency torch appears to be capable of exciting the spectra of an exceptionally large number of elements and compounds with moderate to high intensity. The use of helium or hydrogen atmosphere for sustaining the discharge is advantageous by reason of imposing a minimum background; of these, hydrogen is more effective in volatilizing refractory substances.

The neutral and strongly reducing atmosphere obtained, respectively, with helium and hydrogen, permits compounds to exist in the discharge free of oxidation. The comparatively low neutral gas temperature is advantageous in certain instances in retaining substances in undissociated and un-ionized form, permitting the band spectra of unstable molecules, and atomic first spectra of easily ionized elements, to be strongly excited. However, strong interferences and preferential excitations are also favored by this low neutral gas temperature. The high electron temperature of the discharge provides highly energetic excitation, capable of producing radiation from all elements provided they are vaporized into the flame. The constancy of energy input to the discharge and stability of the flame should lead to constancy of excitation.

ACKNOWLEDGMENT

The various excitation sources employed in this work were constructed by the Amperex Electronic Corporation, Hicksville, Long Island, New York. We acknowledge with pleasure the assistance of C. Roddy in making this equipment available to us. The 30-Mc unit was built by B. Green. The 2450-Mc torch was constructed by W. C. Hickman, who also collaborated in the redesign and carried out the construction of the optical-bench-mounted spectroscopic source of 2450 Mc used throughout this work.

The analytical data presented in this paper were obtained through the assistance of P. Mürau, G. Gundersen, and A. Long, to whom we express our thanks for their competent work.

REFERENCES

1. Y. Asami and T. Hori, Nature 144:981 (1939).
2. E. Bădărău, M. Giurgea, Gh. Giurgea, and A. T. H. Trutia, Spectrochim. Acta (1957), 441.
3. W. E. Bell, A. L. Bloom, and J. Lynch, Rev. Sci. Instrum. 32:688 (1961).
4. L. Bloch and E. Bloch, Ann. Physique 8:397 (1927).
5. H. P. Broida and G. H. Morgan, Anal. Chem. 24:799 (1952).
6. J. van Calker, Proc. IXth Colloquium Spectr. Intern. Lyons, 5-10 June, 1961, p. 48 (Publ. Sept. 1962) G.A.M.S. Publisher.
7. C. L. Chakrabarti, R. J. Magee, and C. L. Wilson, Talanta 9:43 (1962).
8. C. L. Chakrabarti, R. J. Magee, and C. L. Wilson, Talanta 9:639 (1962).
9. J. D. Cobine and D. A. Wilbur, Electronics (June, 1951).
10. J. D. Cobine and D. A. Wilbur, J. Appl. Phys. 22:835 (1951).
11. C. Corliss, W. Bozman, and F. Westfall, J. Opt. Soc. Am. 43:398 (1953).
12. G. D. Cristescu and R. Grigorovici, Bull. Soc. Roum. Phys. 42:3 (1941).
13. H. Dunnken, W. Mikkeleit, and W. Kniesche, Acta Chim. Hung. 33:67 (1962).
14. L. Dunoyer, Compt. rend. 173:472 (1921).
15. L. Dunoyer, J. Physique 3:261 (1922).
16. L. Dunoyer, Compt. rend. 176:953 (1923).
17. E. Fenner, Spectrochim. Acta 1:164 (1941).
18. R. G. Fowler, in "Gas Discharges II," Encyclopedia of Physics, S. Flügge (ed.) (Springer, Berlin, 1956).
19. S. Frisch and E. Schreider, Izvest. Akad. Nauk SSSR, Ser. Fiz. 13:464 (1949).
20. A. Gatterer and V. Frodl, Ric. Spettroscopi. 1:201 (1946).
21. A. Gatterer, Spectrochim. Acta 3:214 (1948).
22. A. Gatterer, Coll. Intern. Spectrographie Strasbourg, 12-14 Oct., 1950, Publ. G.A.M.S. pp. 173-179.
23. A. Gatterer, Mikrochemie and Mikrochim. Acta 36/37:476 (1951).
24. V. B. Gerard, J. Sci. Instrum. 39:217 (1962).
25. W. Gerlach and E. Schweitzer, Z. anorg. allgem. Chem. 195:255 (1931).
26. Goroncy and Urban, Z. anorg. allgem. Chem. 211:28 (1933).
27. R. Grigorovici and G. Cristescu, Optika i Spektroskopiia 6:85 (1959).
28. K. Heinrich, Electrotech. Z. 50:1655 (1929).
29. R. C. Hughes and R. Mavrodineanu, Spectrographic Analysis. U.S.A. Patent Application No. 154-10-468, 1963.

30. R. Ishida, Repts. Govt. Chem. Ind. Research Inst., Tokyo 51:342 (1956).
31. G. Joos, "Excitation of Spectra," in Handbook of Experimental Physics, Vol. 21, Wien and Harms (ed.) (Akademische Verlagsgesellschaft, Leipzig, 1927), p. 197.
32. V. Kapicka, Spisy prirodovedecke fak. Univ. Brno (1961), 269.
33. R. E. Keller and L. Smith, Anal. Chem. 24:796 (1952).
34. F. A. Korolev and Zheenbaev, Izvest. Vysshikh Ucheb. Zabedenii 1959, No. 5, 134.
35. R. Mavrodineanu and H. Boiteux, "L'Analyse spectrale quantitative par la flamme" (Masson, Paris, 1954), p. 26.
36. R. Mavrodineanu and R. C. Hughes, Excitation in Radio-Frequency Discharges, Xth Colloquium Spectroscopicum Internationale, June 18–22, 1962, Washington, D. C. Spectrochimica Acta 19:1309 (1963).
37. R. Mavrodineanu and R. C. Hughes, Excitation of Various Elements by Radio-Frequency Discharges—Potential Analytical Applications. East. Analytical Symposium, Nov. 14–16, 1962, New York, N. Y.
38. W. F. Meggers, J. Opt. Soc. Am. 38:7 (1948).
39. G. D. Morgan, Science Prog. 41:22 (1953).
40. G. Potapenko, Z. anorg. allgem. Chem. 215:44 (1933).
41. C. Roddy and B. Green, Electronics World (1961), 29 and 117.
42. W. Schmidt, Elektronische Rundschau 13:404 (1959).
43. O. Scholz, Chem. Ber. (1959), 497.
44. M. Servigne, P. G. de Montgareuil, and D. Dominé, Compt. rend. 242:2827 (1956).
45. A. L. Stolov, Uchenye Zapiski Kazan. Gosudarst, Univ. im. V. I. Ul'yanova-Lenina, Obscheuniv. Sbornik 116:118 (1956).
46. J. J. Thompson, Phil. Mag. 32:321, 445 (1891).
47. F. S. Tomkins and M. Fred, Report CC-2467, Dec. 18, 1944. Cited after Fred, M., and Scribner, B. F., "Spectrochemical Methods" in Analytical Chemistry of The Manhattan Project, C. J. Rodden (ed.) (McGraw-Hill, New York, 1950), p. 615.
48. F. S. Tomkins and M. Fred, J. Opt. Soc. Am. 47:1087 (1957).
49. G. Vaudet, Compt. rend. 185:1270 (1927).
50. C. S. White, AGARDograph No. 25, 125 (1958).
51. C. S. White and W. R. Lovelace, II, AGARDograph No. 25, 253 (1958).
52. J. G. Winans, Rev. Sci. Instr. 9:203 (1938).
53. M. Yamamoto and S. Murayama, Japan J. Appl. Phys. 2:65 (1958).
54. M. Zelikoff, P. H. Wyckoff, L. M. Aschenbrand, and R. S. Loomis, J. Opt. Soc. Am. 42:818 (1952).
55. Z. Zheenbaev, Inzhenerno-Fizicheskii Zhurnal II:44 (1959).
56. Z. Zheenbaev, Izvest. Vysshikh Ucheb. Zavendenii, Fiz. (1960), 103.

Emission Spectroscopy - Panel Discussion

J. R. Churchill, H. A. Plagge, and J. A. Norris

TRENDS IN THE METAL INDUSTRIES

Emission spectroscopy has shown a rapid and rather consistent growth in the metal industries since 1930 and until rather recently. This is true of both growth as measured by numbers of determinations made and growth in terms of quality of the data produced. The initial and very rapid adoption of emission spectroscopy during the thirties and early forties was primarily due to inherent advantages over wet chemistry that could be realized with relatively little research and with equipment little advanced over that available for many years. By the end of World War II, the development of spectrochemical analysis in the areas of greatest economic importance to the metal industry was approaching a ceiling imposed by inherent limitations of the photographic process. This ceiling was lifted by the advent of the photomultiplier tube which brought the direct reading instruments into being. These have replaced many spectrographs and will replace most but probably not all such instruments in the metal industry. The so-called direct readers improved analysis directly by reducing errors of measurement and indirectly by enabling investigators to measure variables that had been blurred out by photographic variables. Direct readers led to a further and more drastic supplanting of wet methods by emission methods.

The development and availability of vacuum direct readers has created another area of expansion. This development is highly important to the steel industry and other industries in which determinations of certain nonmetallic elements is required routinely. It is relatively unimportant to the aluminum industry.

With the development of vacuum direct readers, the end of an era is in sight. After vacuum instruments are generally accepted and applied, we should expect no further dramatic increase in the statistics of application of spectrochemical analysis, and indeed can anticipate an eventual downward trend.

While quantity of spectrochemical analysis is approaching a ceiling, there remain many interesting and challenging problems and we can expect important progress in three general directions. First there is the improvement and polishing of laboratory equipment and methods to increase accuracy, economy, and scope of application. Achievements in this area are expected to be very important in individual instances, relatively modest in the overall view of the industry. Second and perhaps more important is elimination of laboratory functions by in-line automatic devices, some spectrochemical and some not. The exciting current work on computers in emission spectroscopy is primarily a polishing operation on laboratory analysis but is an important step toward in-line control. Third we can expect the elimination of emission spectroscopy in favor of other techniques in some applications. X-ray emission or fluorescence should be expected to supplant optical emission in many situations where interest is limited to elements heavier than sodium, where higher accuracy is required than is achievable by existing spectrochemical methods or where the traffic will bear both techniques. In trace analysis, the solid state mass spectrometer is likely to prevail over the spectrograph in many cases. Then too, it should be remembered that the trends from one technique to another are two-way streets. Heretical as it may seem, we occasionally go back to wet chemistry as new wet techniques are developed or new requirements imposed. Along this line, there has been a trend from direct readers back to spectrographs resulting from increased needs for sensitivity of detection in certain determinations.

Atomic absorption is another odds on favorite to replace emission spectroscopy in a few specific areas. Its potential in metal analysis seems very limited, but its application of a variety of process controls involved in the metal industries can become very important.

All of the foregoing is stated with considerable assurance. It can be stated with even greater assurance that some of our predictions will look pretty silly a decade from now.

J. R. Churchill
Alcoa Research Laboratories
New Kensington, Pa.

OILS, SLURRIES, AND SOLUTIONS

Mr. Goldblatt mentioned sampling as a major problem. We find that this is also true in the petroleum field. A huge cracking tower filled with tons of catalyst is sampled, the sample is riffled and finally a few milligrams of the catalyst are analyzed by spectroscopy as representative of the whole.

Similarly, the sampling of oils is always a problem. On the subject of spectrographic analysis of oils, solutions, slurries, and soups, I will tell what we are currently doing and in what direction we intend to go.

1. Oils—On the petroleum oils themselves, we do not have any means of exciting the oils directly and obtaining the sensitivity needed. The best sensitivity, directly, is around 1 ppm for Fe, Ni, and V for oils, so a concentration step of some kind is needed. We accomplish this by ashing and analyzing the ash.
2. Soups and slurries—We have no way of continually and reproducibly sampling a thick solution or suspension and usually resort to evaporation and excitation of the residue.
3. We do many types of samples by solution. Catalysts, steels, waters, noble metals, deposits, scales, etc. A metal for trace impurities would be run at about 3% concentrations, but a metal for major components would be run at 1000 ppm or less.

The usual method of our solution work is the rotrode. We have tried the Teflon cap and platrodes but find no advantage in them. Currently we are trying a capillary lower electrode with a metering pump. This device shows much promise. These capillary electrodes could be either metal or graphite. We have also ordered a plasma jet for improved solution work.

I heard so much about sensitivity at yesterday's meeting, that I would like to state what our experience has been. In a dilute HCl solution using a high-intensity spark, we can detect 0.5 ppm of most elements; however, it is usually possible to obtain greater sensitivity for any single element if required. For example, on Zn, a shift from spark to A-C arc excitation will increase sensitivity fivefold.

One of the main trends in solution work is the use of controlled atmospheres. Inert gases are used at UOP for the excitation of hydrocarbons.

Howard A. Plagge
Universal Oil Products Co.
Des Plaines, Illinois

SPECTROSCOPY IN THE ATOMIC ENERGY FIELD

Spectroscopy as practiced in the atomic energy field can be divided into three main areas of activity: (1) radioactive materials, (2) normal materials, and (3) theoretical spectroscopy.

There are two basic approaches used in the analysis of radioactive materials with slight variations due to the type and level of radiation. The first method used is the excitation of the "hot" material in a well-shielded cell from which the light is usually brought outside the cell to the spectrograph. This calls for sources that can be controlled from without the cell through mechanical hands and usually leaves behind radioactive material that must be cleaned up to prevent contamination of the following sample. The inherent difficulties of this type of work combined with the length of time required and the cost of the equipment involved limits, in general, the use of this method to only those samples that must be analyzed under these conditions due to the radioactive hazards involved.

A variation of this technique, however, has been very successful where the amount and type of radioactivity is much less and can be safely handled in small portable enclosed sources that can either be decontaminated after use or so designed that they can be discarded after one use. This type of sample handling can normally be handled on regular out-in-the-room equipment without danger to personnel. Equipment of this type, while not generally available on the commercial market, has been designed by most laboratories handling slightly radioactive materials.

The second approach to the handling of radioactive materials has been the chemical separation of the radioactive material and the analysis of the impurities. This type of analysis works very well where the "hot" material can be readily separated on an ion exchange column, by liquid—liquid extraction techniques, or by classical chemical separations. With this method, the remaining material to be analyzed can normally be handled like any other material, or at least handled with simple protective techniques.

The analysis of normal materials covers a very broad range approaching the scope of the natural and man-made world. A brief review of some of the materials handled within certain fields will give the best picture of the wide interests allied to atomic energy.

Samples taken from the world of nature are mainly of interest to the biologist and health physicist. These samples will include such diverse items as sea water, fresh water, soil, plants, trees, clams, fish, animals, and man.

These studies are carried out to determine the make-up of the natural environment and the effect that radioactive materials have on this environment, both detrimental and beneficial effects being of interest.

The metallurgist and engineer furnishes samples of complex alloys of uranium and other materials used as reactor fuels, and high temperature alloys of many diverse materials used to contain the various reactor fuels under conditions of intense radiation that occur in a working reactor. From the allied field of ceramics and cermets those materials of great interest for high temperature and radiation stability, such as have been proposed for space reactors, are now routine samples. Included in this general area will be metals and compounds of beryllium, boron, carbon, aluminum, titanium, chromium, iron, cobalt, nickel, zirconium, columbium, molybdenum, hafnium, tantalum, tungsten, thorium, uranium, and plutonium. Others of special interest from time to time will also be found on the spectroscopist's desk.

The chemist and physicist also provides samples of a diverse nature. The general interest of the properties of the atomic world in which we live has supported the work of separating the natural stable isotopes of the elements and the determination of the physical properties of these isotopes in pure compounds and metals. This has recently lead to the whole field of high-purity materials, both natural and isotopic, where the total impurity level must be at the ppm level or less. Examples of these materials would include KCl, LiF, and MgO. The analysis of these materials leads the spectroscopist to new areas of problems yet to be satisfactorily solved. The physicist is also interested in unusual alloys of elements in an attempt to better understand the atomic world in which he works. Examples of recent requests include the unusual alloys of Ir-Tb and Ir-Ho.

The theoretical spectroscopist has been responsible for using specialized spectroscopic techniques to provide information for understanding the processes involved in plasmas of interest to the fusion expert. For the field of spectroscopy itself, the theoretical spectroscopist has made use of the many available isotopes to determine the nuclear and electron characteristics necessary for the classification of spectra.

The spectroscopist working in the atomic energy field has a unique opportunity to work with a large number of scientific disciplines and in areas of interest leading to a better understanding of the world in which we live. As are all other spectroscopists, he is faced with solving the hard problems since the easy ones seem to have been finished.

John A. Norris
Oak Ridge National Laboratory
Oak Ridge, Tennessee

Additional Papers on Emission Spectrography

- "Flame Emission Spectrometric Analysis of Rare Earth Mixtures," V. A. Fassel, R. H. Curry, A. P. D'Silva, and R. N. Kniseley, Institute for Atomic Research and Dept. of Chemistry, Iowa State University, Ames, Iowa.

- "Atomic Absorption Determinations Requiring High Sensitivity," W. Slavin, Perkin-Elmer Corp., Norwalk, Connecticut.

- "Observations on the Use of Response Surface Techniques to Optimize Flame Excitation Conditions," R. K. Skogerboe and R. A. Woodriff, Dept. of Chemistry, Montana State College, Bozeman, Montana.

- "Physiochemical Processes in Flame Spectroscopy," W. D. Cooke, Dept. of Chemistry, Cornell University, Ithaca, New York.

- "Flame Emission Detection of Toxic Compounds in Air," R. S. Bramen, W. A. Loseke, and E. S. Gordon, Armour Research Foundation, Chicago, Illinois.

- "Investigations with the Laser Microprobe as a Tool for Spectrochemical Analysis," M. K. Healy and J. J. Devaney, Jarrell-Ash Co., Newtonville, Massachusetts.

- "Spectrochemical Techniques for the Analysis of Wear-Metals in Used Lubricating Oils for the Transportation Industry," F. Leahy, Baird-Atomic, Inc., Cambridge, Massachusetts.

- "A New Vacuum Spectrometer for the Total Analysis of Iron and Steel," A. Davison and F. Leahy, Baird-Atomic, Inc., Cambridge, Massachusetts.

- "The Application of Controlled Atmosphere Spectroscopy to Trace Analysis," W. A. Gordon, National Aeronautics and Space Administration, Lewis Research Center, Cleveland, Ohio.

- "An Application of the Stallwood Jet-D.C. Arc to the Analysis of Air Contamination in Inert Gases," V. Raziunas, W. A. Loseke, and E. L. Grove, IIT Research Foundation, Chicago, Illinois, and B. K. Davis, National Aeronautics and Space Administration, Huntsville, Alabama.

Nuclear Magnetic Resonance
and
Electron Spin Resonance

Application of Electron Spin Resonance Techniques to Biochemical Systems: Current Information on Photosynthetic Systems*

J. J. Heise and R. W. Treharne

Charles F. Kettering Research Laboratory
Yellow Springs, Ohio

ESR has the same advantage of being a nondestructive technique as does optical spectroscopy, and this makes it particularly applicable to biological systems. A review of various applications of ESR to biological systems will be given in such a way as to illustrate certain unique possibilities inherent in the method.

The light-induced ESR signals in photosynthetic plants were among the first biological systems to be studied and are still under intensive investigation. In plants light-energy conversion to useful chemical energy is a central problem. The information obtained by ESR related to this problem will be reviewed and possible interpretations discussed.

Electron spin (or paramagnetic) resonance spectroscopy, since its inception in 1945 by Zavoisky [1], has had a rather slow history of application to biological systems as compared to its use in physics and chemistry. Almost a decade passed before the first biological applications were made [2]. This has been due, in the main, to certain technical difficulties in its application which are inherent with all biological systems. One of them is the spectrometer's low tolerance to water, which is ubiquitous in all biological systems. In addition, only certain limited information can be obtained from compounds and ions which have magnetic properties and are likely to occur in biological systems. Since recent pertinent reviews exist, for example, those by Commoner [3], Androes and Calvin [4], Beinert et al. [5], and Weissman [6], as well as volumes and monographs on the general principles [7-10], only a cursory view of the principles, instruments, and applications will be given. Rather, this paper will review a portion of the experimental information obtained in the last decade about a specific group of complex biological systems and will emphasize the unique types of information which may be obtained by this technique. The biological systems of choice are those capable of photosynthesis. A summary of the contributions which this technique has lent to the understanding of the photosynthetic process also will be given.

Electron spin resonance (ESR) absorption spectroscopy, like other spectroscopic techniques, depends on the interaction of matter, which in this case is an electron, with externally imposed electromagnetic energy. The relative position of the frequency required is outlined in Fig. 1. This region of the electromagnetic spectrum only recently has been open to investigation. The invention of the klystron, which provides the microwave

*Contribution No. 122 of the Charles F. Kettering Research Laboratory.

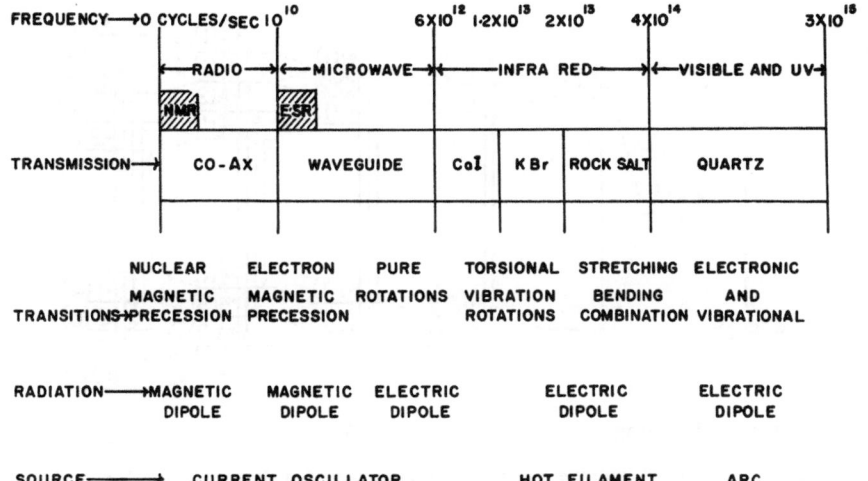

Fig. 1. The electromagnetic spectrum correlating types of transmission, transitions, radiations, and sources. The position in the spectrum of radiation employed for ESR absorption spectroscopy is indicated.

energy, was accomplished by the Varian brothers only 25 years ago [11]. Unlike other forms of spectroscopy, technical considerations dictate that the frequency of the microwave energy (9000 Mcps) must be kept constant rather than varied. Use of the magnetic property of free electrons with unpaired spin states allows microwave energy absorption to be observed when a variable magnetic field is imposed on the sample.

Figure 2 shows that in the idealized case of a single, unpaired free electron, the ground energy state can be split into two levels through perturbation by a magnetic field. Microwave energy of the proper frequency may induce a transition between the two energy levels. Absorption of microwave energy designated by the product of its frequency ν and Planck's constant h in the equation for the resonance condition

$$h\nu = g\beta H \tag{1}$$

is then directly proportional to the applied magnetic field H. Since β, Bohr's magneton, is a constant, g, the "spectroscopic splitting factor," may be calculated, and represents a value for the center of the ESR absorption. The g-value of an ESR absorption is dependent on the intrinsic properties of the sample. For free radicals of the biological systems which will be

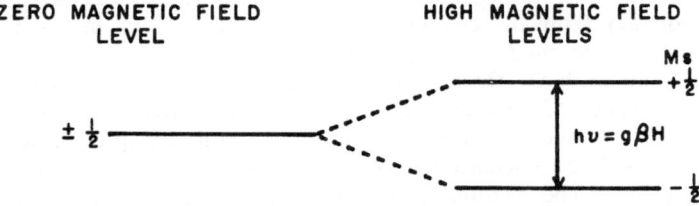

Fig. 2. A diagram of the energy levels of a single, unpaired free electron in the absence and presence of an externally imposed magnetic field.

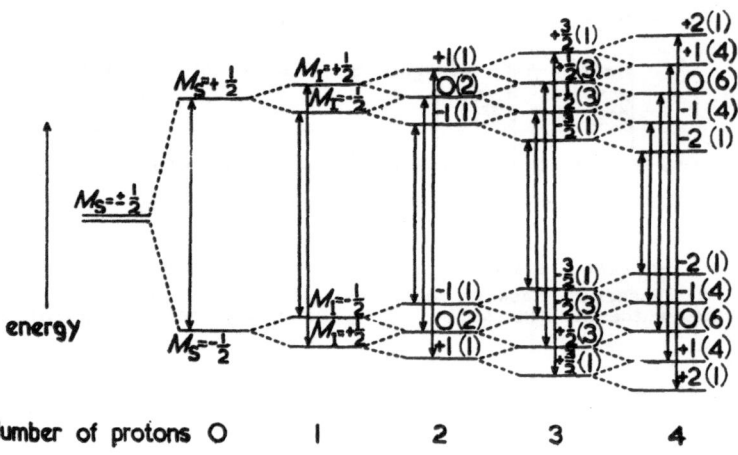

Fig. 3. A diagram of multiple energy levels of an unpaired electron in a molecule with four equivalent protons. (From [12].)

considered, the g-values are all near that of a free electron (2.0023) and vary not more than ±0.003.

Figure 3 illustrates that further complications in the ESR spectrum energy levels may be encountered when magnetic nuclei of a molecule in which the electron is located interact with the electron [12]. In this example the effect of the nuclear magnetic moment is represented by the energy levels dictated by the nuclear spin quantum numbers M_I. The number of absorption lines is given by $2M_I + 1$. This is the diagram of the interaction of four equivalent protons with an electron in a single molecule. The vertical lines indicate the allowed transitions, and the numbers in parentheses at the right indicate the theoretical ratio of the intensity of absorption lines expected in the ESR spectrum which is $1 : 4 : 6 : 4 : 1$. The multiple lines due to interaction of magnetic nuclei with the electron produce what is termed hyperfine structure.

Figure 4 shows the basic components of the Varian instrument most widely used. These include the microwave source produced by the klystron oscillator, a sample cavity, a variable electromagnet, a detector and indicator. In practice the magnetic field is modulated by coils on either side of the sample, and a phase-sensitive detection system of electronics is employed so that the indicator displays the first derivative of microwave absorption with respect to the field. The display is commonly referred to as a "signal" rather than a spectrum. Significant improvement in allowable sample volume and sensitivity have been obtained by increasing the modulation frequency from the conventional 100 kcps to 1 Mcps in a series of ESR spectrometers designed by Dr. J. Townsend [13].

There are at least three important reasons for the inception of studies on the occurrence of unpaired electrons, or free radicals, in photosynthetic systems. First, the existence of the light-excited triplet state of chlorophyll has long been established in simple solutions *in vitro* by optical spectroscopy [14]. The triplet state involves two unpaired electrons having parallel spin states. However, the spin states interact strongly enough to spread the resonance absorption beyond the range of detection of the spectrometers. Photoexcited triplet states have been observed in certain crystals and glasses [15,16].

Although light-induced ESR signals have been found with chlorophyll solutions and crystals, the present interpretation of their occurrence on a molecular level is only provisional. Mention should be made that the photo-reduction of chlorophyll, or the Krasnovsky reaction, which has been hypo-thesized to occur *in vitro* [17] and to form free radicals in solution [18], has not yet been demonstrated to involve free radical intermediates due to chlorophyll itself [19-21]. Single-electron transfer has been implied by the light-induced changes in the equilibrium free-radical concentration of a reductant in the Krasnovsky reaction [21]. Smaller has reported that a given number of chlorophyll molecules excited by intense repetitive light flashes at low temperature have produced practically an equivalent number of un-paired spins [22]. He attributes this ESR signal to the excitation of the pyrrole group [23]. Light-induced transfer of unpaired electrons between solid films of porphyrin analogs of chlorophyll and electron acceptors [24] and between chlorophyll and quinones in solution [25] also is now well established. Since a review of free radicals of chlorophyll and its reactions is perhaps premature, comments will be confined mainly to results obtained either with whole plants, particularly algae or bacterial cells, or with cellular components derived from them.

A second reason for the study of free radicals in photosynthetic systems is suggested by new knowledge about the structure of the photosynthetic apparatus and interest in semiconductor physics. Free electrons may be involved in the poorly understood conversion of light energy into useful

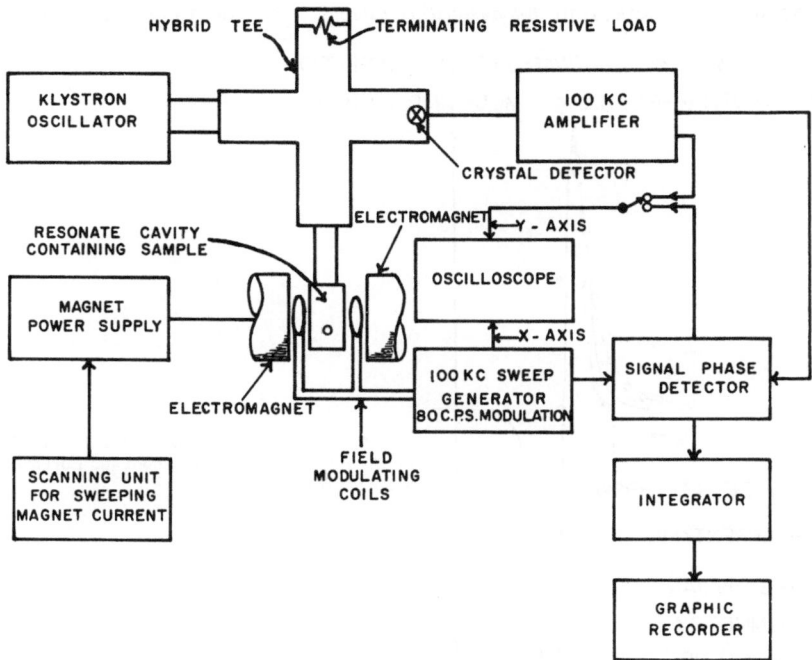

Fig. 4. A block diagram of a typical ESR or EPR spectrometer. Based on [58].

electron potential within the quasi-crystalline structure of the chlorophyll-containing portions of the cell, the chloroplast of green plants or chromato-phore fractions of bacteria [26, 27].

A third reason is that widespread interest in the possibilities of the occurrence of free radicals as intermediates in univalent oxidation–reduction reactions suggest that free radicals occur in the enzyme reactions of chloroplasts [28], which carry out many enzymatic oxidation–reduction reactions.

The use of complex biological material, such as chloroplasts and cells, which have the ability to form many free radicals, may seem impossible. However, even the myriad complexities of a living cell are subject to useful analysis by the ESR techniques. Commoner's pioneering results [29], which showed that green plants exhibit both an ESR signal closely associated with a chlorophyll complex having a photophysical type of response to light and another light-induced ESR signal involving oxidation–reduction reactions, have been substantiated and extended by many laboratories to the extent that they may be placed within the framework of present knowledge of photosynthesis.

Information derived from the characteristics of the light-induced ESR signals themselves, namely, g-value, line width and hyperfine structure, shall be considered in the less complex biological system of photosynthetic bacteria. Then physical parameters, such as the temperature dependence of the kinetics of formation and decay of the signal and effects of wavelength of irradiation, will be presented. Similar considerations of green plants will be given along with additional biological data.

Figure 5 shows a typical light-induced signal from whole cells of the photosynthetic bacterium *Rhodospirillum rubrum*. As the magnetic field is in-

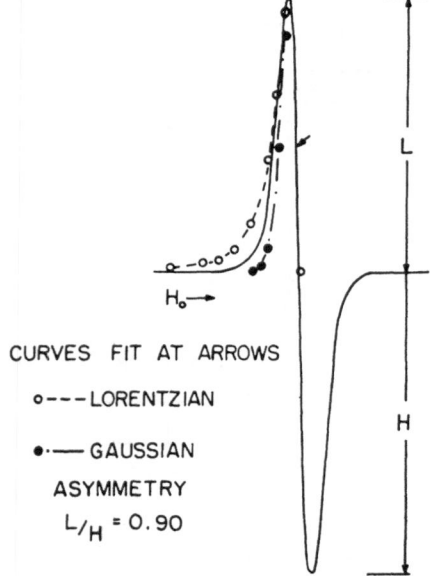

RHODOSPIRILLUM RUBRUM- AQUEOUS SUSPENSION

LINE WIDTH (P-P) = 11.2 g , MOD. AMP. = 0.6 g

$H_o \rightarrow$

CURVES FIT AT ARROWS

o – – – LORENTZIAN

•·—— GAUSSIAN

ASYMMETRY

L/H = 0.90

Fig. 5. A comparison of the light-induced ESR signal of *R. rubrum* cells to Gaussian and Lorentzian line shapes. (From [4].)

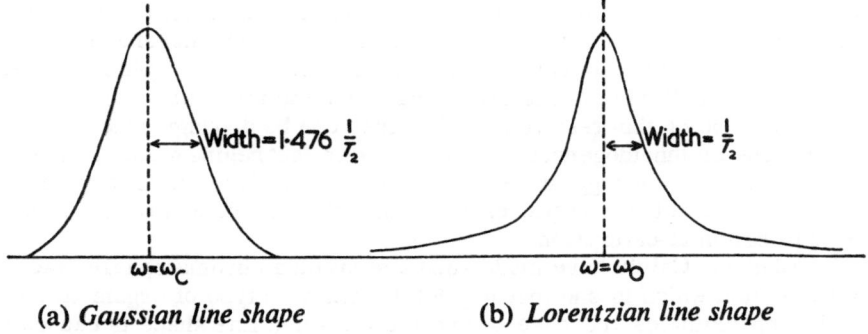

(a) *Gaussian line shape*   (b) *Lorentzian line shape*

Fig. 6. Two theoretical possibilities for the shape of integrated ESR signals. (From [12], cf. [9].)

creased a single, nonstructured ESR signal is obtained, here shown as the first derivative of absorption. It has a g-value between 2.002 and 2.003, which is very near to that of a free electron. The line width ($\Delta H_{\frac{1}{2}}$) is about 11 G. The same light-induced signal has been found in all photosynthetic bacteria, and it has the same spectrum characteristics whether in the whole cell or a chromatophore fraction containing bacteriochlorophyll isolated from the cell [30]. The line shape of the signal is compared in Fig. 5 by Androes, and of the two theoretical possibilities, Lorentzian and Gaussian, it is found to more nearly approximate the Gaussian shape.

Figure 6 shows the two theoretically possible line shapes in what corresponds to the integrated ESR signal. The Gaussian line shape, which has a much broader width at one-half height than the Lorentzian, may indicate line broadening due to electron—electron interaction of neighboring molecules or to electron—nuclear interaction within the same molecule and has the properties of a homogeneous spin system [4]. The Gaussian shape may

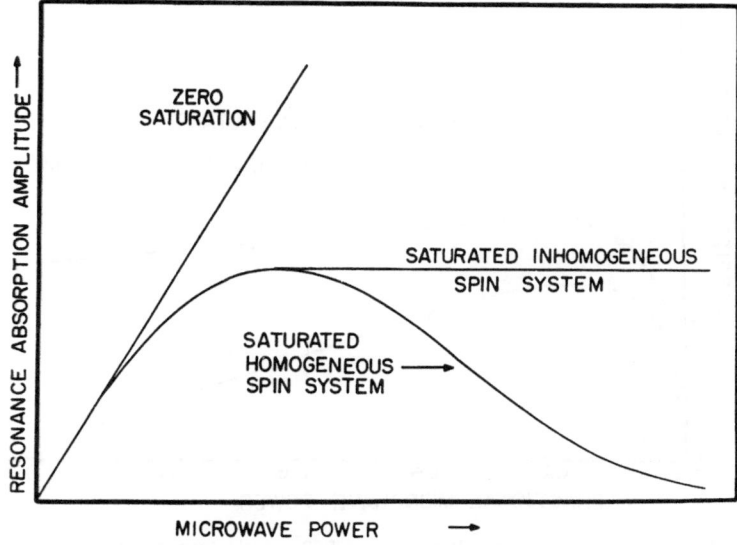

Fig. 7. The idealized power saturation behavior of ESR signals. (From [4], cf. [9].)

also be due to a case where the ESR spectrum is the resulting envelope of a large number of narrower overlapping lines and is inhomogeneous [4].

There exists an experimental method for determining which of these cases exist. In Figure 7 it is shown that when the microwave power is increased a plot of the resulting ESR signal can be distinguished as either homogeneous or inhomogeneous. A saturation effect begins when the number of transitions in energy-state levels induced by the microwave power exceed thermal relaxation processes. The spin system departs from thermal equilibrium at saturation.

Andreos and Calvin have made such a study on a chromatophore fraction of a bacterium which is shown in Fig. 8 [4]. The wild type of organism more closely approximates the case where the Gaussian line shape is caused by inhomogeneous broadening. It therefore is possible that the simple line found in the light-induced ESR signals of bacteria is made up of a number of narrower lines due to the electron interaction with a number of magnetic moments which are randomly orientated. There is no increase in the width of the line as the microwave power is increased. This result further supports the above conclusion. It is interesting to note that the absence of the accessory carotenoid pigments in a mutant organism shows more homogeneous saturation than the wild type. Andreos and Calvin also found the relaxation time to be shorter. These characteristics suggest that fewer varieties of unpaired electrons make up the ESR signal in the low carotenoid-containing bacterium.

Thus, rather sophisticated analysis can be applied even to the study of biological systems by simple methods of observation of line width, shape, and behavior of the signal when the microwave power is increased. In the future it may be possible to determine the specific magnetic moments which bring about this broadening by such techniques as the determination of

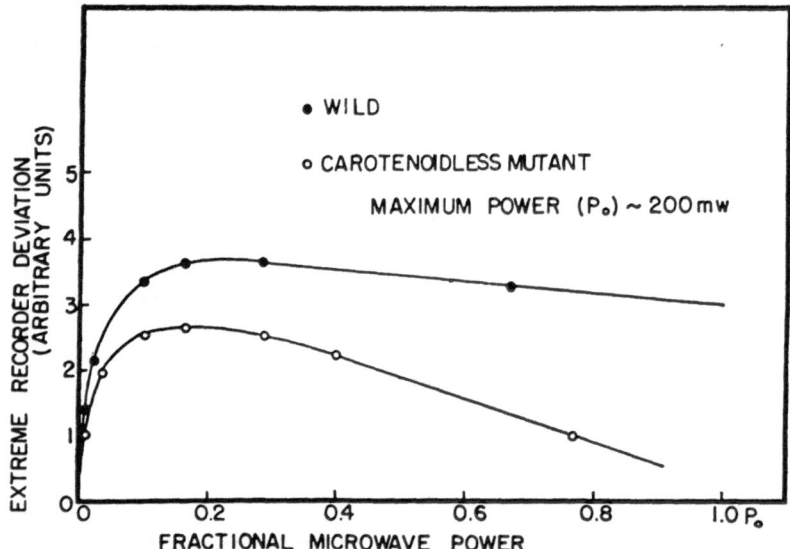

Fig. 8. Power saturation of the light-induced ESR signals from dried films of *Rhodopseudomonas spheroides* chromatophore fractions. (From [4].)

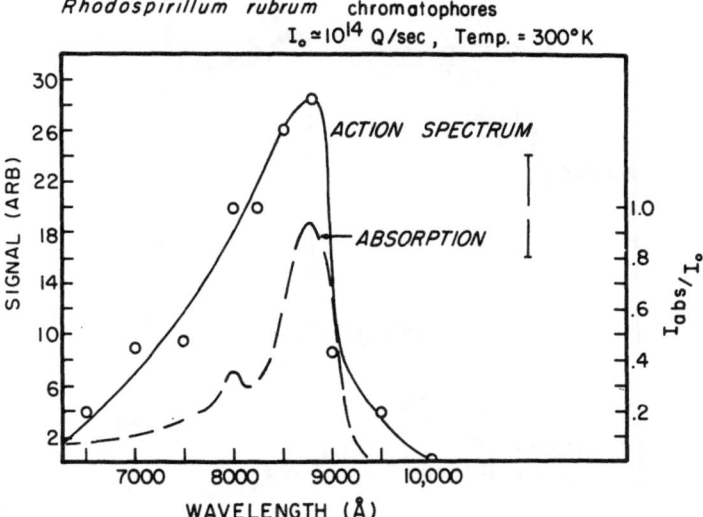

Fig. 9. The action or efficiency spectrum of light-induced ESR signal of a
R. rubrum chromatophore fraction (solid line) compared to optical absorption
spectrum (dashed line). (From [32].)

relaxation times after pulsing with a saturating microwave pulse, and/or
the "electron–nuclear double resonance" techniques [31, cf. 9].

Besides analysis of the ESR signals themselves, other approaches, such
as variation of physical parameters, have recently yielded useful information. Since this discussion is concerned with light-induced signals of photosynthetic systems, the quality of the light inducing the signal is important.
Ruby [32] has recently obtained good correspondence between the optical
absorption spectrum and ESR signal action (or efficiency) spectrum using
relatively low light intensities (Fig. 9). This correspondence indicates
that bacteriochlorophyll is important in the chain of events producing the
ESR signal. However, the quantum efficiency of ESR signal production has
not been measured. This is a rather crucial experiment. The amount of
variation in the production of spins from a quantum yield of one results in
information as to the proximity of the production of spins to the primary
quantum light-conversion step. The existence of rapidly reversible light-
induced ESR signals at −160 C in a chromatophore fraction found by Calvin's
group [33] suggests that at least part of the signal exhibited at room temperature may be concerned with the primary quantum conversion step.

Another way of demonstrating the relationship of signal decay to change
in physical parameters is shown in Fig. 10. In the upper curve the ESR
signal of whole R. rubrum cells forms in the light and decays in the dark in a
fraction of a second. When the chemical processes of the cell are disrupted
by extraction of the chromatophore fraction, the decay is lengthened more
than an order of magnitude in a dried film, as indicated in the lower curve.
Wet chromatophore fractions have decay times similar to the dried films.

Still another method of altering the environment is the biochemical
approach. Vernon has shown that chromatophore fractions catalyze a light-
induced reaction which converts the reduced forms of dyes into the oxidized

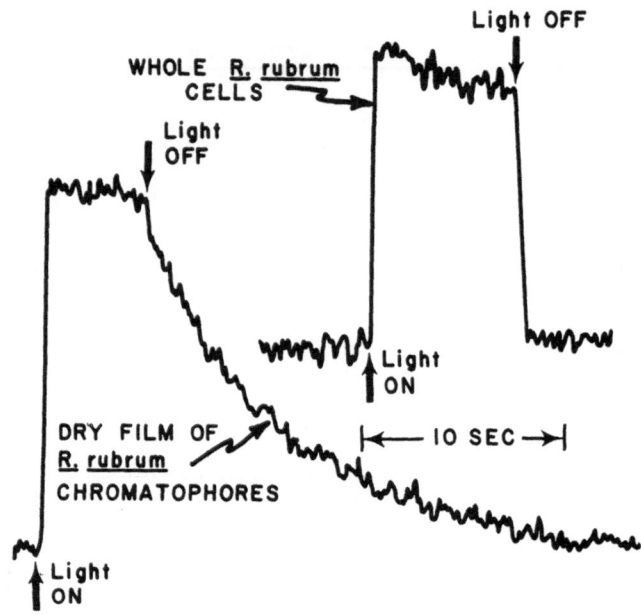

Fig. 10. The light-induced formation and decay of ESR signals with time of whole *R. rubrum* cells compared to a dry film of a chromatophore fraction [54].

forms [34]. Since ESR signal formation in chromatophores in the light indicates an increased capacity of the chromatophores to undergo oxidation-reduction reactions, the ESR signal was examined in the presence of reduced dyes, and is indicated in Fig. 11.

In Fig. 11 the decay of the ESR signal from a chromatophore fraction is plotted in a manner to determine rate constants, viz., the log value of the difference from maximum of ESR signal vs. time. More than one straight line may be drawn through the points. These lines indicate multiple modes of decay. However, since a straight line may be drawn through the majority of points, one decay process apparently predominates.

Table I shows time constants evaluated from these curves. A larger time constant indicates a faster-decaying signal. With reduced indigo carmine it can be seen that a higher final concentration of dye catalyzes the decay of the ESR signal to a greater extent than a lower concentration. Reduced dichlorophenolindophenol accelerated the decay to an even greater extent so that the half-time of decay was about the same as the rise-time (100 msec) [30].

Methylene blue had no effect on the signal from 10^{-6} to 10^{-4} molar concentrations when it was in oxidized form. Before being added to chromatophores, reduction of methylene blue by either dithionite or reduced nicotinamide-adenine dinucleotide (NADH) increased the rate of decay of the ESR signal, even at the lowest concentrations of the dye utilized [35].

Optical absorption studies by Vernon suggest that the photo-oxidation of these reduced dyes is coupled with the photoreduction of chromatophore components [36]. The possibility exists that the chromatophore component

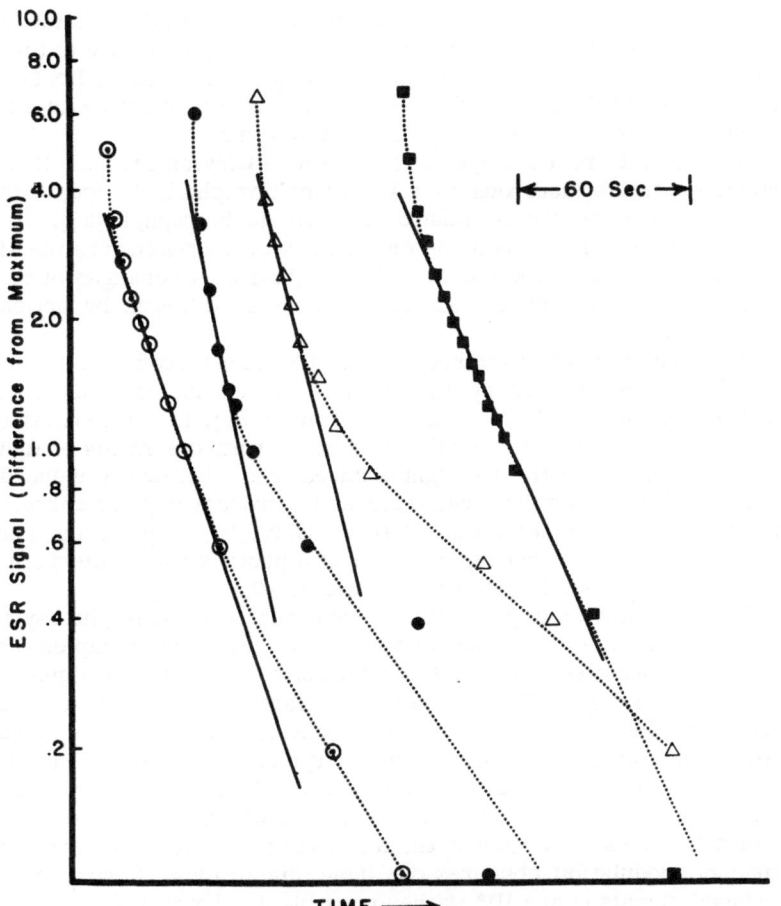

Fig. 11. The effect of dyes on the decay in the dark of light-induced ESR signals of *R. rubrum* chromatophore fraction plotted as the log value of difference from maximum signal. The curves left to right are control with no additions, a lower and higher concentration of reduced indigo carmine, and reduced methyl red. These curves are compared quantitatively in Table I [35].

TABLE I

Time Constants of Decay of the ESR Signal from *R. Rubrum* Chromatophores as Influenced by Addition of Reduced Dyes

Dye	Dye concentration, M	Reducing agent	Time constant, $\Delta \log$ amplitude/sec
None	—	—	0.020
Indigo Carmine	1×10^{-5}	$NaBH_4$	0.029
Indigo Carmine	1×10^{-4}	$NaBH_4$	0.035
Methyl Red	1×10^{-4}	$NaBH_4$	0.031

which is reduced by the dye is bacteriochlorophyll in the oxidized form. One interpretation of the effect of dyes on the signal decay is that light partially oxidizes the bacteriochlorophyll to produce free radicals which give rise to the ESR signal. The reduced dye may then act as an electron coupler and promote a faster decay of the ESR signal.

The results with reduced dyes correlate with Clayton's finding that when ferricyanide extracts electrons from bacteriochlorophyll, the optical density changes correspond to photo-oxidation of bacteriochlorophyll [37]. Experimental confirmation may depend on future measurements such as the simultaneous determination of the kinetics of the light-induced changes of the ESR signal and chlorophyll photo-oxidation effects as followed by optical absorption changes.

Other information about the nature of the ESR signal from chromatophore fractions has been obtained by Loach, Ruby, et al., by manipulation of the chemical environment [38]. This is shown in Fig. 12. A redox titration using a ferri–ferrocyanide couple was made with the ratios of light and dark ESR signals. Both the light-induced and chemically induced ESR signals in the dark reach their half-values at a crossover point corresponding to a redox potential of about ± 0.46 V. Calvin's group has been able to infer from these results that the acceptor of a photoextracted electron from chlorophyll has a rather low redox potential of −0.1 V [39].

The ESR signals from green plants, which in contrast to photosynthetic bacteria have the capacity to photoevolve oxygen, are more complex. Figure 13 shows it is possible to obtain four distinctly different ESR signals from *Chlorella pyrenoidosa* cells. Treharne and associates have found that the paramagnetic ions Fe^{+++} and Mn^{++}, whose ESR signals arise from unpaired spins of uncompleted shells in d-orbitals typical of transition metal ions, can also be found in washed algae cells [40]. These signals occur even in the dark. Light induces a free-radical signal which is narrower than that of the metallic ions. Narrowing the magnetic field sweep and reducing its amplitude of modulation obscures the signal due to ions. Resolution of two light-induced signals (I and II)* is then possible as shown in Fig. 13.

Figure 14 shows the effect of light on the integrated signals of the green alga *C. pyrenoidosa*. The residual signal in the dark having a time constant of decay of several hours has hyperfine peaks and will be referred to as signal II. The ESR signal in white light, composed of both signals I and II, shows marked increase in both amplitude and asymmetry compared to the signal in the dark. The asymmetry is explained by signals I and II having different g-values. Signal I occurs at a higher magnetic field position than II.

In green plants as in bacteria, the line shape of signal I, which more nearly approaches the Gaussian distribution of resonance absorption, indicates interaction with other magnetic moments [41]. Not only line shape, but g-value, response to light frequency, temperature, and the kinetic behavior of signal I from green plants is so similar that it may be equated in an operational way with the ESR signal exhibited by photosynthetic bacteria.

Figure 15 shows that signal II differs from signal I at the point where it saturates with increasing microwave power. Treharne and Johnson have shown that signal II exhibits saturation at a lower microwave power value and has a longer relaxation time [52]. Therefore, the electrons giving rise

*Since the original designation [42], signal I has also been designated: f, n [39]; R [44, 46, 50]; LS [43]; II [47, 49]. Signal II has been designated: s, b [39]; S [44, 46, 50]; DS [43]; I [47, 49].

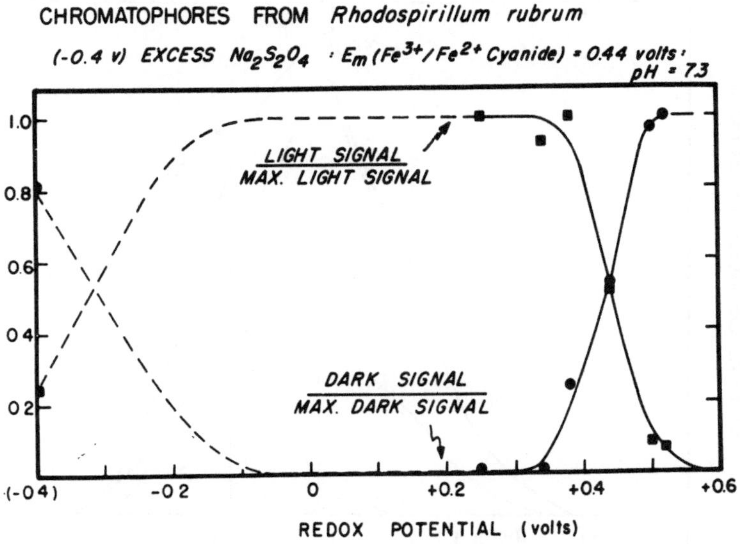

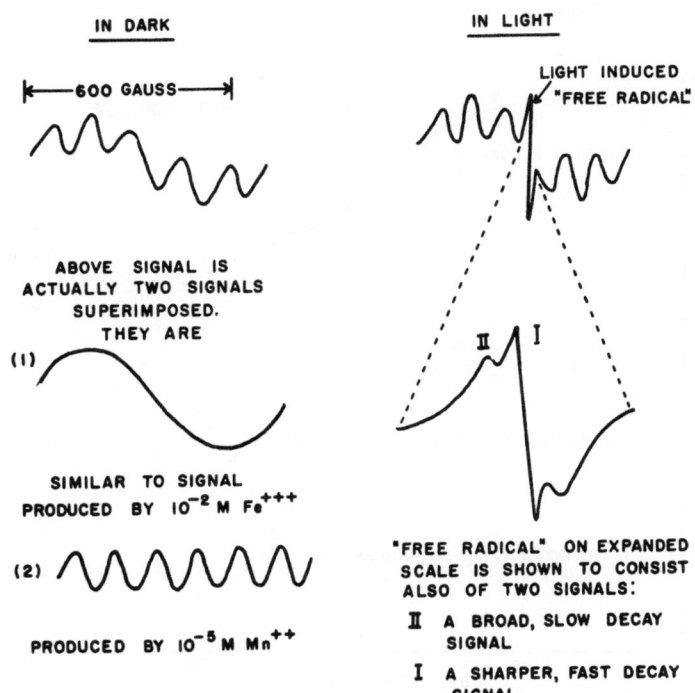

Fig. 12. Light-induced and chemically induced (dark) ESR signals of a *R. rubrum* chromatophore fraction shown as a redox titration using a ferri-ferrocyanide couple. (From [38].)

Fig. 13. A diagrammatic representation of the ESR signals (1) Fe^{+++}; (2) Mn^{++}; and light-induced signals I and II in *C. pyrenoidosa* [40].

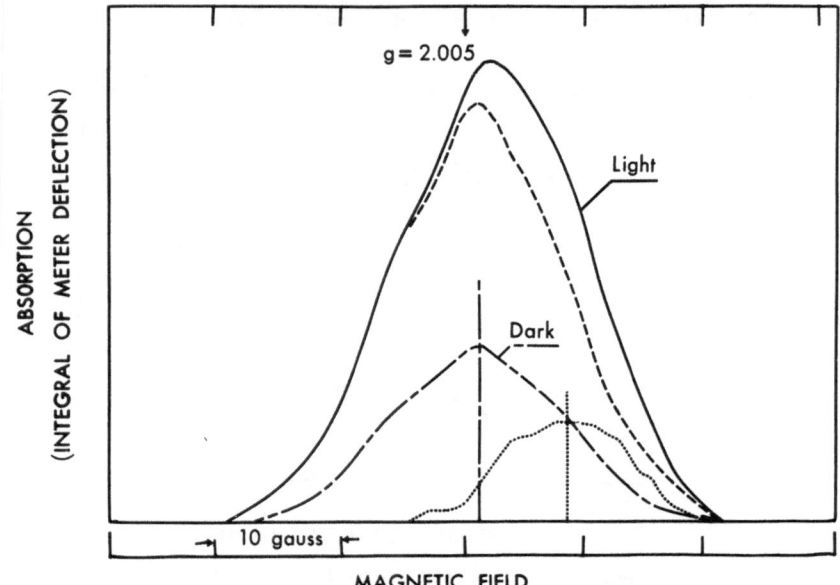

Fig. 14. Integrated ESR signals of *C. pyrenoidosa* in the light and dark at 5C. The segmented curve represents an average of residual light-induced ESR signals in the dark, and the solid curve represents ESR signals in the light. The dashed curve is a representation of the increase in signal II in the light, and the dotted curve is the light-induced signal I obtained by subtraction of the dashed from the solid curve. (From [21].)

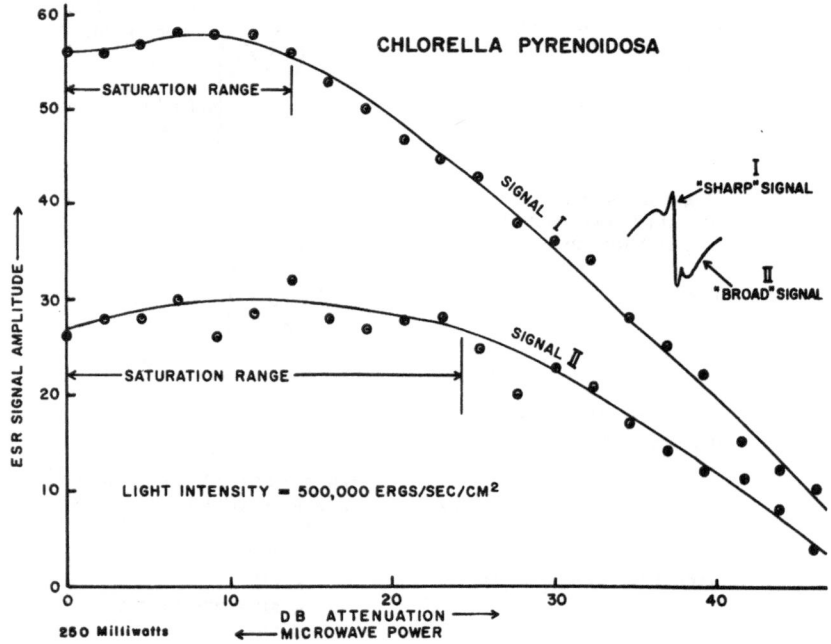

Fig. 15. The saturation effect of the light-induced ESR signals I and II of *C. pyrenoidosa* cells as the microwave power is increased from right to left [52].

to a signal II may be less tightly bound than electrons giving rise to signal I. This conclusion appears to be in conflict with the suggestion that signal I has properties of a semiconductor electron [24] and signal II is due to a free radical in a specific molecular complex [29]. Since one report has given the opposite result [47], further data are needed to resolve this inconsistency.

Since signal II has hyperfine structure, the presence of a nuclear magnetic moment which can influence the electron is implied. The possible types of atoms with magnetic moments found in large quantity in biological materials are hydrogen, nitrogen, and phosphorus. Substitution of such atoms with their isotopes of different magnetic moment is a difficult task for the chemist, even in simple molecules. However, since there are autotrophic organisms which can grow on inorganic nutrients, water, carbon dioxide, and light, isotope incorporation by the plant is possible. Through the extensive program at Argonne National Laboratory, Katz, Commoner, and Heise were able to show an effect on the signals by growth of algae in nearly pure deuterium oxide [21,42]. This effect is indicated in Table II as a narrowing of the widths of the ESR signals. Both of the signals are influenced by protons because they are both narrowed. In Table II the ratio of magnetic moments of protons to deuterons is 2.6 for signal II. A comparison of this ratio to the theoretical value (3.3) indicates that all the hyperfine structure is due to protons. The intensities of the hyperfine peaks most closely approximate the ratio $1:4:6:4:1$. In the complicated energy diagram presented in Fig. 3, such a ratio is characteristic of an organic free radical in which the unpaired electron is associated with a resonance system composed of four chemically equivalent carbons with a proton attached to each. The structural possibilities which could give rise to this spectrum are shown in the following structures:

The unpaired spin which gives rise to signal II is probably either a portion of a linear array of atoms or a ring structure. This moiety is probably located in the matrix of a larger structure, the nature of which awaits the isolation and biochemical characterization of the cellular component. The properties of the two ESR signals so far considered are summarized in Table III.

TABLE II
ESR Signal Widths $\Delta H_{1/2}$ (Gauss)

	H_2O	D_2O	H_2O/D_2O
Signal II	17	6.6	2.6
Signal I	13	3.0	4.3

TABLE III

Properties of the Light-Induced ESR Signals of Photosynthetic Plants*

Property	Signal I	Signal II	Reference
Center g-value	2.002-2.003	2.004-2.005	[29, 44]
Width of signal	8-9 G	19-21 G	[21]
Hyperfine splitting values	–	5-7 G	[21]
Shape	Nearer Gaussian than Lorentzian	Nearer Gaussian than Lorentzian	[4, 44]
Dependence on microwave power	Saturates at higher value (but see [47])	Saturates at lower value (but see [47])	[21, 44, 55]
Type of broadening	Inhomogeneous	Inhomogeneous	[4]
Reason for broadening	Many magnetic moments	Principally protons	[21, 42]
Electron environment	Randomly orientated moments	Probably four equivalent protons	[21, 41, 42]
Hyperfine structure	Broadened to single line	Line intensity ratio 1:4:6:4:1	[21]
Narrowing by deuterium substitution	H/D about 4.3	H/D about 2.6	[21]

*This table is a compilation of presently available data and is subject to modification. No distinction is made between the signal in bacteria and signal I of green plants or between whole and disrupted cells.

Considerable information has been obtained about the ESR signals by variation in physical parameters on the photosynthetic organisms. Change in the wavelength of light has been mentioned. Table IV is a summary of the effects of certain variations. The following effects are worth emphasizing. The light quality and efficiency spectra effects follow chlorophyll absorption and are similar for both signals. However, signal II saturates at lower light-intensity values comparable to those of the photosynthetic process. The kinetics of signal II formation and decay are slower and vary in different ways within the biokinetic range of the organism (5 to 30 C). The rate constants of formation correspond to no single order of reaction over the biokinetic range. Calculations of activation energy from the rate constants indicate either that the process of formation of the ESR signals has no constant activation energy or that there are a number of competing processes involved [21].

Perhaps the most important result of kinetic analysis of the ESR signals in whole cells made by Heise and Commoner [21,42] is shown in the proposed scheme of Fig. 16. Light activates processes which give rise to signal I. Some of these electrons are quickly transferred to free radical component II, and electron flow may continue into photosynthetic electron transport. Temperatures in the lower portion of the biokinetic range which suppress the biochemistry of photosynthesis reduce electron turnover rate. The reduced requirement of electrons manifests itself in increased ESR signals. Naturally occurring components of the electron chain, or ones artificially introduced into the system, would be expected to change the steady-state levels of the signals. When flavinmononucleotide, ascorbic acid, nicotinamide-adenine dinucleotide phosphate, dyes, and CO_2 were added to chloroplasts and to whole cells, there occurred a reduction in the steady-state concentration of ESR signal II in the light [21]. The majority of these

TABLE IV

Kinetics and Effect of Variation of Physical Parameters on the ESR Signals
of Photosynthetic Plants

Physical parameter	Signal I	Signal II	Reference
λ for maximum efficiency	Greatest efficiency at absorption λ_{max} of chlorophyll	Same as I—Possibility of two efficiency peaks	[21, 32, 44]
λ for maximum signal	λ independent about 400 to 759 mμ at high intensity	Same as I	[44]
λ for upper limit	Signal at 754 but not 760 mμ	Same as I	[44]
Intensity	Saturates at high intensity	Saturates at low intensity comparable to values for photosynthesis	[21, 44 40, 45]
Above biokinetic range	60 C (5 min) enhances signal	60 C (5 min) destroys signal	[45]
Biokinetic range (5-30 C)	Greater increase below 20 C than II	Increases below 20 C	[21]
Below 0 C*	Larger—55 C than 35 C Smaller—160 C	—	[4, 21]
Decay time	Fractions of a second	Initial decay—seconds	[21, 42]
		Complete decay—hours	[21, 42]
Formation time	Fractions of a second	Fractions of a second	[21, 42]
Decay time (below 0 C)	−55 C to 30 C multiple decay rates		[33, 39]
	−160 C single, fast decay rate		[33, 39]
Rate Constants	Rate constants larger than for free radicals in solution		[21, 42]
Whole cells (biokinetic range)	Not same for decay and formation; therefore different processes probably involved		[21]
	Formation has multiple-order rate constants		[21]
	Decrease at lower temperatures		[21]
Dry film (−160 C)	Similar for decay and formation;* Chloroplasts have large time constant for decay		[33, 39]

*Data obtained with bacteria. All other data are obtained with green algae.

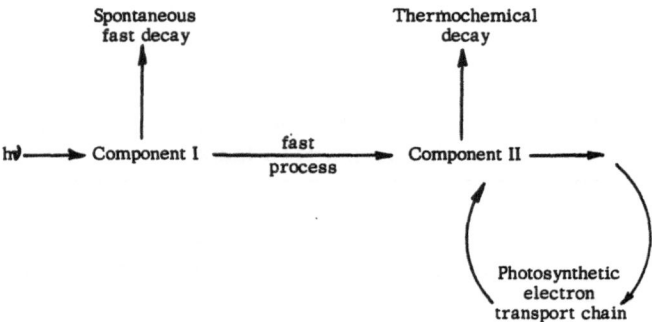

Fig. 16. One possible scheme of transfer of light energy to useful electron potential in the living cell based on kinetic analysis of the light-induced ESR signals I and II over the biokinetic range of the organism. (From [21].)

agents are reversibly oxidizable intermediates and cofactors which act to transfer electrons in the photosynthetic process.

The effect of gases other than CO_2 have been noted. Storey et al. found that O_2, which may act as a substrate to produce more photosynthetically derived O_2, increased signal II and promoted its decay [43]. Weaver found that lack of O_2 brought about by respiration of cells in the dark over a long period of time increased the capacity of the cells to exhibit signal I [44]. Clearly, a more thorough investigation of the effect of variation in the biochemical parameters on these signals is warranted. It is presently one of the largest gaps in our knowledge about the signals.

Agents which impair the photosynthetic process reduce signal II. Treharne [45], Weaver [46], Allen [47], and Heise [21] have shown that inhibitors of photosynthesis, here defined as O_2 evolution, reduce signal II. The powerful inhibitor of O_2 evolution 3-(3,4-dichlorophenyl)1,1-dimethylurea enhances the ratio of signal I to signal II. It is evident that the free radical which gives rise to signal II is an important part of the O_2 evolution process of photosynthesis.

Variation of biological parameters has given support to this assertion, as is illustrated in Table V. Cell disruption, whether by sonicating, heating, freezing, or drying, or subsequent treatments to the chloroplast, reduce signal II, while only extreme treatment eliminates the photophysical response of signal I to light. A specific binding of component II, the molecule which gives rise to signal II, is necessary in order to observe the ESR signal. With the loss of signal II there is a concomitant reduction in the capacity of cell fragments to photoevolve O_2. Oxygen evolution with pigment complexes derived from cells requires special oxidants and does not exhibit signal II.

Selection of specialized photosynthetic cells is another biological parameter which suggests signal II is intimately associated with O_2 evolution (Table V). As mentioned before, bacteria do not evolve O_2, and there is no ESR signal II.

Mutant cells in which part of the photosynthetic process is no longer operative have proved to be the most powerful biological parameters yet found for elucidation of the role of signal II in the photosynthetic process. Certain mutant cells when grown in the dark have incomplete development

TABLE V
Effect of Variation of Biological Parameters on the ESR
Signals of Photosynthetic Plants

Biological parameter	Signal I	Signal II	O_2 evolution	References
Selection of Genus:				
Green plants	Present	Present	Normal	[21, 45]
Algae	Present	Present	Normal	[21, 45]
Bacteria	Present	Absent	None	[21]
Cell treatment:				
By sonication	Present*	Absent	None	[45, 55]
By heating				
60 C 5 min	Enhanced	Absent	None	[45]
100 C 5 min	Absent	Absent	None	[45]
	Signal I/II ratio			
By freezing	Enhanced		None	[45]
By washing				
chloroplasts	Enhanced	Reduced	Small	[29]
By extraction	Present in			
of pigment	P700 and			
complexes	P672	Absent	Small	[56, 57]
Mutants:				
Lacking caro-				
tenoids				
(bacterium)	Present	Absent	None	[4]
Etoliated	Absent	Absent	None	∟ [2, 4]
Etoliated cells	Increases		Corre-	
exposed to	with	—	lates	
light	"greening"		with	
			greening	
Lacking green				
pigments	Absent	Absent	None	[21]
Requiring dye				
and reductant				
to reduce NADP	Present	Absent	Some	[49]
Having normal				
pigments but				
no O_2 evolution	Present	Absent	None	[50]
Having normal				
pigments but	Absent #8	Present	Normal	[50]
no CO_2 evolu-	Small #17			
tion	Normal #18			

*Ochromonas danica grown in organic medium loses signal I with this treat-
ment [45].

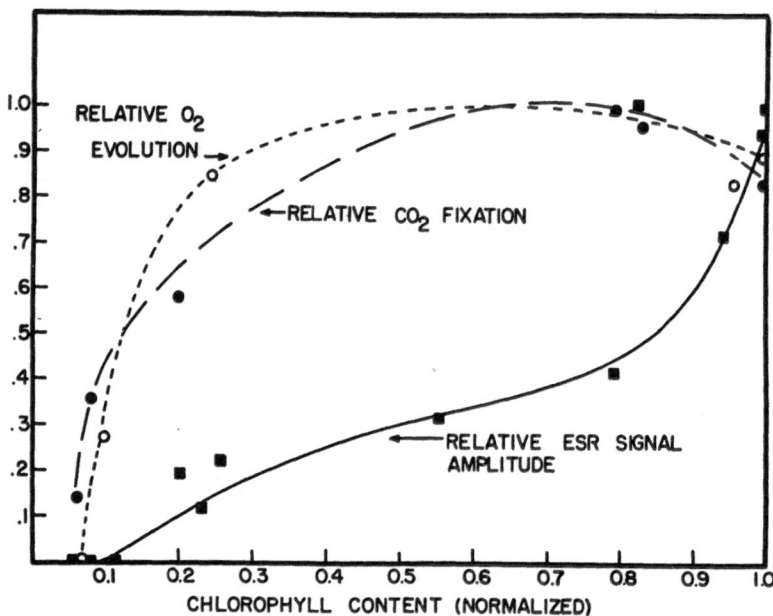

Fig. 17. A comparison of relative O_2 evolution and CO_2 fixation to the ESR signal amplitude of dark-grown cells (*Chlamydomonas* mutant) upon reversion in the light to the normal green state. (From [4].)

of chlorophyll and give no ESR signal. In Fig. 17 Androes et al. [48] show that the ESR signals begin to increase in the light as photosynthesis increases, although not in direct proportion.

Levine and Piette have shown that mutants which require special oxidants and reductants to evolve O_2 have no signal II [49] (Table V). The most convincing evidence that the presence of signal II is associated with an unimpaired capacity for photosynthetic O_2 evolution was obtained by Weaver and Bishop [50]. A normal pigment-containing mutant of the green alga *Scenedesmus* which lost its capacity to photoevolve O_2 was isolated and characterized by Bishop [51]. The organism had no indication of signal II.

The above compilation of ESR data hopefully will serve to stimulate further interest and investigation into the process of photosynthesis with this form of spectroscopy. It is clear that a complete description of the ESR signals on a molecular or biological structural level has not been accomplished. Similarly, the proximity of component I to the primary energy conversion act is not yet determined, nor have the signals from green plants been placed within the framework of biochemical knowledge of the photosynthetic processes such as O_2 evolution, CO_2 assimilation, phosphorylation, or NAD reduction. When these and other gaps in our knowledge are filled, it is probable that ESR spectroscopy may make a considerable contribution to the understanding of the transfer of light energy into useful chemical potential in photosynthesis.

In summary, analysis of the light-induced ESR signals in photosynthetic systems indicates there are two, as yet unidentified, light-induced unpaired electron components. Component I which gives rise to signal I: (1) is found in all normal photosynthetic organisms (bacteria, algae, and higher plants); (2) is nearer to the primary light-conversion step than component II; and (3) appears to transfer unpaired electrons to component II. Components I and II are affected differentially by variation in physical, chemical, and biological parameters. Presently available experimental evidence indicates that component II: (1) exists only in green plants and algae; (2) is a free-radical molecule exhibiting hyperfine structure; and (3) correlates in some manner with the normal processes of oxygen evolution in photosynthesis.

REFERENCES

1. E. Zavoisky, J. Phys. SSSR 9:211 (1945).
2. B. Commoner, J. Townsend, and G. E. Pake, Nature 174:689 (1954).
3. B. Commoner, Acad. Roy. Belg., Classe Sci., Mem., Collection in 8°, 33:114 (1961).
4. G. M. Androes and M. Calvin, Biophys. J. 2:217 (1962).
5. H. Beinert, W. Heinen, and G. Palmer, Enzyme Models and Enzyme Structure, Brookhaven Symposia in Biology No. 15 (1962), p. 229.
6. S. I. Weissman, Comprehensive Biochemistry, M. Florkin and E. H. Stotz (eds.) Vol. 3 (Elsevier Publ. Co., Amsterdam, 1962), p. 189.
7. G. E. Pake, Paramagnetic Resonance (W. A. Benjamin, Inc., New York, 1962).
8. M. S. Blois, Jr., H. W. Brown, R. M. Lemmon, R. O. Lindblom, and M. Weissbluth (eds.), Free Radicals in Biological Systems (Academic Press, New York, 1961).
9. D. J. E. Ingram, Free Radicals as Studied by Electron Spin Resonance (Butterworths, London, 1958).
10. D. J. E. Ingram, Spectroscopy at Radio and Microwave Frequencies (Butterworths, London, 1955).
11. R. H. Varian and S. F. Varian, J. Appl. Phys. 10:321 (1939).
12. A. Carrington, Quart. Rev. (London) 17:67 (1963).
13. T. S. Hahs, Studies in Electron Spin Resonance: A Free Radical Formed from Ascorbic Acid. Thesis, Washington University, St. Louis, 1960.
14. R. Livingston, Handbuch der Pflanzenphysiologie, W. Ruhland (ed.) Vol. 5, Pt. 1 (Springer-Verlag, Berlin, 1960), p. 830.
15. C. A. Hutchinson and B. W. Mangum, J. Chem. Phys. 29:952 (1958).
16. J. H. Van der Waals and M. S. de Groot, Mol. Phys. 3:190 (1960).
17. J. W. Coleman and E. Rabinowitch, J. Phys. Chem. 63:30 (1959).
18. A. A. Krasnovsky and A. V. Umrikhina, Dokl. Akad. Nauk SSSR 66:663 (1955).
19. H. Linschitz and S. I. Weissmann, Arch. Biochem. Biophys. 67, 491 (1957).
20. S. S. Brody, G. Newell, and T. Castner, J. Phys. Chem. 64:554 (1960).
21. J. J. Heise, Electron Spin Resonance Studies of Free Radicals in Photosynthetic Systems. Dissertation, Washington University, St. Louis, 1962.
22. B. Smaller, Free Radicals in Biological Systems, M. S. Blois, Jr., et al. (eds.) Academic Press, New York, 1960), p. 315.
23. B. Smaller, Personal Communication.
24. M. Calvin, J. Theoret. Biol. 2:258 (1961).
25. G. Tollin and G. Green, Biochim. Biophys. Acta 60:524 (1962).
26. E. Katz, Photosynthesis in Plants, J. Frank and W. E. Loomis (eds.) (Iowa State College Press, Ames, Iowa, 1949), p. 287.
27. D. F. Bradley and M. Calvin, Proc. Natl. Acad. Sci. U.S. 41:563 (1955).
28. B. Commoner, J. J. Heise, and J. Townsend, Proc. Natl. Acad. Sci. U.S. 42:710 (1956).
29. B. Commoner, J. J. Heise, B. B. Lippincott, R. E. Norberg, J. V. Passonneau, and J. Townsend, Science 126:57 (1957).
30. J. J. Heise and L. P. Vernon, Bacterial Photosynthesis, H. Gest, A. San Pietro, and L. P. Vernon (eds.) (Antioch Press, Yellow Springs, Ohio, 1963).
31. G. Feher, Phys. Rev. 103:500 (1956).
32. R. H. Ruby and M. Calvin, Bacterial Photosynthesis, H. Gest, A. San Pietro, and L. P. Vernon (eds.) (Antioch Press, Yellow Springs, Ohio, 1963).
33. M. Calvin, A Symposium on Light and Life, W. D. McElroy and B. Glass (eds.) (Johns Hopkins Press, Baltimore, 1961), p. 317.
34. L. P. Vernon, J. Biol. Chem. 234:1883 (1959).
35. J. J. Heise and L. P. Vernon, Abstracts 7th Biophysical Society Meeting, New York, Section ME 9 (1963).
36. L. P. Vernon, Bacterial Photosynthesis, H. Gest, A. San Pietro, and L. P. Vernon (eds.) (Antioch Press, Yellow Springs, Ohio, 1963).
37. R. K. Clayton, Photochem. Photobiol. 1:201 (1962).
38. P. A. Loach, G. M. Androes, G. M. Maskin, and M. Calvin, manuscript in preparation; cf. [32].
39. M. Calvin and G. M. Androes, Science 138:867 (1962).
40. R. W. Treharne, T. E. Brown, H. C. Eyster, H. A. Tanner, Biochem. Biophys. Res. Commun. 3:119 (1960). Treharne and Eyster, ibid. 8:477 (1962).
41. E. C. Weaver and H. E. Weaver, Photochem. Photobiol., in press.

42. B. Commoner, A Symposium on Light and Life, W. D. McElroy and B. Glass (eds.) (Johns Hopkins Press, Baltimore, 1961), p. 356.
43. W. H. Storey, C. M. Monita, and D. G. Cadena, Jr., Nature 195:963 (1962).
44. E. C. Weaver, Carnegie Inst. Wash. Year Book 61 (1962), p. 353.
45. R. W. Treharne, T. E. Brown, and L. P. Vernon, Biochim. Biophys. Acta (in press).
46. E. C. Weaver, Arch. Biochem. Biophys. 99:193 (1962).
47. M. B. Allen, L. R. Piette, and J. C. Murchio, Biochim. Biophys. Acta 60:539 (1962).
48. G. M. Androes, M. F. Singleton, J. Biggins, and M. Calvin, Biochim. Biophys. Acta 66:180 (1963).
49. R. P. Levine and L. R. Piette, Biophys. J. 2:369 (1962).
50. E. C. Weaver and N. I. Bishop, Science 140:1095 (1963).
51. N. I. Bishop, Nature 195:55 (1962).
52. R. W. Treharne and C. E. Johnson, unpublished results.
53. E. H. Rogers, NMR and EPR Spectroscopy, 3rd workshop on NMR and EPR sponsored by Varian Associates (Pergamon Press, New York, 1960), p. 1.
54. J. J. Heise, unpublished results.
55. R. W. Treharne and L. P. Vernon, Biochem. Biophys. Res. Commun. 8:481 (1962).
56. H. Beinert, B. Kok, and G. Hoch, Biochem. Biophys. Res. Commun. 7:209 (1962).
57. M. B. Allen and J. C. Murchio, Biochem. Biophys. Res. Commun. 11:115 (1963).
58. W. C. Lockhart and R. C. Jones, 1960. NMR and EPR Spectroscopy, 3rd workshop on NMR and EPR sponsored by Varian Associates (Pergamon Press, New York, 1960), p. 65.

Electron Spin Resonance Spectra of Gamma Radiolysis Products of Solid Acetonitrile

D. Dunbar, D. Hale, L. Harrah, R. Rondeau, and S. Zakanycz

Wright-Patterson Air Force Base
Dayton, Ohio

The electron spin resonance spectra of radicals formed in solid acetonitrile at 77 K under the action of Co ^{60} γ-radiation were observed to guide in the interpretation of the γ-radiolysis of this material. The hyperfine pattern is centered at a g-value of 2.0025 ± 0.0004, and exhibits eleven lines symmetrically situated about this value. Analysis of this hyperfine structure indicates the probable presence of at least two different species. The radical present in the largest concentration is formed from acetonitrile by removal of a hydrogen atom and is responsible for about four-fifths of the observed signal. The remaining intensity is thought to be due to radicals resulting from the addition of a single hydrogen atom to the parent substance. The decay of these radicals has been measured at various temperatures and the resulting material analyzed by mass spectrometry and gas chromatography. The 100-eV yields for these radicals were estimated using a diphenyl picryl hydrazyl solution standard.

INTRODUCTION

The radiation chemistry of triply bonded organic compounds is currently under investigation in our laboratory. Although the radiation chemistry of the simplest compounds containing the triple bond has been studied, little attention has been given to more complex molecules containing this linkage, and hence no generalizations as to the effect of this moiety on the radiation chemistry of organic systems have been made.

The simplest analogs, acetylene and hydrogen cyanide, have been studied extensively [1-14] and acetylene has found uses as a vapor-phase chemical dosimeter [4, 5]. Only recently has the radiation chemistry of the next number of the C≡N series (acetonitrile) been studied [15, 16], and little work has been reported on the radiation chemistry of propyne, the next number of the acetylene series.

We are currently investigating the radiation chemistry of these materials in the vapor, liquid, and solid phases, and the results for the solid phase radiolysis of acetonitrile are compared here with the results and interpretation of the ESR spectra of the irradiated solid.

EXPERIMENTAL

Sample Purification

The samples of CH_3CN were prepared from Eastman spectro grade acetonitrile which was subjected to gas chromatographic analysis and found

361

to contain traces of propionitrile and water, together with some light
hydrocarbons. (These were found to be present whenever this material was
stored for any length of time after opening.) These impurities were removed in three steps.

1. The acetonitrile was distilled through a Todd column at a reflux
   ratio of 5:1.
2. Returned to the pot and distilled from P_2O_5 into anhydrous $NaHSO_4$.
3. Introduced into a vacuum system together with P_2O_5 and degassed
   by alternate freeze–thaw–pump cycles until the pressure observed,
   when opening the solid to the vacuum system, did not exceed 0.1
   μ at −195 C. These samples were then distilled directly into the irradiation vessels and sealed at −195 C. A vacuum of less than 0.1
   μ was maintained throughout the sealing operation.

Analyses performed on the samples purified before radiolysis indicated
only water was present to the extent of approximately 100 ppm. This analysis, since it involved a short exposure to the atmosphere, may not be reliable for the water impurity, and levels less than 100 ppm are probably
attained.

The vacuum system used for sample purification and handling is shown
in Fig. 1.

Radiolysis Vessels

Figure 2 shows the vessels used in our radiolysis experiments. The
vessels are of three types. The largest is used exclusively for gas-phase
experiments and has a sample volume of about 200 cc. The intermediate
size is used both for liquid and solid phase work with a volume of about
5 cc. The small tubing is of fused quartz and is used for the ESR studies.
About 2 cc when full, the tubing is normally filled only to about 0.5 cc so
that it may be annealed to remove the color centers produced in the quartz
by the radiolysis when repeated radiolyses were done on the same sample.

Spectrometer

The spectra were obtained on a Varian Associates V-4502 ESR spectrometer equipped with a multipurpose cavity, a variable temperature accessory including a liquid-nitrogen dewar, and a 100-kc modulation unit.
The recorder used was a Mosley X-Y, the x-direction being driven by the
magnet field exciting current. This modification was made to simplify data
handling. Our quantitative work was done by double integration, and this
modification gave a chart display that was directly transferable to a computer for the integration.

Frequency Measurement

The operating frequency of the microwave bridge, as measured with a
cavity wavemeter, was found to be 9.116 kMc with the quartz dewar, sample,
and liquid nitrogen inserted.

Fig. 1. Vacuum purification system.

Field Measurement

The field was measured at the sample position with a proton resonance gaussmeter and the span of the recorder, in field, for several sensitivities and positions was noted. In practice the field was monitored with solid 1,1-diphenyl 2-picryl hydrazyl as a primary standard and the quartz color center as a secondary "internal" standard for the radiolysis samples. Field differences, as obtained from the calibrated recorder span, were used to calculate the g-values given. All CH_3CN spectra were obtained with a recorder span of 100 G in 38 cm.

Calibration Standard

The quantitative measurement of radical yield was performed by double integration of the spectra and comparison of these areas with a 1,1-diphenyl 2-picryl hydrazyl standard prepared in the following way. A large single crystal of DPPH was weighed and dissolved in a known volume of purified carbon tetrachloride. This sample was transferred and sealed in a quartz tube of the same dimensions as those containing the CH_3CN. Since the surface area to volume ratio of this crystal is relatively small, it was assumed to contain a stoichiometric quantity of radicals. This standard was then inserted in the cavity and the resulting spectra integrated in the same manner as the sample spectra.

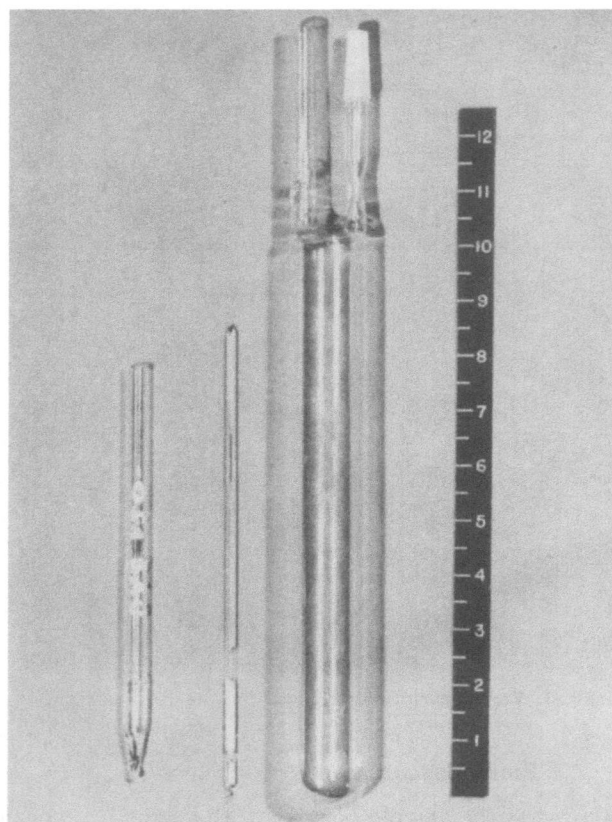

Fig. 2. Sample vial.

Carbon tetrachloride was chosen as a solvent since the sample thus prepared has the same resonant frequency and apparent cavity Q as that of the frozen CH_3CN and presumably the same "filling factor." Errors introduced by this method are probably smaller than those inherently produced by voids appearing in the frozen samples and inaccuracies in dose determinations. The sample density was assumed to be that of the liquid at its freezing point.

Radiolysis Sources

The radiolyses were carried out in two Co^{60} facilities. A small water-cooled source of 1500 Ci of Co^{60} was used for the liquid samples. This is diagrammed in Fig. 3.

For the solid-phase experiments a large-volume source of 25,000 Ci having a cylindrical sample space 6 in. × 24 in. long was used. The dose rate in the large source was 26.8×10^{19} eV/g-hr and the small source, $\frac{1}{6}$ of this value. Dosimetry was accomplished with the ceric–cerous dosimeter.

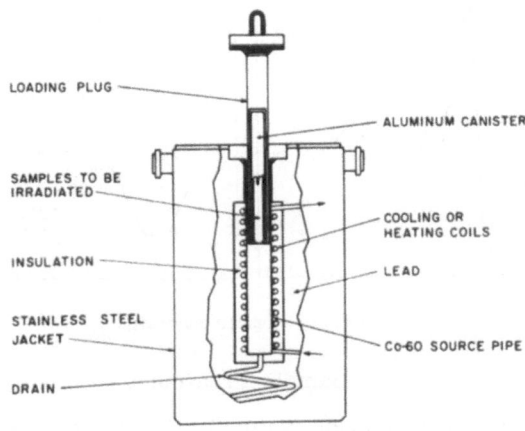

Fig. 3. Fifteen hundred-curie source.

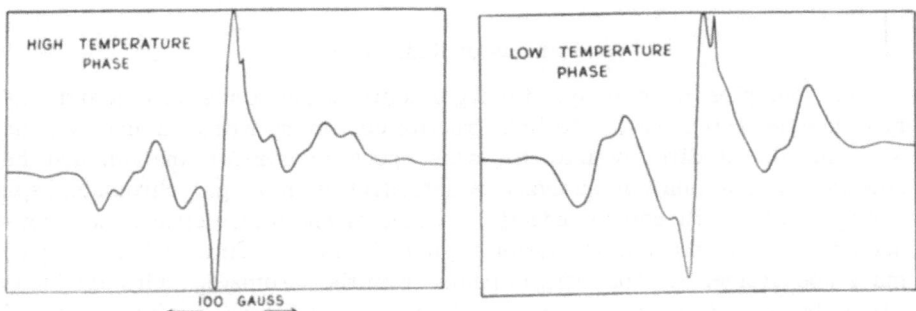

Fig. 4. ESR spectra of irradiated CH₃CN.

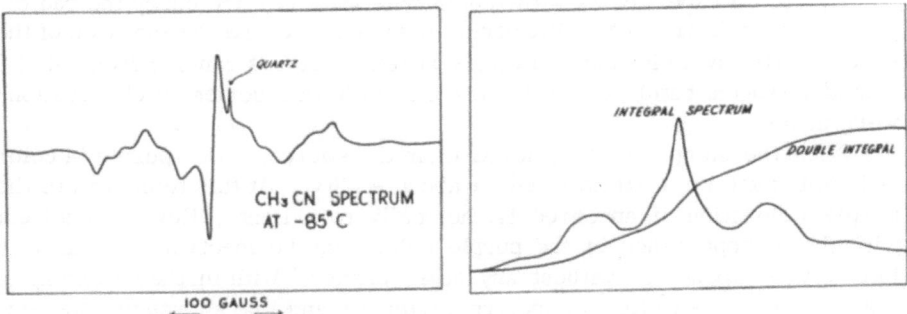

Fig. 5. Integral spectra.

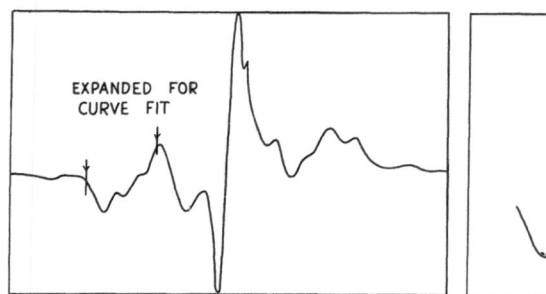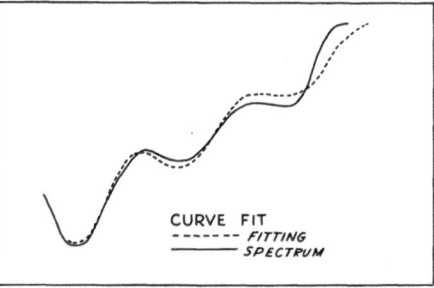

Fig. 6. Computer curve fit.

Sample Treatment

The samples were frozen prior to the radiolysis at liquid-nitrogen temperature in the following two ways: (1) insertion of the room-temperature quartz tubing directly into liquid nitrogen to give a "quick freeze" and (2) insertion into baths at various temperatures from the freezing point of CH_3CN, −42 C, to −95 C (toluene slush) and from these baths into the liquid nitrogen for radiolysis. All radiolyses were carried out in liquid nitrogen.

Periods of Radiolysis

The samples were analyzed using a combination of gas chromatography and mass spectrometry. The light fraction containing hydrogen and methane was introduced directly into the mass spectrometer for analysis and the condensable fraction was analyzed by introduction into a gas chromatograph equipped with a thermal conductivity detector and whose effluent was connected directly into the ionization region of a Bendix Time of Flight (TOF) mass spectrometer. The effluent peaks from the chromatograph were identified by retention times and by their mass spectra, obtained by rapid scanning of the Bendix TOF during their appearance.

RESULTS

The initial experiments with acetonitrile were carried out by the radiolysis of "quick frozen" acetonitrile samples and the observation of the spectra directly following radiolysis at temperatures ranging from −195 C (liquid-nitrogen bath) to −70 C, and the following series of observations were made.

When the sample was removed from the source, it was purple in color and remained so until warmed to about −126 C. At this temperature the purple coloration disappeared isothermally to a light yellow. Coincident with the disappearance of the purple coloration the spectrum increased in intensity by about 40% without any other change. With further warming to −80 C a sharp reduction in spectrum intensity and line sharpening was observed. This behavior was interpreted as indications of phase changes and subsequently the X-ray powder diffraction patterns of acetonitrile at −130,

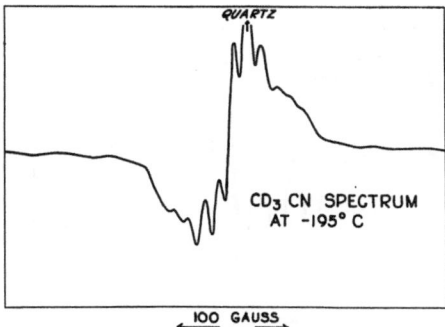

Fig. 7. Spectrum of irradiated CD$_3$CN.

−95, and −78 C were observed. These patterns did indeed indicate the presence of two distinct phases, at −130 C and below and above −78 C with a region of apparent mixed phase appearing between these two temperatures. A re-examination of the spectrum with heat treatment before radiolysis gave the two spectra shown in Fig. 4.

Spectra

The spectrum in Fig. 4a was obtained by irradiating "quick frozen" acetonitrile and scanning at liquid-nitrogen temperature while the spectrum in Fig. 4b is that of acetonitrile which was slowly cooled to liquid-nitrogen temperatures and stored at −195 C for about 50 hr before the radiolysis.

The sharp line appearing at the right of center of these spectra is due to the quartz color center and is present to varying degrees in all our spectra. It provides little interference and is in fact useful as a secondary field standard in our measurements.

Figure 5 shows a spectrum of the radical formed in the high-temperature phase obtained at −85 C, together with the integral and double-integral spectra. In the integral spectrum the central maximum is observed to be not coincident with the major triplet structure. Estimates of intensities of

TABLE I
Spectral Data

CH$_2$CN $\{$	Major triplet low-temperature phase.	23.2 G
	Major triplet high-temperature phase	22.9 G
N	Hyperfine triplet high temperature phase.	4.9 G
N	Hyperfine triplet half-width, high-temperature phase	4.2 G
CD$_2$CN	Pentet.	3.3 G
CN	Triplet.	9.0 G
CCl$_4$ $\{$	Solution of DPPH $\{$ Separation	9.1 G
	Half width	8.4 G
	Benzene solution of DPPH $\{$ Separation.	10.0 G
	Half width	10.9 G

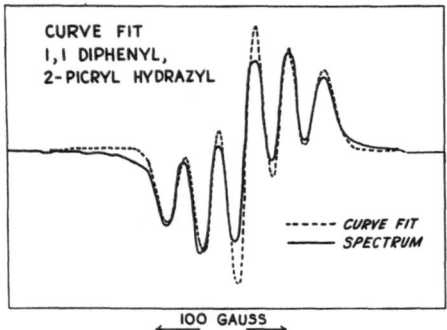

Fig. 8. DPPH spectrum in CCl₄.

the major triplet, excluding the sharp central maximum and portions of the wings of the central peak, are in the ratio 1:2:1. From these observations we conclude that this spectrum results from at least two radical species with the small peaks in the wings probably attributable to a third. From this information, together with analyses showing succinonitrile as one of the major radiolysis products, we deduced that the major triplet structure is consistent with the formation of $\cdot CH_2CN$ radicals making up the greater part of the intensity. If these radicals have a small electron density at the nitrogen atom, one would expect a slight N hyperfine splitting superimposed on the hyperfine splitting of two equivalent protons. On this assumption we made a curve fit shown in Fig. 6.

If it is assumed that the results of this partial curve fit the postulated radical structure, the pattern can be fitted with a $\cdot CH_2CN$ spectrum and superimposed on this a second pattern of three lines of equal intensity and splitting of about 9 G. This second radical probably shows only N hyperfine splitting and may be due to CN radicals or radicals obtained by the addition of an H atom to the central carbon in CH_3CN.

The spectrum of the deuterated acetonitrile in the low-temperature phase was run to allow us to decide on this second radical structure. If coincidence exists between the two spectra, it must arise from radical species not containing hydrogen or not exhibiting hydrogen hyperfine splitting.

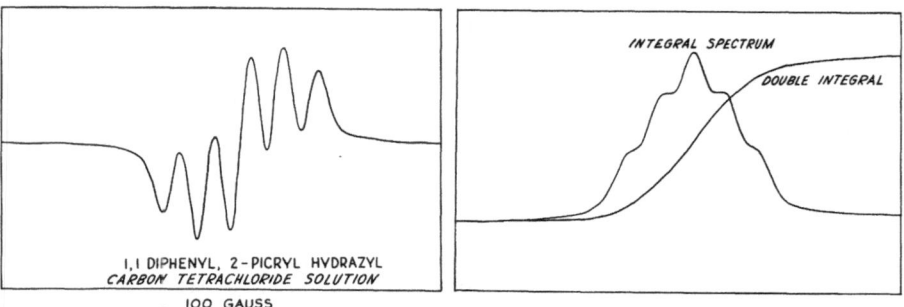

Fig. 9. Integration of DPPH spectrum.

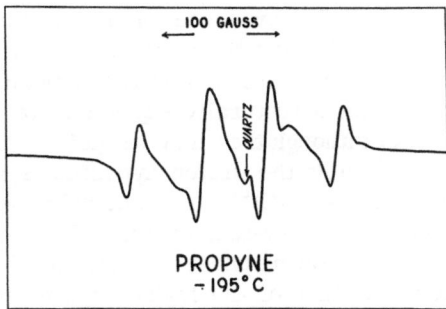

Fig. 10. ESR spectrum of irradiated CH₃C₂H.

The spectrum of irradiated CD_3CN is shown in Fig. 7. Three peaks, including the central maximum, are coincident with those in the CH_3CN spectrum and indicate the presence of a radical exhibiting only nitrogen hyperfine splitting. This radical has a hyperfine splitting of 9 G and is postulated to be the cyanogen radical.

Table I summarizes the spectral data. Included in this table are the results for 1,1-diphenyl 2-picryl hydrazyl–carbon tetrachloride solution used as a standard in the quantitative estimation of radical content. The spectrum of this solution is shown in Fig. 8 together with a curve fit for this spectrum using Gaussian line shapes. Figure 9 shows this spectrum together with the integral and double-integral spectra.

CONCLUSIONS

Radical Structures

Warming of the irradiated samples, whose spectra are shown in Figs. 4a and b, does not cause either spectra to go over to the other, thereby indicating that the structure of the radical is affected by the environment of the parent molecule during radiolysis.

Both spectra show the same coarse features and differ only in the

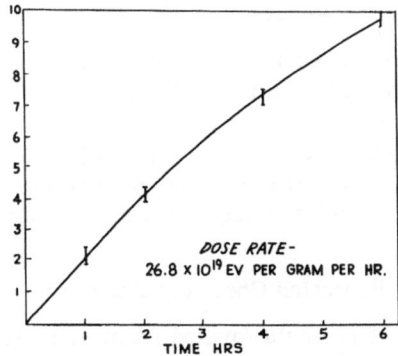

Fig. 11. Relative yield as a function of total radiation time.

breadth of the lines and the presence of resolvable nitrogen hyperfine splitting for radicals formed in the "quick frozen" samples.

We conclude from the observations that the radicals formed in the low-temperature phase are not structurally equivalent to those formed in the high-temperature phase, although they may be quite similar.

Since both spectra exhibit the proton hyperfine splitting that would be expected for the structure $\cdot CH_2C\equiv N$, (two equivalent protons), it is postulated that both the high- and low-temperature phase radicals contain this basic unit. In the low-temperature phase the distance of approach of nearest neighbors is shorter than that of the high-temperature phase and the observed broadening of the lines is expected. If this were the only origin of the spectral differences, the spectra would be convertible by proper heat treatment. That they are not may be due to the formation of a local unit with the structure of the low-temperature phase, which is stable even after the bulk has undergone a phase change to the higher-temperature form.

We postulate that this local stability arises from the coupling of an electronically excited parent molecule with the radical $\cdot CH_2CN$ to form the metastable entity

$$CH_3-C\equiv N\equiv CH_2-C\equiv N$$

in the low-temperature phase.

The high-temperature phase radicals are then the unreacted $\cdot CH_2CN$ species. The formation of the radical dimer only in the low-temperature modification may be a result of closer proximity or more favorable orientation, so that the electronic excitation results in bonding before it can be dissipated as heat or fluorescence.

The simple monomeric radical can be described by the following two formal representations:

$$(I) \quad \cdot CH_2-C\equiv N \qquad (II) \quad CH_2=C=N\cdot$$

The presence of nitrogen as well as proton hyperfine splitting indicates that both forms contribute to the radical structure. The similar molecule propyne (methyl acetylene) should give radicals exhibiting much the same type of resonance, the forms for this radical being:

$$(III) \quad \cdot CH_2-C\equiv C-H \qquad (V) \quad CH_2=C=\overset{\cdot}{C}H$$

The acetylenic proton should give rise to a doubling of each of the methylene proton lines rather than a trebling as in the case of $CH_2C\equiv N$. A spectrum obtained from irradiated propyne is shown in Fig. 10. The set of four resolvable lines indicates that the splitting for the acetylenic proton is nearly the same as that for the methylene protons and does support the postulated structures for both the acetonitrile and propyne radicals.

Radiation Chemical Behavior

The concentration of radicals formed from solid acetonitrile irradiated for 1, 2, 4, and 6 hr in a Co^{60} gamma source giving a dose rate of 26.8×10^{19} eV/g-hr were measured by the technique described earlier. These samples

TABLE II*

	At 4 hr	Extrapolated to zero date
Total radical	0.49	0.57
Prompt radical	0.35	0.41

Yield of Major Products		
	Solid	Liquid
H_2	0.91	0.75
CH_4	0.65	0.75
$(CH_2CN)_2$	0.32	0.52

*Yields are given in molecules per 100 eV.

were then allowed to warm to −126 C and the radical concentration again estimated. Ratios of initial radical signal to the signal after warming to −126 C were constant at about 1.4 : 1 for each sample. The results for total radical yield vs. radiation time are shown in Fig. 11.

Table II summarizes the observed radical yield at four hours and the extrapolated yield at zero dose. The 4-hr value is given since it represents the dose-time unit used for the samples which were chemically analyzed. Also included in Table II are the measured yields for three major products found in both the liquid- and solid-phase radiolyses.

Postirradiation radical production, as observed in our samples, may arise from a metastable excited electron state which undergoes relaxation to the ground state by dissociation or atom abstraction to give rise to additional radical signals. An excited triplet configuration of CH_3CN is postulated to account for this postirradiation radical production.

The following sequence of reactions is suggested to explain the observed radical production, where the left-hand terms represent the initial events:

$$CH_3-C{\equiv}N \longrightarrow H + CH_2C{\equiv}\overset{*}{N} \tag{1a}$$

$$CH_3-C{\equiv}N \longrightarrow CH_3\cdot + \cdot CN \tag{1b}$$

$$CH_3-C{\equiv}N \longrightarrow CH_3-\dot{C}{\equiv}\dot{N} \text{ (triplet state)} \tag{1c}$$

Abstraction reactions at −195 C

$$H + CH_3CN \longrightarrow \cdot CH_2CN + H_2 \tag{2a}$$

$$CH_3 + CH_3CN \longrightarrow \cdot CH_2CN + CH_4 \tag{2b}$$

$$CN + CH_3CN \longrightarrow \cdot CH_2CN + HCN \tag{2c}$$

Postirradiation reactions at −126 C

$$CH_3-\ddot{C}N \longrightarrow CH_3 + \cdot CN \tag{3a}$$

$$CH_3 + CH_3CN \longrightarrow CH_4 + \cdot CH_2CN \tag{3b}$$

$$CN + CH_3CN \longrightarrow HCN + \cdot CH_2CN \tag{3c}$$

In the low-temperature modification

$$CH_2CN^* + CH_3CN \longrightarrow CH_3CN-CH_2CN^* \tag{4}$$

In the high-temperature modification

$$CH_2CN^* + CH_3CN \longrightarrow CH_2CN + CH_3CN + E \tag{5}$$

It is evident from our results that reactions (2c) and (3c), although energetically possible, do not occur with a high rate at the temperatures of our observations even though the slightly larger CH_3 radical does react. This may well be due to "solvation" of the more polar CN radical causing preferential stabilization.

With the above restriction, the observed spectra are consistent with this reaction sequence.

Upon warming the liquid, the remaining radicals will decay in the following manner:

$$CN + CH_3CN \longrightarrow CH_2CN + HCN \tag{6a}$$

$$2CH_2CN \longrightarrow CN-(CH_2)_2-CN \text{ High temp. phase} \tag{6b}$$

$$2CH_3CNCH_2CN \longrightarrow (CH_2CN)_2 + 2CH_3CN \text{ Low temp.} \tag{6c}$$

$$CH_3CNCH_2CN + CH_3CN \longrightarrow \text{Possible polymerization route} \tag{6d}$$

These reactions imply the following relationships of the final products:

(A) $2(CH_2CN)_2 = CH_4 + H_2 + HCN$

(B) $CH_4 = HCN$

and the relation to radical yield

(C) $(CH_2CN)_2 = \text{Total Radical}$

From our estimate of the final radical ratio as observed in the spectrum (i.e., that CH_2CN made up $4/5$ of the intensity) we estimate the ratio of products from reaction (1a) to products from the sum of reactions (1b) and (1c) as $1.5:1$, so that we have the relation

(D) $H_2/CH_4 = 1.5$

and finally

(E) Total Radical $= 2(CH_2CN)_2 = 3.5 \, CH_4$

We find the radical yield, as measured by our techniques, to be consistent with the concentration of succinonitrile found in the analysis; but both hydrogen and methane are formed in large excess of the predicted amounts although they are in approximately the correct ratio. Hydrogen cyanide is not found in our reaction products but radiolysis of mixtures of hydrogen cyanide and acetonitrile indicate that hydrogen cyanide polymerization is catalyzed by some of the minor radiolysis products and may disappear before the analysis can be performed.

The inconsistency in the concentrations of hydrogen and methane probably implies that production of the greater part of these materials takes place by a process not involving radical intermediates nor resulting in the production of succinonitrile.

Our experiments do, however, indicate that succinonitrile is produced almost exclusively by a radical process, the dimerization of the radical CH_2CN.

REFERENCES

1. W. Mund and W. Koch, Bull. soc. chim. Belg. 34:125, 241 (1925); J. Phys. Chem. 30:389 (1926).
2. W. Mund, "L'action chimique des rayons alpha en phase gazeuse," (Hermann et Cie., Paris, 1935).
3. S. C. Lind, D. C. Bardwell, and J. H. Perry, J. Am. Chem. Soc. 48:1556 (1926).
4. L. M. Dorfman and F. J. Shipko, J. Am. Chem. Soc. 77:4723 (1955).
5. L. M. Dorfman and F. J. Shipko, J. Phys. Chem. 41:479, 651 (1937); 52:474 (1948).
6. C. Rosenblum, J. Phys. Chem. 41:651 (1937); 52:574 (1948).
7. W. Mund and C. Rosenblum, J. Phys. Chem. 41:469 (1937).
8. W. Mund and E. Bogaert, Bull. soc. chim. Belg. 34:410 (1925).
9. S. C. Lind and P. S. Rudolphe, J. Chem. Phys. 26:1768 (1957).
10. L. M. Dorfman and A. C. Wahl, Radiation Research 10:680 (1959).
11. W. Mund and J. C. Jungers, Bull. soc. chim. Belg. 40:158 (1931).
12. A. Russell Jones, J. Chem. Phys. 32:953 (1960).
13. S. C. Lind and D. C. Bardwell, J. Am. Chem. Soc. 48:1562 (1926).
14. D. C. Bardwell and D. K. Naylor, Radiation Research 11:432 (1959).
15. ASD-TDR-63-144, "Radiation Chemistry of Triple Bonds: Part I: Acetonitrile and Butyne-2, Liquid Phase."
16. S. Okamura et al., J. Chem. Soc. Japan 65:728 (1962).

Additional Papers on Nuclear Magnetic Resonance

● "An Empirical Approach to the Interpretation of NMR Spectra," R. Bible, G. D. Searle, and Co., Chicago, Illinois.*

● "Proton Magnetic Resonance Spectra of Substituted Pyridines and the Effect of Protonation," C. Bell, R. S. Egan, and L. Bauer, University of Illinois, Chicago, Illinois.

● "Hydrogen-Deuterium Exchanges in Organic Molecules by NMR Spectroscopy," R. W. Mattoon and R. Kriese, Abbott Laboratories, North Chicago, Illinois.

---

*To be published in expanded form as a monograph entitled "Interpretation of NMR Spectra—an Empirical Approach" by Plenum Press, New York, 1964.

Ultraviolet Spectroscopy

Absorption Spectra of Diatomic Molecules in Liquid and Crystalline Rare Gases

Gerald L. Pollack and Antoinette A. Pollack

National Bureau of Standards
Washington, D. C.

Examination of absorption spectra of molecules in liquid and crystalline matrices gives information on two fundamental kinds of problems: perturbations of molecules by surrounding fields, and effects of impurities on crystals and crystal growth. Rare-gas solid and liquid matrices are ideally suited to such problems since the forces between rare-gas atoms are especially weak, symmetric, short-range, and well known. Experimental studies of growth and observation of pure rare-gas crystals, transparent in a broad range from the infrared to the vacuum ultraviolet, and of low concentrations of diatomic molecules in solid solution with rare gases are described. As an example, the system NO in liquid and crystalline Kr was studied in the range from 2000-5000 A. Even at the smallest concentrations of NO no trace of discrete spectra of the NO γ system was found. However, a cutoff at short wavelength was observed in the liquid, increasing with increasing NO concentration. This absorption cutoff may be due to the transition from the ground state $X^2\pi$ to the theoretically predicted $^4\pi$ state. Both spectroscopic and solid state conclusions are drawn from these experiments and experiments on other systems of the same kind.

INTRODUCTION

The idea of studying absorption spectra of atoms and molecules in condensed matrices and the theory behind it are rather old. The techniques have been extensively applied to molecules such as methane and ammonia; one examines the infrared absorption spectra and determines, for example, whether the molecules' rotation is free or hindered.

More recently, theories have been worked up and experiments carried out on the problem of electronic absorption spectra of molecules in matrices. The principle here is to put molecules at appropriate sites in appropriate matrices and to examine the resultant electronic absorption spectra as a function of temperature, matrix, and concentration, and deduce whatever one can. The heart of the matter is to choose a suitably coupled pair of matrix and molecule, in general diatomic molecules are best since their electronic spectra are best understood. The molecule should have well-understood, conveniently observable, and highly probable electronic transitions in regions where the matrix is transparent. For its part the matrix should be as transparent as possible over a broad wavelength range, and one should be able to form mixtures of the matrix and molecule as solid and liquid solutions and vary the concentrations widely. Also the matrix and diatomic molecule must both be available very pure to be sure that the absorption spectra are those of the system of interest and not the result of some impurity. Most important of all, the matrix must be as symmetric as possible and the fields inside the matrix must be particularly simple, weak, and well known. Ideal matrices for these criteria are solid rare gases and to a slightly lesser extent liquid rare gases. We shall here discuss this type of experiment in general, the techniques and applications, and then describe some specific experiments and the results.

The kind of information obtainable about the diatomic molecule from these experiments stems from the opportunity of seeing the electronic levels perturbed principally by weak van der Waals forces. Theories lead to expectations of seeing otherwise forbidden transitions and in general transition probabilities change sometimes dramatically, weak lines become strong and strong lines weak, etc. This kind of phenomenon also happens in rotational and vibrational spectra of molecules in matrices. Not only can the molecule be thus studied weakly perturbed, but the perturbations can be controlled to some quantitative extent, e.g., lowering the temperature or raising the pressure increases the density of the matrix and hence the perturbing field.

Experiments of this kind are also useful for studying problems of the solid state of the rare gases or other matrices, especially for obtaining information about the local microscopic fields in the liquids and solids. The intermolecular fields in rare-gas matrices are rather well known and simple, so that if the diatomic molecules can be localized on symmetrical sites in the crystal, or if all the diatomic molecules appear on the same kind of site then from knowledge of the weak field acting on the molecule, at least in principle, the matrix elements may be constructed which give the transition probabilities between the perturbed energy levels. Conversely, although this is much more difficult, from the change in absorption spectra as the matrix transforms, e.g., from liquid to crystal, one may learn how the fields and hence the molecular environment in the matrix changes. We thus have a technique for investigating the mechanisms of crystal growth. If one uses matrices that are somewhat more complex, the change in spectrum as the matrix goes through a solid–solid phase transition may give information about this mechanism, too. Rare-gas solids, in particular, neon, argon, krypton, and xenon, have been intensively studied recently [1, 2] and we shall very briefly discuss them here.

RARE-GAS SOLIDS

Solids may be more or less arbitrarily divided into classes [3]: metals, ionic crystals, hydrogen-bonded crystals, covalent crystals, and molecular crystals. Molecular crystals are characterized as follows: on every lattice point identical molecules are situated and the principal forces between the molecules are van der Waals forces. For rare gases the intermolecular potentials have been much studied since it is just their details which account for the slight departures from ideality of the gases and these potentials have been extensively applied to studying the solid and liquid rare gases too. If some solid state property is to be calculated, for example, specific heat, surface tension, or even structure, one sets up a partition function or other relevant thermodynamic function for which on each molecule only the short-range forces of its 12 nearest neighbors (for the face-centered cubic lattice) need be considered. To a good approximation these are central, two-body forces.

Although many simplifying approximations may be made in calculating solid state properties for rare-gas solids, measuring them is rather more difficult than usual due partly to the low temperatures necessary. In particular many properties can only be measured on samples of known crystal size and shape, preferably even on single crystals. In this class are, e.g., molecular diffusion, thermal conductivity, ultrasonic attenuation, and bulk elastic properties. In addition rare-gas crystals are important because they may serve as prototype systems for studying crystal growth.

Crystalline rare-gas matrices suitable for such studies are most conveniently grown near their triple points; the temperatures are: for Ar, 83.8 K; for Kr, 116.0 K; for Ne, 24.6 K; and for Xe, 161.3 K. The specific experiments we shall discuss concern only the first two of these. The solids are best grown from the melt and their grain sizes may be conveniently studied from thermal etch patterns which form on the surfaces of polycrystalline samples as a result of judicious manipulation of thermal gradients [4]. Etch patterns of this kind on the surface of pure Ar grown from the melt are shown in Fig. 1. It is worth noting that no etch patterns have yet been reported on the solid solutions, but it may be expected that the relatively small concentrations of diatomic molecules present in the solutions do not critically affect the crystallization habits. The critical condition is that the spectroscopic absorption be due to diatomic molecules in the ordered crystal rather than to those in the disordered intercrystal environment.

NITRIC OXIDE IN KRYPTON [5]

The system NO in Kr is a potentially useful one for several reasons. At about 2265 A, NO shows in absorption the γ system, which is conveniently observed through fused silica windows and easily identified by its double double-headed bands. The transition is between the $X^2\pi$ and $A^2\Sigma^+$ states. Furthermore, the triple point of NO is at 109.6 K, close to the Kr triple point temperature, and the approximate length of an NO molecule is 2.5 A, compared to a diameter of 3.4 A for Kr molecules. The lattice parameter for Kr at its triple point is about 5.8 A so that although an NO molecule is too large to fit into an interstitial position in the crystal, it could fit on a lattice site. One thus expects that these two molecules are an appropriate pair. In order to minimize the effects of NO–NO interactions and particularly of dimerization, it is necessary to keep the concentrations low. Special dewar vessels must be used for obtaining these spectra so that the source does not pass through the coolant but only through the crystalline solid. The details of apparatus and technique have been described elsewhere.

Figure 2 shows the absorption spectra obtained with a xenon-arc source. The spectrum labeled (f) is the copper-arc comparison spectrum from which the wavelength scale is established. The five spectra above it show the absorption of liquid solutions in Kr at different NO concentrations in the vapor from which the liquid is condensed: (e) 0% NO, (d) 0.0177% NO, (c) 0.0609% NO, (b) 0.197% NO, and (a) 0.646% NO. The range is thus from no NO up to one NO molecule per 150 Kr molecules. There are several things to be determined from these spectra.

The most prominent feature of the spectra (a) through (e) is that a cutoff due to the NO is present and this cutoff moves to longer wavelengths and

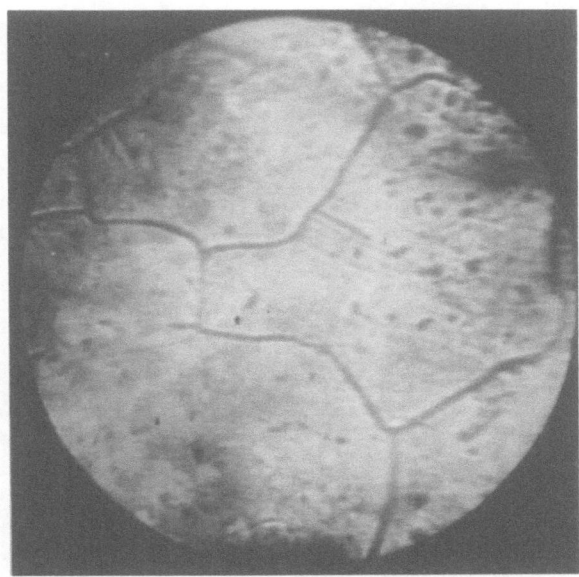

Fig. 1. Thermal etch patterns on the surface of polycrystalline argon at 77 K. Grain boundaries, enclosing single crystals
in the polycrystalline block, appear as dark heavy lines. The
finer striations may be coherent twin boundaries or slip lines.
The diameter of the field shown is about 2 mm.

becomes sharper as the NO concentration is increased. The cutoff stops
transmission in the 2265 A region for most of the spectra, but for case (d)
there is sufficient density on the plate to determine that the NO γ band heads
are missing. This is surprising since ideally no cutoff would be observed
but instead the γ band heads would appear shifted in frequency or with
changed intensity or modified in some other way which would make possible
quantitative determination of just where the NO molecules are and what
fields act on them. The disappearance of the NO γ system and the behavior
of the cutoff may be due [5, 6] to transitions between the $X^2\pi$ and a shallow
$^4\pi$ state predicted [7] by Mulliken in 1932. Although such spin-forbidden
transitions are forbidden in the gas, they might be allowed in the weak
fields obtaining in rare-gas matrices since π states may be relatively
easily perturbed. Spin-forbidden transitions of this same general kind can
be observed in CO and N_2 even in the gas.

The spectrum (g) in Fig. 2 shows the absorption of the solid obtained
from the most concentrated NO solution. All the other solids showed
neither a cutoff nor any γ system, just as if they were pure Kr, and this is
the only solid solution that showed any spectroscopic activity. Note that the
cutoff in the solid looks much like the one in the corresponding liquid (a)
but it occurs at shorter λ; this means that the NO concentration is less
in the solid than in the liquid but that the local environment around the NO is
probably the same. Furthermore, the NO that was in the solid disappeared
in about an hour so that this solid also transmitted as well as pure Kr; the
NO diffuses from the solid into the liquid and does it rapidly.

If the cutoff wavelength in the melt is plotted as a function of NO concentration in the vapor the resultant curve is smooth, as may be seen on

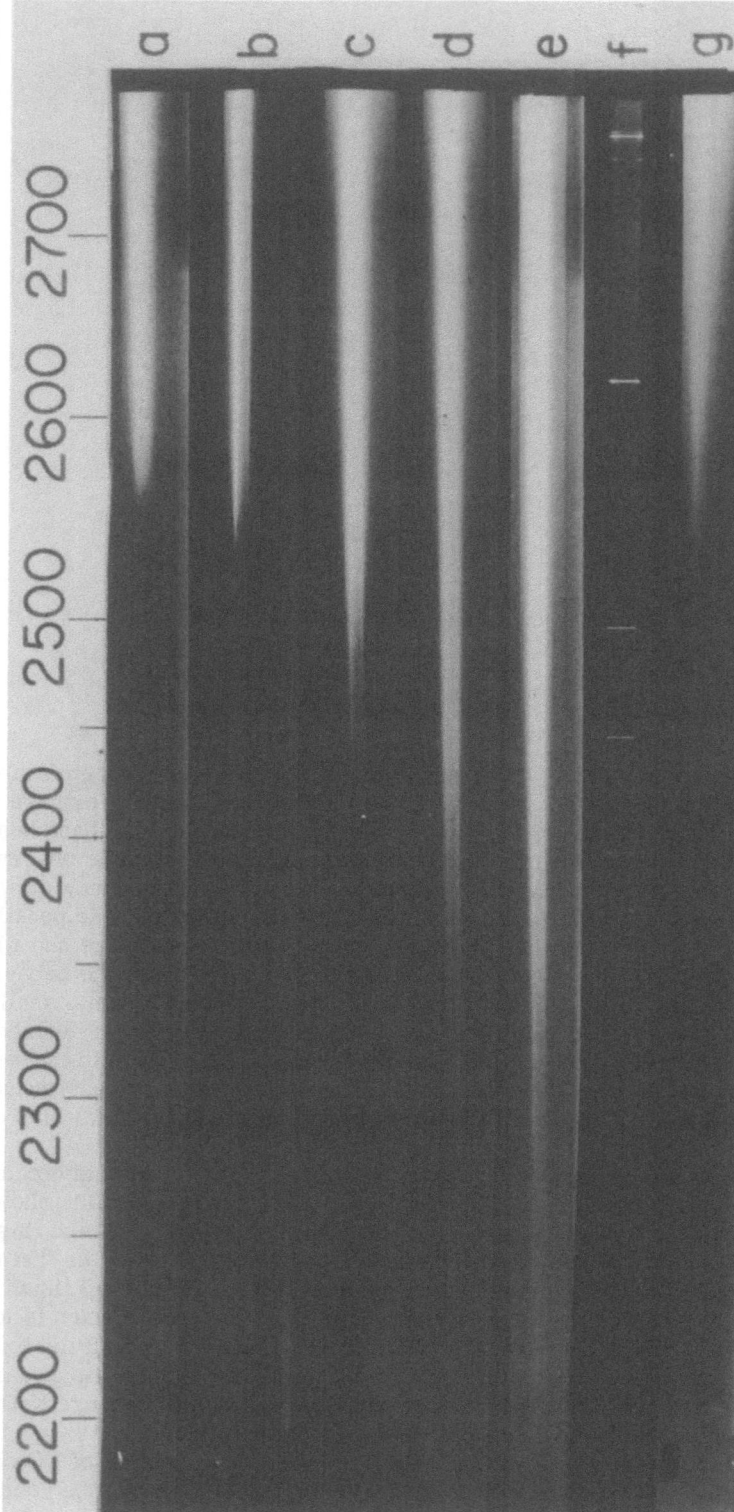

Fig. 2. NO absorption in condensed krypton. Absorption of continuous Xe-arc source in liquid condensed from vapor of (a) 0.646% NO concentration in Kr, (b) 0.197% NO in Kr, (c) 0.0609% NO in Kr, (d) 0.0177% NO in Kr, (e) pure Kr. (f) Cu-arc comparison spectrum. (g) Absorption in the crystalline mixture grown from the liquid of (a). (Courtesy of J. Chem. Phys.)

Fig. 3. This indicates that in principle the NO concentration in the melt and solid could be determined from the cutoff wavelength. The technique could thus be used to determine the phase diagram of the binary system Kr + NO for low NO concentrations. One difficulty is that the NO concentration has been determined only in the vapor and not in the liquid or solid. However, since the total amount of NO in the system is known and since the system is closed, the ordinate axis on Fig. 3 may be recalibrated to read % NO in liquid by measurements of NO concentration remaining in the vapor from which the liquid or solid is grown.

The point on Fig. 3 which indicates the cutoff in the solid is worth special consideration. Since it lies to the left of the curve for liquid cutoff it indicates that the NO distribution favors higher concentrations in the liquid. This is just the principle involved in zone refining, much used to purify crystals. The preference of impurities to concentrate in the liquid has also been recently observed in studying crystallization of methane [8].

OTHER SYSTEMS

In general homonuclear diatomic molecules are more easily treated theoretically than heteronuclear ones and absorption spectra for O_2 (triple

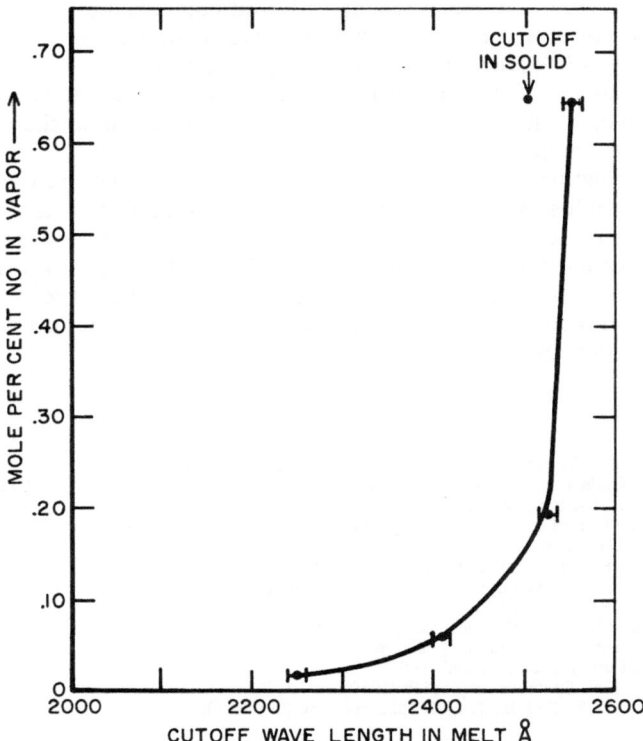

Fig. 3. Wavelength of transmission cutoff in condensed phases as a function of NO concentration in vapor. Note that all solids transmitted without a cutoff except the one from the 0.646% NO vapor mixture. (Courtesy of J. Chem. Phys.)

point 55 K) have been much observed in pure condensed phases and in matrices.

In studying the absorption spectrum of impure liquid CH_4 at its boiling point (111.7 K) weak and broad absorption bands have recently been observed [9] at 2483, 2535, 2598, 2609, and 2675 A (respectively, 40,270; 39,450; 38,490; 38,330; and 37,380 cm^{-1}). These are probably due to liquid-oxygen impurities and may be the result of the forbidden transition $A^3\Sigma_u^+ - X^3\Sigma_g^-$ reported and thoroughly investigated by Herzberg [10] in oxygen gas. Several bands at wavelengths very close to these have also been reported in the spectrum of solid α-oxygen by Hörl [11]. No attempt has as yet been made to follow the oxygen absorption in methane with changing phase or temperature. This would be of much interest especially in the region of the solid–solid phase transition at 20.4 K for pure methane. Solid methane is unfortunately not quite so ideal a molecular solid as the rare-gas solids but it is probably the simplest molecular solid with such a phase transition.

A thorough study of the vacuum ultraviolet absorption spectra of oxygen in liquid and crystalline Ar and N_2 has recently been reported [12]. This work extended over a broad range of oxygen concentrations (from 0.01% to 10%), temperature (from 60-90 K), and wavelength (from 1600-3100 A), so that the behavior of the Schumann–Runge bands as well as the Herzberg bands under different perturbations were observed. The Schumann–Runge bands in the condensed matrices are shifted compared to their position in the gas and the shifts are slightly temperature-dependent. There is remarkably little difference in the bands upon change from liquid to solid of the matrix, and spectra in the nitrogen and argon matrices are also remarkably alike. The Herzberg bands were observed to be the same under variation of the temperature and the matrix.

We may conclude then that the observation of electronic transitions of diatomic molecules in liquid and solid rare-gas matrices is a potentially powerful and interesting way of studying both spectroscopy and the solid state. So far each of the systems studied has shown its own informative surprises and much knowledge can be expected from the study of such molecules as H_2 and I_2 or Cl_2, HI, CO, etc. in liquid and crystalline matrices of He, Ne, Ar, Kr, Xe, CH_4, and others.

REFERENCES

1. E. R. Dobbs and G. O. Jones, Repts. Progr. Phys. 20:516 (1957).
2. A. C. Hollis Hallett in Argon, Helium, and the Rare Gases, Volume 1, G. A. Cook, (ed.) (Interscience Publishers, Inc., New York, 1961).
3. C. Kittel, Introduction to Solid State Physics (John Wiley and Sons, Inc., New York, 1953).
4. D. Stansfield, Phil. Mag. 1:934 (1956).
5. G. L. Pollack and H. P. Broida, J. Chem. Phys. 38:2012 (1963).
6. H. P. Broida and M. Peyron, J. Chem. Phys. 32:1068 (1960).
7. R. S. Mulliken, Revs. Mod. Phys. 4:1 (1932).
8. G. L. Pollack and R. S. Williams (to be published).
9. G. L. Pollack and H. P. Broida (unpublished).
10. G. Herzberg, Can. J. Phys. 30:185 (1952).
11. E. M. Hörl, J. Mol. Spectroscopy 3:548 (1959).
12. A. M. Bass and H. P. Broida, J. Mol. Spectroscopy (in press).

Additional Papers on Visible–Ultraviolet Spectroscopy

- "What We Don't Know About Electronic Spectra," D. S. McClure, University of Chicago, Chicago, Illinois.

- "The Optical Spectrum of $HI_3 \cdot 2C_6H_5\text{-}CONH_2$: A Model of the Starch-Iodine Complex," M. B. Robin, Bell Telephone Laboratories, Murray Hill, New Jersey.

- "Absorption in Chromium (VI) Oxide," J. Anysas and A. Companion, Illinois Institute of Technology, Chicago, Illinois.

- "Attenuated Total Reflection in the Visible Spectrum," W. N. Hansen, North American Aviation Science Center, Canoga Park, California.

- "Absorption and Emission of Broad Bands in Solids," J. J. Markham, Illinois Institute of Technology, Chicago, Illinois.

- "An Indirect Spectrophotometric Determination of Metals in Their Metal Oxides by the Reduction of the Nitrite Ion," H. F. Combs and E. L. Grove, ITT Research Foundation, Chicago, Illinois.

Gas Chromatography

Applications of Gas Chromatography to Phosphorus-Containing Compounds*

A. Davis, A. Roaldi, J. G. Michalovic, and H. M. Joseph

Research Department
Hooker Chemical Corporation
Niagara Falls, New York

Very little has been published on the use of gas chromatography for phosphorus-containing compounds. We have found it to be an extremely valuable aid in both the research and production of organic and inorganic phosphorus compounds. Methods have been developed for assaying finished products and for control of many of the intermediate streams for a number of processes producing phosphorus compounds. Gas chromatographic analysis of inorganic phosphorus compounds—PCl_3, $POCl_3$, and $(PNCl_2)_3$ to $(PNCl_2)_4$—have appeared in the literature. Gas chromatographic methods developed in our laboratory for these mixtures are also given in this paper. Several useful suggestions on gas chromatographic procedure when analyzing for phosphorus compounds are direct injection on the column, the use of glass columns, careful drying of the carrier gas, special treatment of the support to reduce tailing and rearrangement on the column, and the use of temperature programming.

INTRODUCTION

The trend in analytical chemistry has been more and more towards gas chromatographic analysis for both organic [5] and inorganic [3, 15] compounds. While very little has been published on gas chromatography for the analysis of phosphorus compounds, we have found this to be a very fruitful field for gas chromatography. A brief review of the literature of the technique applied to phosphorus-containing compounds is included.

PYROLYSES AND DEGRADATION

Many phosphorus compounds have very low vapor pressure. Several authors have avoided this problem by use of pyrolysis followed by gas chromatographic analyses of the resulting volatiles. Legate [12] used infrared and mass spectroscopy, after gas chromatographic separation to identify the organic pyrolysis products for a complex zinc, lead, or potassium phosphate or thiophosphate. Szymanski [19] used an induction furnace for pyrolysis. The organo-phosphorus compound was mixed with iron and a high-powered induction coil was used to rapidly heat the sample to 900C. Figure 1 shows several compounds used in this study, together with their retention times. Dulon [8] also used gas chromatography to identify products of pyrolysis of organo-phosphorus compounds. This method certainly shows good promise for qualitative analysis, and may also be of value in quantitative analysis for high-boiling phosphorus compounds. Williams [21] irradiated

*Reprinted by permission from J. Gas Chromatography 1:23 (August, 1963).

Compound

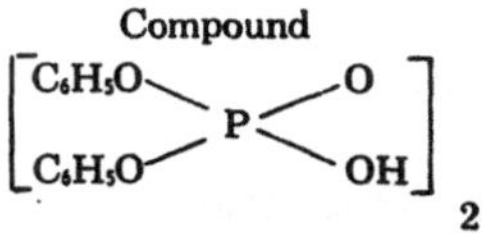

$$\left[\begin{array}{c} C_6H_5O \\ C_6H_5O \end{array} P \begin{array}{c} O \\ OH \end{array} \right]_2 \qquad 5.9 \text{ and } 9.4$$

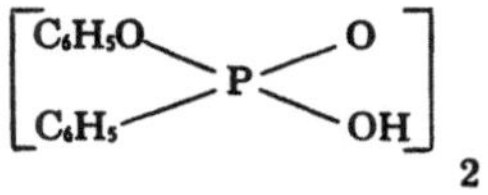

$$\left[\begin{array}{c} C_6H_5O \\ C_6H_5 \end{array} P \begin{array}{c} O \\ OH \end{array} \right]_2 \qquad 3.0$$

$$\left[\begin{array}{c} C_8H_{17}O \\ C_8H_{17} \end{array} P \begin{array}{c} O \\ OH \end{array} \right]_2 \qquad 2.1, \ 4.0, \ 4.1 \text{ and } 6.0$$

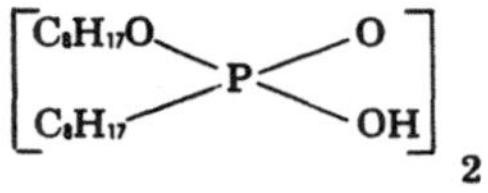

Fig. 1. Retention time in minutes of pyrolysis product.

tributyl phosphite saturated with water with a Van de Graaff generator and studied the resulting products using gas chromatography.

CHEMICAL REACTIONS

Sauers [16] used gas chromatography to identify the products of re-arrangement during phosphoryl chloride-pyridine dehydration. No phosphorus compound actually passed through the gas chromatograph. Walling [20] used gas chromatography to identify and estimate the product of the reaction of thiol or alkoxy radicals with trialkyl phosphites. No details as to gas chromatographic conditions are given in his paper. Denney [6] investigated the mechanism of the reaction of tributyl phosphines with episulfides. The butene generated was analyzed by gas chromatography. DeRose [7] used gas chromatography to investigate ester interchange in dialkyl hydrogen phosphites. The column used was either 5% Apiezon or 5% squalane on

TABLE I
Dialkyl Hydrogen Phosphites

$$(RO)_2P \begin{array}{c} O \\ H \end{array}$$

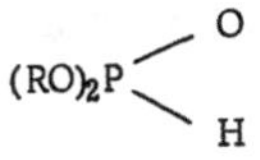

R	Boiling point C/mm Hg	5% Apiezon	5% Squalane
Me	59-60/14	0.67	0.53
Et	77-79/16	0.10	1.13
Pr^n	92/11	2.19	3.07
Pr^i	78-80/14	1.42	1.32
Bu^n	126-127/16		8.04
Bu^i	118/14	4.35	6.00
Methyl- propyl			1.25
Napthal- ene		1.00	1.00

TABLE II
Separation of PCl_3, $POCl_3$, and $PSCl_3$

Instrument	Burrell K-2
Column	Silicone grease (Burrell Cut #341-136)
	250-cm glass
Carrier gas	Helium
Initial temperature	60 C
Program rate	3 V/min
Final temperature	250 C
Flow rate	87 ml at 4 psi
Sample size............	5 µl
Detector	Thermoconductivity

80-100 mesh Celite at 100 C. Table I shows the retention time for a series of secondary phosphites.

INORGANIC PHOSPHORUS COMPOUNDS

Several authors have published methods for the separation of PCl_3 mixtures. Shipotofsky [17] used Kel F-90 on Fluoropack 80 to separate $PSCl_3$–PCl_3 and $POCl_3$–PCl_3 mixtures. Separation of $PSCl_3$ from $POCl_3$ was not achieved. Staforel [18] also separated PCl_3 from $POCl_3$, but he used silicone elastomer E-301 (SE-30) on Celite 545 to do the separation. Abe [1] used DC-703 for best results in separating PCl_3, HCl, $SiHCl_3$, $SiCl_4$, and BCl_3. He also investigated several other substrates, some of which caused hydrolysis and degradation on the column.

Table II gives the conditions used in our laboratory to separate PCl_3, $POCl_3$, and $PSCl_3$. Figure 2 shows the chromatogram of this separation. Several authors have published methods for analyzing cyclic phosphonitrilic halides. Mou [14] and Chapman [4] both have published a method for phosphonitrilic fluorides. Chapman used silicone elastomer E-301 (SE-30) on Celite 545. Gimblett [10] used the same substrate and support to separate trimer and tetramer of cyclic phosphonitrilic chloride. Table III gives the conditions for analysis of the cyclic trimer to heptamer used in our laboratory. It was found that the addition of a small amount of carbowax 20 M

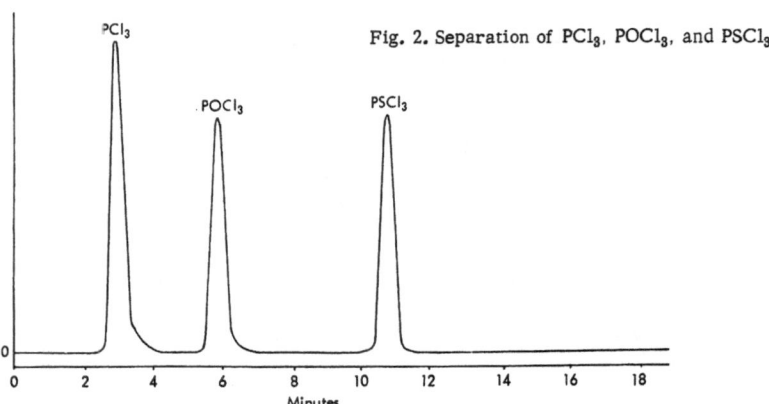

Fig. 2. Separation of PCl_3, $POCl_3$, and $PSCl_3$.

TABLE III
Separation of Phosphonitrilic Chlorides

Instrument . . . . Burrell K-2
Column . . . . 3% Dow 11 plus 0.15% Carbowax 20 M on
Columpak 80/100 mesh 100-cm glass U
Carrier gas . . . . Helium
Initial temperature . . . . 30 C
Program rate . . . . 2 V/min
Final temperature . . . . 230 C
Flow rate . . . . 50 ml/min
Sample size . . . . 20 μl
Detector . . . . Thermoconductivity
Solvent . . . . CH_2Cl_2

markedly reduced tailing.   Figure 3 is a chromatogram of a typical crude mixture of cyclic phosphonitrilic chlorides.

ORGANIC ANALYSIS

Lewis [13] in his article on analysis of ester-type plasticizers gives conditions for the gas chromatographic separation of triethyl phosphate, triphenyl phosphate and tris(m-tolyl)phosphate on a 60-cm column of Apiezon oil K on Celite 545 at a column temperature at 283C.

Gudzinowicz [11] on a 5% SE-30 silicone gum rubber achieved separation for the compound shown in Table IV.   The sample was programmed from 240 to 350C.

Buckler [3] used gas chromatography to separate trialkyl phosphines and their oxidation products. Conditions for separating the tributyl phosphine and the cyclohexyl phosphines are given in [9].   Feinland used silicone grease (Dow Corning high vacuum) and Reoplex 400 on 60 to 80 mesh Chromosorb W.   Shipotofsky [17] separated dimethyl phosphite from diethyl phosphite on di-n-butyl phthalate on Fluoropak 80 at 107 C.

The conditions for separation of crude trimethyl phosphite used in our laboratory are listed in Table V.   This same column, with slightly different temperature programs, also serves for analysis of triallyl phosphite, and

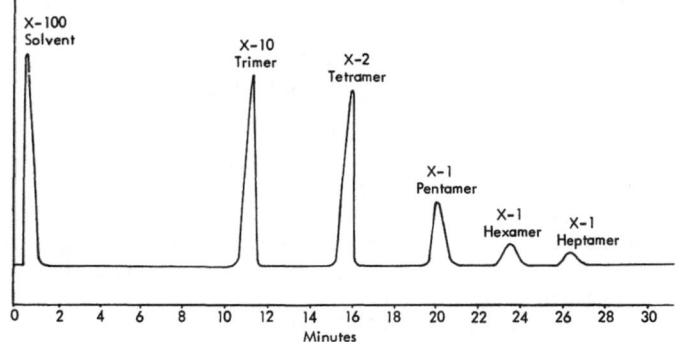

Fig. 3. Gas chromatogram of cyclic phosphonitrilic chlorides—$(PNCl_2)_x$.

TABLE IV
Aryl Phosphines and Phosphine Oxides

Compound	Structure	Molecular weight	Melting point, C
Triphenylphosphine	$(\bigcirc)_3 P$	262	79
p-Hydroxyphenyldiphenylphosphine	$HO-\bigcirc-P-(\bigcirc)_2$	278	105-106
p-Methoxyphenyldiphenylphosphine	$CH_3O-\bigcirc-P-(\bigcirc)_2$	292	78-79
Di(p-methoxyphenyl)phenylphosphine	$(CH_3O-\bigcirc-)_2 P-\bigcirc$	322	87-89
Tri(p-methoxyphenyl)-phosphine	$(CH_3O-\bigcirc)_3 P$	352	129-131
Triphenylphosphine oxide	$(\bigcirc-)_3 P = O$	278	153
p-Methoxyphenyldiphenylphosphine oxide	$CH_3O-\bigcirc-P-(\bigcirc)_2$	308	116-118
Tri(p-methoxyphenyl)phosphine oxide	$(CH_3O-\bigcirc-)_3 P = O$	368	144-145

for the separation of dimethyl fumarate, trimethyl phosphonopropionate, and tetramethyl phosphonosuccinate. For triethyl phosphate, in order to achieve separation of a particular impurity it was found necessary to use 10% diethylene glycol adipate in place of 20% SE-30. In all cases, separations were comparable to those shown in the previous figures, and there were no difficulties with tailing.

DISCUSSION

Several points should be emphasized in the gas chromatographic analysis of phosphorus compounds. All water should be kept out of the system. The carrier gas should be dried very carefully. We normally pass the carrier gas through a trap packed with 5A molecular sieve. This trap is periodically dried at 300–400C. Glass columns and direct injection on the column have been found to reduce decomposition and tailing. We were unable to reproduce many of the over fifteen different gas chromatographic analyses of phosphorus compounds when metal columns or a metal inlet were substituted for glass.

Presently, we are using only Chromosorb W acid base, washed and silanized for the column support. We have not yet found any case where silanizing hurt the analysis, and there are several instances where this has reduced tailing. Program temperature is required to handle the wide-boil-

TABLE V
Analysis for Trimethyl Phosphite

Instrument . . . . Burrell K-2
Column . . . . 20% SE-30 on Chromosorb W 250-cm glass U
Carrier gas . . . . Helium
Initial temperature . . . . 90 C
Program rate . . . . Isothermal for 30 min then raise the temperature as rapidly as possible to 250 C
Final temperature . . . . 250 C
Flow . . . . 40 ml/min
Sample size . . . . 5 μl
Detector . . . . Thermoconductivity

ing-range mixture without serious reduction of resolution. But of even greater importance is that programmed temperature allows for the removal of volatile compounds, often temperature-sensitive, without subjecting them to the higher temperatures needed for the remaining components in the sample. This is a great aid in the handling of mixed phosphites.

Gas chromatography has served as a very valuable analytical tool in the analysis of phosphorus compounds in our laboratory, and we strongly believe that in the future many more will use gas chromatography as a major analytical control in the analysis of phosphorus compounds.

ACKNOWLEDGMENT

The authors would like to gratefully thank the members of the organic and inorganic phosphorus research groups of the Hooker Chemical Corp. for supplying the many compounds used for these analyses. We would also like to acknowledge the work of D. W. Eastman, J. A. Kaplan, C. R. Kregg, E. J. Lobbett, and N. J. Roller, who all helped in the development of gas chromatographic methods for phosphorus compounds.

REFERENCES

1. Y. Abe, Banseki Kaguka 9:795 (1960).
2. E. W. Abel, G. Nickless, and F. H. Pollard, Proc. Chem. Soc. (London) 288 (1960).
3. S. A. Buckler, J. Am. Chem. Soc. 84:3093 (1962).
4. A. C. Chapman, N. L. Paddock, H. D. Paine, H. T. Searle, and D. R. Smith, J. Chem. Soc. (1960) p. 3608.
5. S. Dal Nogare and R. S. Juvet, Gas—Liquid Chromatography (John Wiley and Son, New York, 1962), p. 383.
6. D. B. Denney and M. J. Boskin, J. Am. Chem. Soc. 82:4736 (1960).
7. A. DeRose, W. Gerrad, and E. F. Mooney, Chem. and Ind. 36:1449 (1961).
8. R. Dulon, G. Quesnel, and M. deBottom, Bull. Soc. Chim. (France) 9:1340 (1959).
9. R. Feinland, J. Sass, and S. A. Buckler, Anal. Chem. 35:920 (1963).
10. F. G. R. Gimblett, Chem. and Ind. 12:365 (1958).
11. B. J. Gudzinowicz and R. H. Campbell, Anal. Chem. 33:1510 (1961).
12. C. E. Legate and H. D. Burnham, Anal. Chem. 32:1042 (1960).
13. J. S. Lewis and H. W. Patton, in Gas Chromatography, V. J. Coates et al. (eds.) (Academic Press, New York, 1958), p. 149.
14. T. J. Mou, R. D. Dresdner, and J. A. Young, J. Am. Chem. Soc. 81:1020 (1959).
15. C. S. G. Phillips and P. L. Timms, Anal. Chem. 35:505 (1963).
16. R. R. Sauers and J. M. Landesberg, J. Org. Chem. 26:964 (1961).
17. S. H. Shipotofsky and H. C. Moser, Anal. Chem. 33:521 (1961).
18. F. G. Staforel, J. Chromatog. 4:419 (1960).
19. H. A. Szymanski, Lectures on Gas Chromatography, Szymanski (ed.) (Plenum Press, New York, 1963).
20. C. Walling and R. Rabinowitz, J. Am. Chem. Soc. 81:1243 (1959).
21. T. F. Williams, R. W. Wilkinson, and T. Rigg, Nature 179:540 (1957).

Some Experiments in Gas Chromatography

H. A. Szymanski, C. McMenamy, J. Kuczkowski, K. Broda, and J. May

Department of Chemistry
Canisius College
Buffalo, New York

The authors have been experimenting with very fast and slow pyrolysis, and report preliminary results on this work. Some work has also been in progress using fast induction heating to change column temperatures, and is also presented in this paper.

INTRODUCTION

This paper is divided into three general topics. The first concerns carrier-gas purity for trace analysis; the second is a report on our work using molecular sieves as solid supports; and the third topic concerns very fast temperature changes for sample vaporization and/or column heatings.

CARRIER-GAS PURITY FOR TRACE ANALYSIS

Recently in the Russian literature a new approach to gas chromatography was suggested [1], which roughly translated can be termed "blank chromatography." A simple example can be used to illustrate this new field. The example chosen is based on some fairly old unpublished work in the United States and serves to illustrate that "blank chromatography" has been utilized by other workers, who neglected to publish their technique [4]. This unpublished work was not done at our laboratory, but we feel it is of general interest to call it to the reader's attention.

Consider a carrier gas such as hydrogen, which has traces of oxygen and nitrogen. If this carrier gas passes through both the reference and sample side of the detector, no signal will be observed to indicate the presence of the oxygen and nitrogen. Now if an ultrapure sample of hydrogen gas having no oxygen or nitrogen is introduced into the sample stream, negative peaks will appear for the oxygen and nitrogen present in the carrier gas. Quantitative analysis can be done using such negative peaks. Blank chromatography is based on detecting any sample using negative peaks in a manner similar to that described above. The example illustrates another point. Unless the carrier gas is free of the component which is to be detected, accurate quantitative analysis, especially in the trace region, is very difficult. For trace analysis, we would like to make some suggestions concerning the carrier gas and its accompanying pressure regulators and pressure lines.

It is possible to obtain very pure specimens of carrier gases. The analysis of the gas is also available from the manufacturer. For example, nitrogen can be obtained as 99.995% N_2, the impurities being oxygen and the rare gases. A trace of water is also present. Impurities, including water, can be picked up from rubber hoses, needle valves, rubber diaphragms in regulators and inboard leakage in regulators, but most of these sources of impurities can be eliminated. For example, stainless steel tubing welded together to eliminate fittings can be used to connect to the pressure regulators. The Linde Co. recently announced two new regulators which can be utilized to eliminate diaphragm contamination and inboard leakage. The R-2130 rare-gas regulator has a Teflon diaphragm to reduce rubber contamination. This regulator has an inboard leakage of $6.5 \cdot 10^{-5}$ cc/sec. The second regulator R-1500 has no diaphragm and an inboard leakage of $3 \cdot 10^{-9}$ cc/sec. This latter regulator, while more expensive, appears quite promising for trace analysis work.

Discussions with a number of workers as to the value of inserting a drying tube between the carrier gas tank and the chromatograph indicated the consensus of opinion to be that the purifiers do more harm than good, especially if the analysis of the carrier gas is available, since it can be used to identify and control the trace impurities present.

MOLECULAR SIEVES AS SOLID SUPPORTS

Our interest in molecular sieves as solid supports was generated by the concept that column bleeding could possibly be reduced if the liquid partitioning agent could be strongly adsorbed on the solid support. However, it is well known that a highly active solid support produces tailing as well as irreversible adsorption of the sample. Even a fairly thick layer of liquid partitioning agent on the solid support may not cover highly active sites. Deactivation of solid supports has been extensively studied. Materials such as hexamethyldisilazane and dimethyldichlorosilane have been used to reduce the tailing effect found for oxygenated compounds when fire brick is used as the solid support. While it appeared that molecular sieves would be particularly poor for solid supports since they have very active sites, several aspects appeared worthy of investigation. First, synthetic sieves are fairly homogeneous and the capillary openings have been extensively studied. The adsorption characteristics are therefore well known and could presumably be controlled. Also the high selectivity of sieves for certain substances appeared especially promising. For example, certain branched hydrocarbons easily pass through a sieve column but the straight-chain analogs of these branched hydrocarbons are irreversibly absorbed. Further, we felt that if a series of molecules could be absorbed on sieves with their polar ends "sticking out" of the sieve capillaries, unusual separations could be obtained. Experimentation was therefore begun to verify our general ideas. Some general conclusions can be drawn from the following known facts concerning sieves. With dry 5A sieves, the order of elution of gases is oxygen, nitrogen, methane, and carbon monoxide. For wet sieves the order is oxygen, nitrogen, carbon monoxide, and methane. It is obvious the water is acting as a partitioning agent.

While our work was underway, workers in Europe reported the separation of isoprene in the presence of other C_5 hydrocarbons [2]. Our early work was similar to this report in that we utilized polar materials such as acetone to cover the active sites of molecular sieves. However, we found that a number of factors must be considered in utilizing sieves as solid supports. In

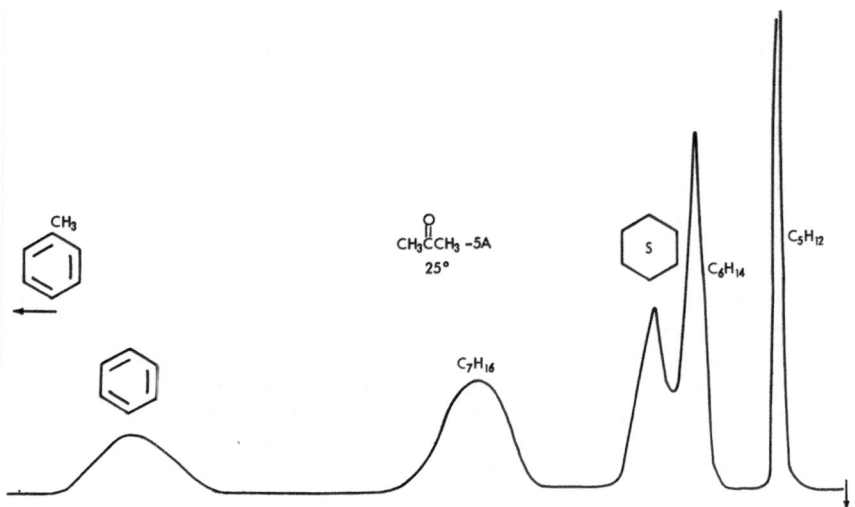

Fig. 1. A chromatogram obtained on an acetone—5A sieve column at 25C. Components separate as marked.

our work we examined the properties of three kinds of sieves. These are Linde 4A, 5A, and 13X. The 13X was very highly adsorbing even when coated with a high percent of liquid substrate. It did have some unusual properties which we shall describe later. In general it can be stated that none of the liquid partitioning agents we utilized completely covered the surface activity of 13X, 5A, or 4A sieves play a part in the observed separations in all cases, the order of decreasing surface activity being 13X, 5A, and 4A. Liquid partitioning agents utilized include organo-phosphorus acids, phosphates, formic acid, water, acetone, duodecane, and several conventional liquid agents. Some typical results are as follows: For a hydrocarbon sample consisting of normal pentane, normal hexane, normal heptane, benzene, and toluene, all the

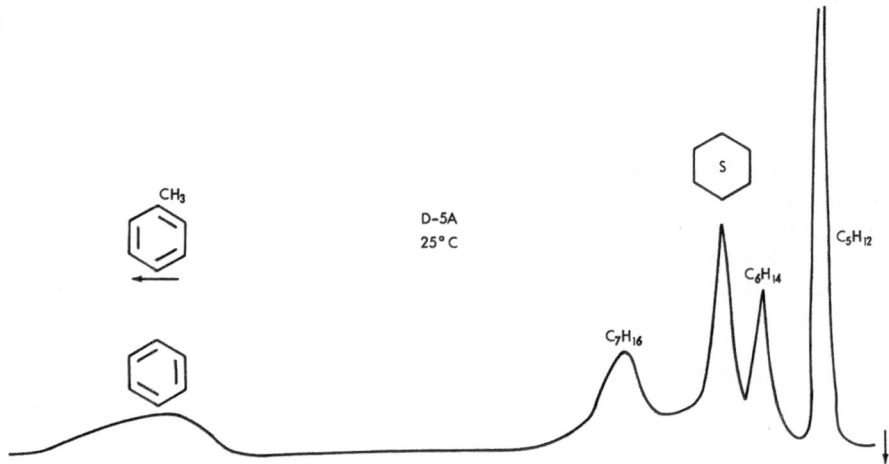

Fig. 2. A chromatogram obtained in a duodecane—5A sieve column at 25C. Components separate as marked.

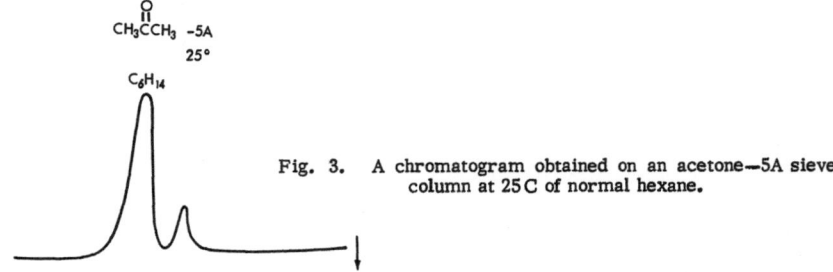

Fig. 3. A chromatogram obtained on an acetone—5A sieve column at 25 C of normal hexane.

components could be separated on 13X, 5A, and 4A sieves regardless of the partitioning agent used on the sieves. This is illustrated in Figs. 1 and 2 where a column of acetone on 5A sieves and one of duodecane on 5A sieves were used to obtain the chromatograms shown. The activity of the sieve is important only when two closely related compounds are to be separated. For example, in Fig. 3 a chromatogram of normal hexane is shown. This was obtained with the acetone 5A column, and it can be seen that an impurity was separated from the main component of the hexane.

Upon examining a large number of chromatograms, the following general conclusion was reached. The highly adsorbing 13X sieve when used as a solid support held up ketones, halides, and hydrocarbons at temperatures well above the boiling points of these components, but when a temperature was reached where these would be released off the column, excellent separations were obtained. Thus a highly adsorbing substrate can be useful. Apparently because the materials are strongly held, slight differences in properties become accentuated and separations can be obtained that could not be achieved by using a liquid—gas partitioning effect alone. This is illustrated in Figs. 4, 5, 6, and 7, where chromatograms are presented for a column of triethylhexyl phosphate (TEHP) on 13X sieve. This column appeared particularly selective for aliphatic halides although it was necessary to operate it near 300 C to obtain the separations. Examining the data indicates that impurities in these compounds can be separated. The same com-

Fig. 4. A chromatogram of ethylene bromide on a triethyl-hexyl phosphate—13X column at 300 C.

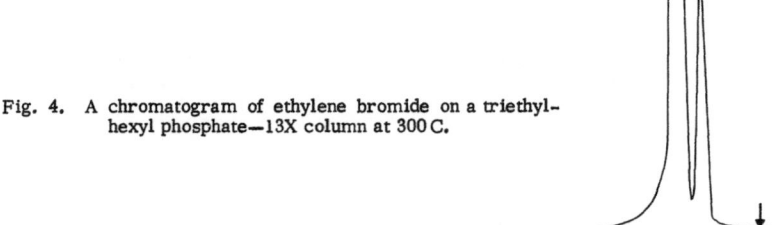

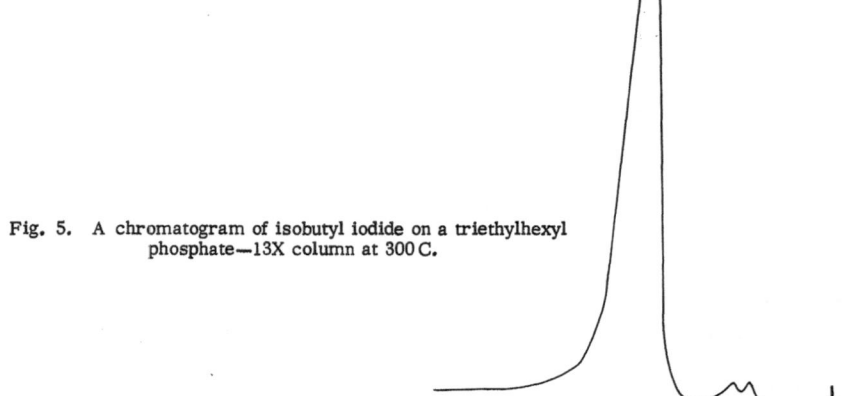

Fig. 5.   A chromatogram of isobutyl iodide on a triethylhexyl
phosphate—13X column at 300 C.

pounds were checked on 5A and 4A sieve columns having the same partitioning agent and on conventional columns, and the separations were not as good as those found with the 13X column. The one disadvantage of the 13X column was the high operating temperature required.

Further disadvantages of all sieve columns tested here are the difficulties of covering adsorption sites which hold oxygenated compounds, such as ketones. While hydrocarbons and aliphatic halides could be separated on most columns we investigated, ketones (and aromatic halides) could not be satisfactorily separated regardless of the type or amount of liquid agent placed on the sieve. None of the above agents tested proved satisfactory. It is obvious there are adsorption sites on sieves which cannot be easily covered even with such polar materials as water and acetone or mixtures of polar and nonpolar materials, such as acetone and duodecane. Unless these sites can be covered, oxygenated compounds will show tails or be adsorbed.

Aromatic halides are highly adsorbed on all columns using sieves as solid supports, even if large amounts of liquid partitioning agents are used.

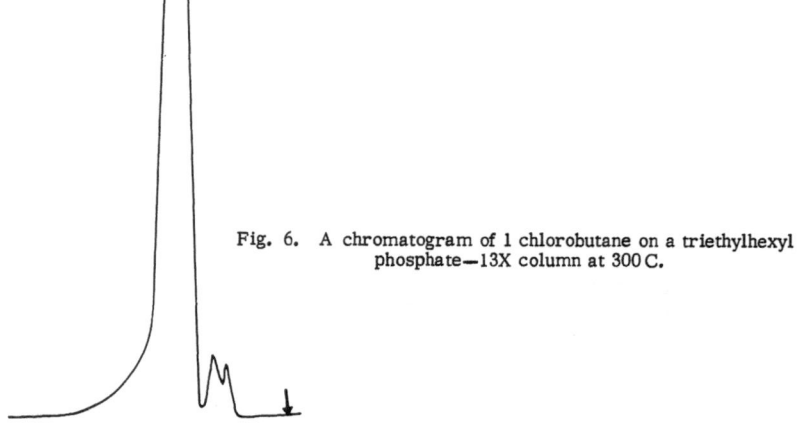

Fig. 6.   A chromatogram of 1 chlorobutane on a triethylhexyl
phosphate—13X column at 300 C.

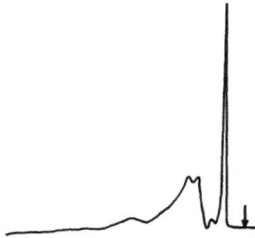

Fig. 7. A chromatogram of carbon tetrachloride on a triethylhexyl phosphate—13X column at 300 C.

The adsorption of these halides, as well as the general behavior of sieves as solid supports, suggests that short precolumns of coated sieves could be used to remove certain groups of compounds before the remainder of the sample enters the column. Thus a precolumn of 13X sieves with TEHP passes low-molecular-weight hydrocarbons, ketones, and aliphatic halides, but holds up aromatic halides. At a lower temperature, ketones or aliphatic halides can selectively be held up. We are examining such systems at present.

FAST TEMPERATURE CHANGES FOR SAMPLE VAPORIZATION AND/OR COLUMN HEATING

We recently reported a technique of using induction heating to obtain very fast column and precolumn temperature changes [3]. Since this work was reported we have redesigned our apparatus to utilize smaller-diameter columns and a flame ionization detector. Further, we have experimented with column packings utilizing metals dispersed in molecular sieves. Induction heating requires a metal to be dispersed in the column packing which is capable of being heated by high-frequency induced currents. In our work we examined silver- and platinum-loaded sieves to see if the fine dispersions of metal in these sieves can be heated by induction. These metal-loaded sieves were originally introduced for catalytic work and would appear to have some value as column-packing materials in gas chromatography. We found these metal-loaded sieves could not be heated by induction and therefore we concluded the metal is in a state which is not truly metallic; rather, since they are prepared by reducing the metal oxide with hydrogen, they are probably in a reduced oxide state.

We are working, at present, on thermally unstable compounds to see if these can be chromatographed by utilizing fast temperature changes. Included in the list of classes of compounds under study are steroids and other biomedical compounds which decompose at high temperatures. This work is in a preliminary stage and no results will be presented here.

REFERENCES

1. A. A. Zhuhovitskii and N. M. Turkeltaub, Dokl. Akad. Nauk SSSR 143:3, 646 (1962).
2. O. Grubner, M. Ralek, and J. Svoboda, Institute Physical Chemistry, Czechoslov Academy of Science, Prague, CSSR.
3. H. A. Szymanski, in Lectures on Gas Chromatography, 1962 (Plenum Press, New York, N. Y., 1963), p. 237.
4. Private communication.

A Broad-Range Ionization Detector

W. M. Barbour and D. R. Rushneck

Barber–Colman Company
Rockford, Illinois

In the course of designing an electron attachment detector, a study of design parameters led to some rather interesting results. The effect of interelectrode distance was of particular interest. This variable has a pronounced effect on attachment behavior at spacings less than 8 mm, the extent of the attachment effect decreasing abruptly as spacing is reduced below this value. At very narrow spacings virtually no attachment occurs, resulting in high sensitivity to permanent gases. Thus, a detector with adjustable electrode spacing, as described, can be used for both fixed gases and electronegative materials. Applications are presented which illustrate the detector's versatility and sensitivity in various operating modes.

INTRODUCTION

In the course of developing an electron attachment detector the effects of several design parameters were studied. Of these parameters, electrode spacing was one of the most significant and its pronounced effect suggested that control of spacing might permit use of a single detector in at least two different modes. By making spacing easily adjustable, the user would be provided with means of optimizing performance in various modes. Further investigation has revealed that a detector with adjustable electrode spacing can be used in the electron attachment, cross-section, helium diode, and Shahin–Lipsky modes.

DETECTOR DESIGN

Description

A sectional view of the detector is shown in Fig. 1. The detector is cylindrical, with an inner diameter of 1 cm. Parallel-plate electrodes ensure a uniform field within the detector. A thin disc of metal foil containing 300 Mc of tritium on one side side serves as the cathode and is perforated to allow passage of gas. The anode is made of stainless steel and is adjustable to permit use of electrode spacing from 1-25 mm. The body of the detector is glass-supported Teflon, and O-rings are used to ensure gas-tight seals. A fitting is provided for introduction of scavenge gas when the detector is used with capillary columns. This fitting is normally capped for packed-column work.

Effect of Spacing on Electron Attachment Phenomenon

The dependence of detector response on electrode spacing was reported earlier [1]. Figure 2 illustrates the effect of this variable in response to

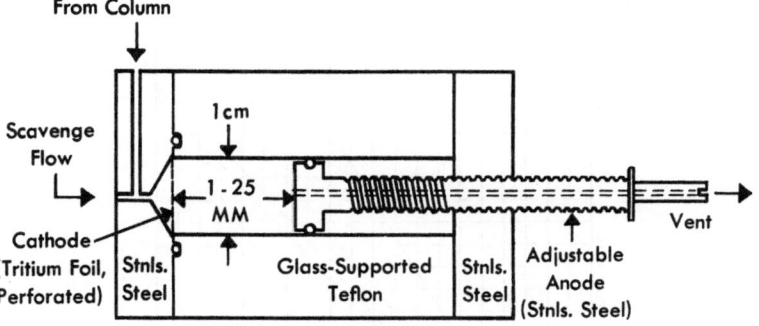

Fig. 1. Adjustable anode detector.

1,4-dichlorobenzene.   At spacings of 8 mm or greater, the detector gives excellent response to electronegative compounds.   The essentially complete elimination of attachment behavior at narrow spacing apparently contributes to the high sensitivity of the detector in other modes of operation.   For example, at narrow spacings the same detector can be used for permanent gases and other compounds of high ionization potentials.   The effect of narrow spacings has also been reported by Lovelock et al. [6] and by Shahin and Lipsky [9].   Their detectors are similar in two respects—an interelectrode spacing of 1 mm, and good sensitivity to permanent gases.

RESPONSE IN VARIOUS MODES

Cross Section

The cross-section detector is the simplest of the ionization types employing radioactive sources.   The theoretical aspects of cross-section response were best described by Otvos and Stevenson [7], who have shown that the total ion concentration of an element or compound exposed to a source of ionization is approximately proportional to the ionization cross section of the ionized material.   Ionization cross section can be visualized as the physical size of the element or compound.   In gas chromatography applications, the response should be proportional to the difference between cross sections of carrier and eluted constituent.   At present, however, there is insufficient experimental data to prove the validity of this assumption.

Hydrogen has a very low cross section and therefore should serve as an excellent carrier for cross-section detection.   Figure 3 presents results

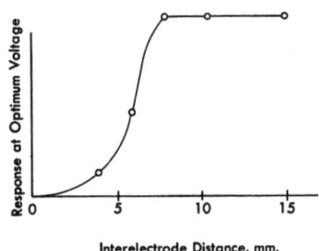

Fig. 2. Effect of electrode spacing on response to 1,4-dichlorobenzene.

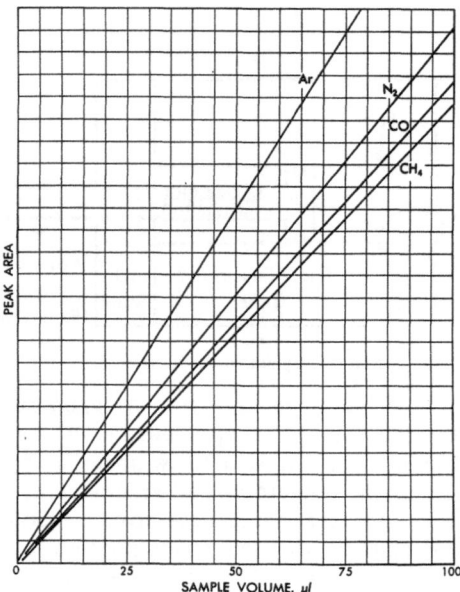

Fig. 3. Response in cross-section mode (1 mm
spacing, hydrogen carrier, 300 V).

obtained with the detector of Fig. 1 operated at a 1-mm electrode spacing
and an applied potential of 300 V. Response is linear and slightly greater
than that obtained with Gow-Mac W-2 filaments at 300 mA.

Table I shows a comparison of the slopes of the response curves for the
various sample gases and the relative sensitivities predicted on the basis
of theoretical cross sections as presented by Otvos and Stevenson [7]. From
a consideration of the theoretical cross sections, methane should have a
greater response than nitrogen or carbon monoxide, but the experimental
response to methane was lower than to both of these gases. With nitrogen
as a reference the difference between the experimental and theoretical responses becomes apparent. These data indicate that the response is not

TABLE I

Comparison of Experimental and Theoretical Cross-Section Response Factors (Hydrogen carrier, 1 mm spacing, 300V)

Compound (x)	Peak area slope	Q_x^* molecular cross section	$Q_x\text{-}Q_{H_2}$	$\dfrac{Q_x\text{-}Q_{H_2}}{Q_{N_2}\text{-}Q_{H_2}}$	$\dfrac{\text{Slope x}}{\text{Slope } N_2}$
H_2	—	2.0	—	—	—
N_2	4.8	7.70	5.70	1.0	1.0
CO	4.7	7.45	5.45	0.96	0.98
CH_4	4.4	8.16	6.16	1.08	0.92
Ar	6.3	10.9	8.9	1.56	1.31

*Theoretical molecular cross section (Otvos and Stevenson).

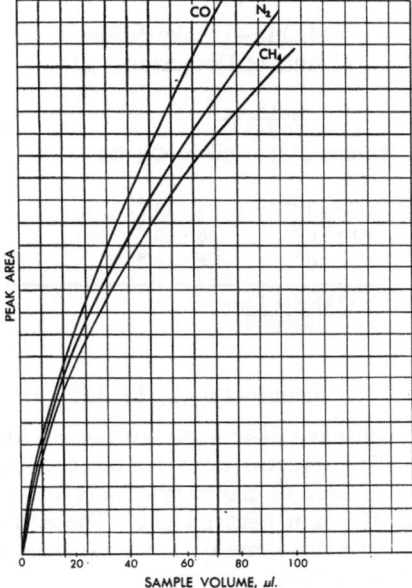

Fig. 4. Response with helium carrier (1 mm
spacing, 300 V).

proportional to cross section alone, but since only four sample gases were
used, more data are needed to properly evaluate the response.

Helium Diode

Since helium has a lower cross section than hydrogen, the cross-section
response when helium is used as the carrier gas should be greater than with
hydrogen, and the standing current should be one-third as great. However,
when helium was used as the carrier with 1 mm spacing and 300 V applied
potential, eluted constituents caused decreases in detector current, whereas
current increases are predictable on the basis of cross-section response.
One suggested explanation of this type of response has been called "electron
mobility detection" by Lovelock [4], who describes response in this mode as
follows: metastable helium atoms ionize the carrier contaminants to cause
a current in the detector; when a sample of permanent gas enters the detector, there is a decrease in the number of metastable helium atoms and
therefore a decrease in the number of contaminant atoms ionized, causing a
decrease in current.

This theory, however, does not adequately explain results obtained over
a wide range of voltages. Standing current remains constant from 5 to 300 V,
indicating a maximum concentration of ions from carrier contaminants over
this voltage range. At low voltages (5-20 V) increases in current were observed for samples of permanent gases, whereas decreases in current were
measured at 300 V.

Figure 4 shows calibration curves for several permanent gases and the
response is slightly nonlinear. The response of the detector to hydrogen is
similar up to about 0.04 ml, after which it reaches a plateau.

TABLE II
Shahin–Lipsky Mode
(Limit of Detection,* μl)

| | Shahin–Lipsky | | Adjustable anode detector |
Sample	Reference [8]	Barber-Colman	
O_2	$2.8 \cdot 10^{-14}$	$6 \cdot 10^{-4}$	$1.2 \cdot 10^{-3}$
N_2	$1 \cdot 10^{-3}$	$8 \cdot 10^{-4}$	$1 \cdot 10^{-3}$
CO	$4 \cdot 10^{-4}$	$7.8 \cdot 10^{-4}$	$1 \cdot 10^{-3}$

*Limit of detection = $\dfrac{(2 \times \text{Noise Level})}{\text{Response}}$

Shahin–Lipsky Mode

Shahin and Lipsky [9] have described a detector which is very sensitive to permanent gases. The detector consists of concentric electrodes with a 1 mm spacing. Argon is used as the carrier gas and a potential of 1-2 V is used for the analysis of permanent gases. The adjustable anode detector with a 1 mm spacing gave the same type of response, but with slightly less response than the Shahin–Lipsky detector.

Table II presents a comparison of the results reported by Shahin and Lipsky and those obtained with a Barber-Colman detector of the Shahin–Lipsky design and the adjustable anode detector.

Figure 5 presents a comparison of the response vs. voltage curves for the Shahin–Lipsky detector and the adjustable anode detector with a 0.02 ml air sample. At the 1-2 V recommended for this mode of operation there is a sharp maximum in response which may be too sharp for practical operation but the response even at 10 V is still good. Figure 6 shows typical calibra-

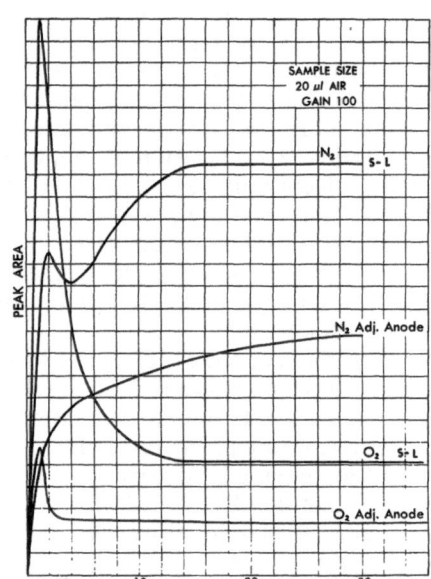

Fig. 5. Response vs. voltage for Shahin-Lipsky detector and adjustable anode detector at 1 mm spacing.

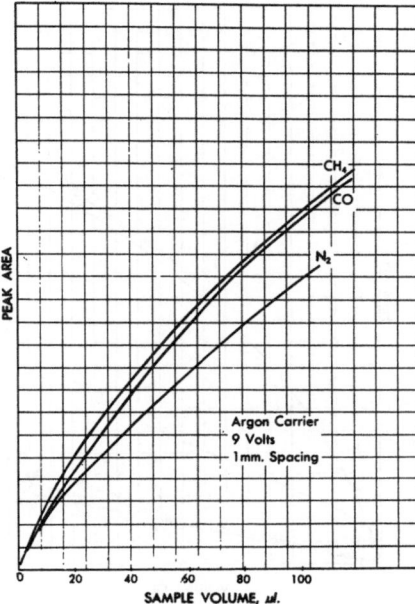

PEAK AREA

Argon Carrier
9 Volts
1mm. Spacing

SAMPLE VOLUME, μl.

Fig. 6. Response of adjustable anode Detector
in Shahin-Lipsky mode.

tion curves for the adjustable anode detector and shows that the response is nonlinear, as is the Shahin–Lipsky detector.

The Effect of Temperature on Response

Figure 7 shows the effect of temperature on response of the adjustable anode detector when it is operated in the cross-section, electron mobility, or Shahin–Lipsky modes of detection. For maximum response in these modes the detector temperature should be between 170-220 C. This must be emphasized because detectors for the analysis of permanent gases are normally used at much lower temperatures.

Comparison of Response to Permanent Gases in Various Modes

Figure 8 compares the response to nitrogen in the various modes with thermal conductivity response. Table III summarizes the limits of detection of several permanent gases in the various modes of detection.

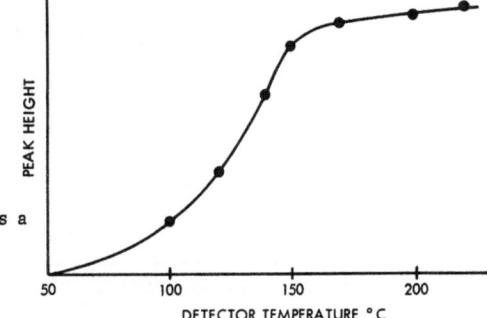

PEAK HEIGHT

DETECTOR TEMPERATURE ° C

Fig. 7. Response to permanent gases as a
function of detector temperature.

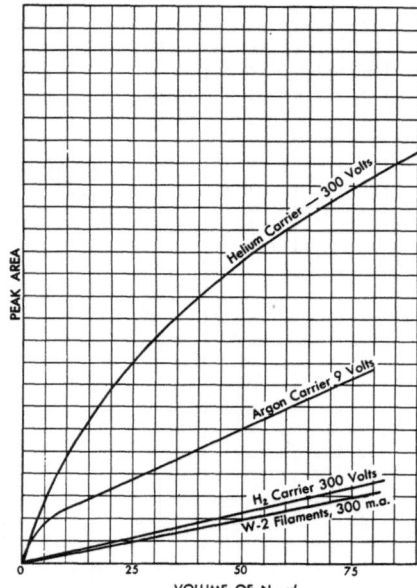

Fig. 8. Relative responses of thermal conductivity and adjustable anode detectors.

VOLUME OF N₂, μl

Electron Attachment Mode

In the electron attachment mode the carrier gas ionized by the radioactive
source yields electrons and gaseous ions. As a result of multiple collisions
with gas molecules, or because of their method of formation, electrons are
present at low energy levels. Accelerated by a weak electric field, they
attain relatively low velocities in traveling to the anode. At low velocity and
low energy the probability of their attachment to electronegative material
entering the detector is quite high. As these electrons are removed by attachment the detector current decreases, providing a measure of concentration
of electronegative material.

Argon, nitrogen, and 5-10% methane in argon are the commonly used
carrier gases. With nitrogen alone in the detector the standing current is
approximately 10^{-8} A. In practice reliably measurable decreases in current
range in magnitude from approximately 10^{-12} to $3 \cdot 10^{-9}$ A.

Because the electron affinity of different compounds varies, sensitivities
will vary greatly depending on the electronegative group and its position in

TABLE III

Carrier	Approximate limit of detection,* μl			
	N_2	O_2	CO	CH_4
Ar	$1 \cdot 10^{-3}$	$1.2 \cdot 10^{-3}$	$8 \cdot 10^{-4}$	$9 \cdot 10^{-4}$
H_2	$6 \cdot 10^{-3}$	$7 \cdot 10^{-3}$	$7 \cdot 10^{-3}$	$7.8 \cdot 10^{-3}$
H_e	$1 \cdot 10^{-3}$	$1 \cdot 10^{-3}$	$8 \cdot 10^{-4}$	$2 \cdot 10^{-3}$

*Limit of detection = $\dfrac{(2 \times \text{Noise Level})}{\text{Response}}$

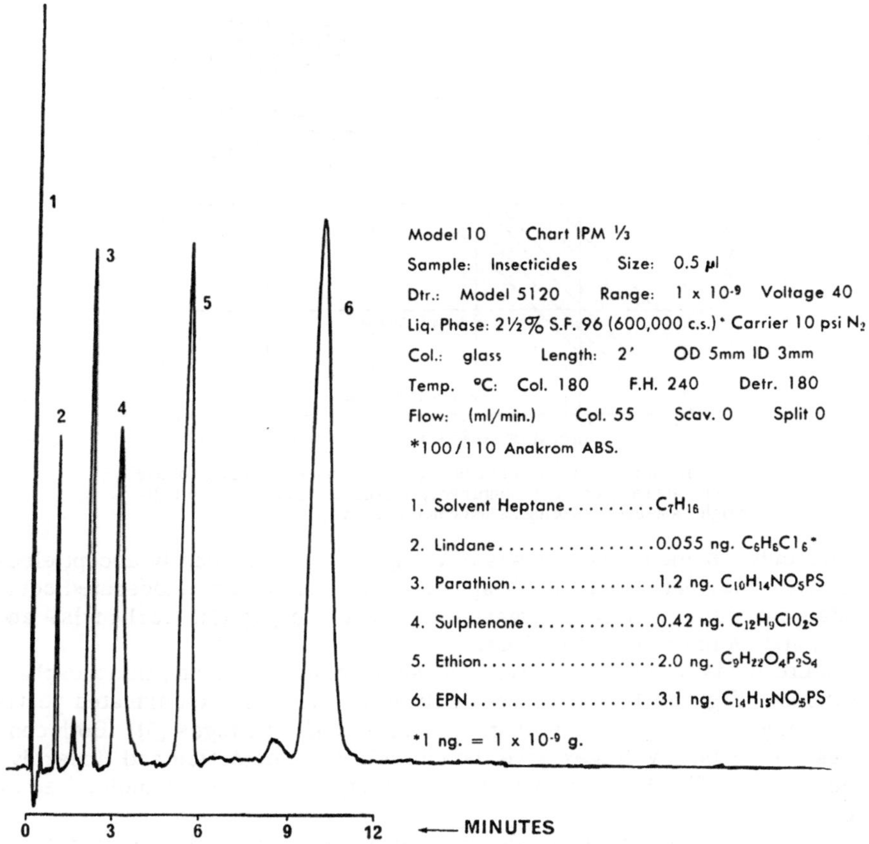

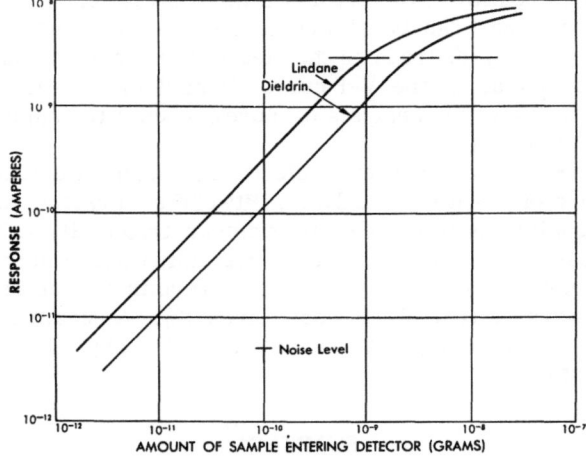

1
2
3
4
5
6
Model 10 Chart IPM 1/3
Sample: Insecticides Size: 0.5 μl
Dtr.: Model 5120 Range: 1 x 10^{-9} Voltage 40
Liq. Phase: 2½% S.F. 96 (600,000 c.s.)* Carrier 10 psi N_2
Col.: glass Length: 2' OD 5mm ID 3mm
Temp. °C: Col. 180 F.H. 240 Detr. 180
Flow: (ml/min.) Col. 55 Scav. 0 Split 0
*100/110 Anakrom ABS.
1. Solvent Heptane........C_7H_{16}
2. Lindane...............0.055 ng. $C_6H_6Cl_6$*
3. Parathion..............1.2 ng. $C_{10}H_{14}NO_5PS$
4. Sulphenone............0.42 ng. $C_{12}H_9ClO_2S$
5. Ethion.................2.0 ng. $C_9H_{22}O_4P_2S_4$
6. EPN..................3.1 ng. $C_{14}H_{15}NO_5PS$
*1 ng. = 1 x 10^{-9} g.
0
3
6
9
12
MINUTES

Fig. 9. Chromatogram of insecticides.

10^{-8}
10^{-9}
10^{-10}
10^{-11}
10^{-12}
RESPONSE (AMPERES)
Lindane
Dieldrin
Noise Level
10^{-12}
10^{-11}
10^{-10}
10^{-9}
10^{-8}
10^{-7}
AMOUNT OF SAMPLE ENTERING DETECTOR (GRAMS)

Fig. 10. Electron attachment response vs. concentration, using DC voltage.

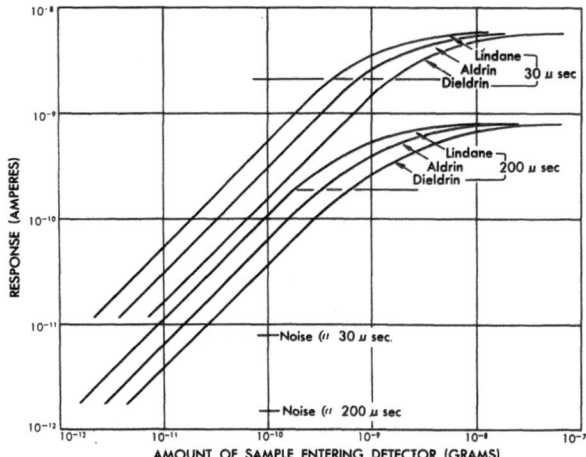

Fig. 11. Electron attachment response vs. concentration using voltage
pulses (pulses of 80-V amplitude, 0.4-μsec pulse width at half-
height, and pulse spacing as indicated in figure).

the molecule. Some of the classes of compounds which readily accept electrons are chlorinated and thiophosphate pesticides, other halogenated compounds, nitriles, nitrates, organo-metallics, conjugated carbonyls, and
steroid and amino acid derivatives.

Figure 9 is a chromatogram of some chlorinated and thiophosphate
pesticides. Figures 10 and 11 are calibration curves for chlorinated pesticides using continuous DC voltage and pulsed DC voltages [5]. Both continuous and pulsed voltages gave the same limit of detection and linear dynamic range. The limit of detection in many cases is 10^{-15} moles/sec or
better.

From the study of electrode spacing it was found that a spacing of
8-15 mm gave optimum response (see Fig. 2), and a spacing of 10 mm was
used to obtain the results in Figs. 9-11. Analysis of more concentrated
samples is possible with this detector by simple alteration of conditions. The
linear dynamic range can be shifted to higher concentrations of electronegative materials by decreasing the electrode spacing and using a voltage
that is on the plateau of the V-I curve. This decreases the sensitivity of
the detector but allows the analysis of more concentrated samples while retaining linearity of response.

The most popular application of electron attachment detectors has been
in determination of pesticides, and the utility of the adjustable anode detector
in this type of application has already been demonstrated [8]. Other potentially useful areas of application are in determination of steriods via their
monochloroacetate derivatives [3], and in determination of amino acids via
the methyl esters of their DNP derivatives [2]. An example of the latter
application is shown in Fig. 12, which reveals the high sensitivity of electron attachment to these compounds.

SUMMARY

From these examples of the use of the detector in various modes of operation it has been shown that this detector has a broad range of application.

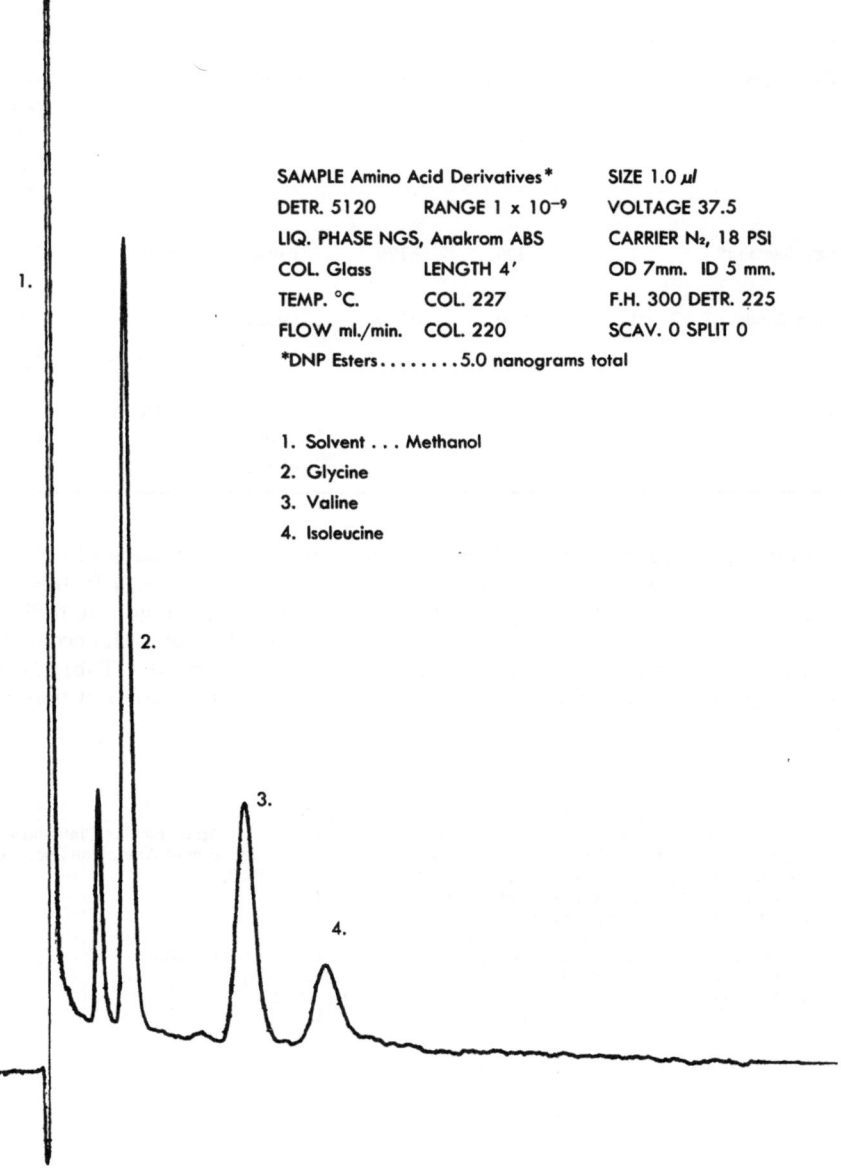

Fig. 12. Electron attachment detection of amino acid derivatives.

TABLE IV

Versatility of Adjustable Anode Detector

Mode of operation	Spacing, mm	Voltage	Temperature, C	Carrier	Optimum noise level, A	Compounds detected
Shahin–Lipsky	1	1-15	>170	Argon	$3 \cdot 10^{-13}$	Permanent gases and others of high ionization potential
"Cross-Section"	1	300	>170	Hydrogen	$3 \cdot 10^{-13}$	All volatile compounds
Helium Diode	1	300	>170	Helium	$2 \cdot 10^{-12}$	Permanent gases
Electron Attachment	8-15	10-100		Nitrogen or 5% CH_4 in argon	$5 \cdot 10^{-12}$	Electronegative compounds

Permanent gases can be analyzed in any one of three modes of detection. With hydrogen as a carrier gas, the response (cross section) is linear and slightly greater than thermal conductivity response. With argon or helium as carrier gàses the response is greater but is nonlinear. Electronegative compounds are detected using the electron attachment mode. Table IV lists the conditions for operation of the detector in the various modes of detection.

REFERENCES

1. Barber-Colman Quarterly Chromatogram, Vol. 2, No. 2, 1962.
2. R. A. Landowne and S. R. Lipsky, "Ultrasensitive Analysis of Amino Acids by Gas Chromatography and Electron Capture Spectrometry," 47th Annual Meeting of Federation of American Societies for Experimental Biology, Atlantic City, New Jersey, April 16-20, 1963.
3. R. A. Landowne and S. R. Lipsky, Anal. Chem. 35:532 (1963).
4. J. E. Lovelock, Anal. Chem. 33:162 (1961).
5. J. E. Lovelock, Anal. Chem. 35:474 (1963).
6. J. E. Lovelock, G. R. Shoemake, and A. Zlatkis, Anal. Chem. 35:460 (1963).
7. J. W. Otvos and D. P. Stevenson, J. Am. Chem. Soc. 78:546 (1956).
8. D. L. Petitjean and C. D. Lantz, J. Gas Chromatog. 23: (Feb. 1963).
9. M. M. Shahin and S. R. Lipsky, Anal. Chem. 35:467 (1963).

Additional Papers on Gas Chromatography

- "High-Temperature Gas Chromatographic Separation of Inorganic Compounds," R. S. Juvet, Jr. and F. Tivin, University of Illinois, Urbana, Illinois.

- "Factors Affecting the Efficiency of Gas Chromatography of Steroids," M. D. Whittier, L. Mikkelson, and N. Armstrong, F and M Scientific Corporation, Avondale, Pennsylvania.

- "Recent Advances in Preparative Chromatography," K. P. Dimick, Wilkens Instrument and Research, Inc., Walnut Creek, California.

- "Review of Advances in Column Theory," D. D. DeFord, Northwestern University, Evanston, Illinois.

- "An Application of Gas Chromatography to Solid State Physics: Determination of the Gases Evolved when Colored Potassium Chloride Dissolves," F. T. Phelps, Argonne National Laboratory, Argonne, Illinois.

- "Studies on the Detection of Narcotics in Human Urine by Gas Liquid Chromatography," J. Reffner, McCrone Associates, Chicago, Illinois.